BORDERING

Bordering
Identity Processes between the National and Personal

ANDERS LINDE-LAURSEN
Marshall University, USA

ASHGATE

Published by
Ashgate Publishing Limited
Wey Court East
Union Road
Farnham
Surrey, GU9 7PT
England

Ashgate Publishing Company
Suite 420
101 Cherry Street
Burlington
VT 05401-4405
USA

www.ashgate.com

British Library Cataloguing in Publication Data
Linde-Laursen, Anders.
  Bordering : identity processes between the national and
  personal.
  1. Boundaries. 2. Boundaries--Social aspects.
  3. Boundaries--Political aspects. 4. Human territoriality.
  5. Nationalism. 6. Denmark--Boundaries. 7. Sweden--
  Boundaries. 8. National characteristics, Danish--History.
  9. National characteristics, Swedish--History.
  I. Title
  304.2'3-dc22

Library of Congress Cataloging-in-Publication Data
Linde-Laursen, Anders.
  Bordering : identity processes between the national and personal / by Anders Linde-Laursen.
     p. cm.
  Includes bibliographical references and index.
  ISBN 978-0-7546-7905-9 (hardback) -- ISBN 978-0-7546-9790-9
(ebook) 1. Human geography--Denmark. 2. Human geography--Sweden. 3.
Denmark--Boundaries--Sweden. 4. Sweden--Boundaries--Denmark. 5. National
characteristics, Danish. 6. National characteristics, Swedish. I. Title.

  GF616.L56 2009
  304.2'3--dc22
                                                                        2009038672

ISBN   9780754679059 (hbk)
ISBN   9780754697909 (ebk)

Mixed Sources
Product group from well-managed
forests and other controlled sources
www.fsc.org Cert no. SGS-COC-2482
© 1996 Forest Stewardship Council

Printed and bound in Great Britain by
TJ International Ltd, Padstow, Cornwall

Contents

List of Figures[1]

1   Every effort has been made to find and contact owners of copyrights for all illustrations reproduced in this book. In a few cases such efforts have remained fruitless. In these cases, if on a later date rights can be documented, the owners will be treated on par with already identified and contacted copyright holders.

Preface

This book summarizes my research on the Danish-Swedish border over the last two decades. During those years, my interests have wandered widely but have constantly been anchored in my curiosity for what a border was/is and how people (myself included) affect and are affected by this geographical phenomenon. In this book I have tried to keep a focus on that central concern.

Over the years, my work has been supported by different grants, mostly obtained with colleagues at Lund University, in particular Orvar Löfgren. Grants that have supported this work include Swedish Research Council for the Humanities and Social Sciences (1991-1996), a position as a Mellon Foundation Postdoctoral Fellow at Duke University (1996-1997), Swedish and Danish municipalities and the European Regional Development Fund (1998-1999), The European Regional Development Fund (1999-2000), and The Bank of Sweden Tercentenary Foundation (2000-2003). I thank all the granting agencies and their administrators for their generosity and support.

Many individuals have over the years generously offered comments on ideas and drafts or otherwise let me take advantage of their time and knowledge: Orvar Löfgren, Jan Olof Nilsson, Fredrik Nilsson, Birgitta Svensson, Tom O'Dell, Gunnar Alsmark in Lund, Ulf Hannerz and Helena Wulff in Stockholm, Christian Tangkjær and Bjarne Stoklund in Copenhagen, Rick Wilk at Indiana University, John Richards and Edward Tiryakian at Duke University, Eve Darian-Smith at University of California Santa Barbara, Tim Hulsey at Virginia Commonwealth University, Gustav Peebles at The New School, and Kateryna Schray at Marshall University. None of them, of course, should be held accountable for any part of this book or my use of their advices. I also must thank the people in libraries, collections or elsewhere who have helped me find at times obscure information including help with locating illustrations and allowing me to print the same. In particular, I would like to thank Benny Andersen for his permission to print his poem "*Closet Swedes*" as part of the postscript, Ola Terje for finding pictures of his dioramas and allowing me to use the same, and Björn Andersson, Historiska Media, for his extensive work with photographing many of the illustrations.

Last but not least I thank my wife, anthropologist Elizabeth Faier. Over the years, her insight in the topic discussed, her anthropological curiosity, her encouragement and generous help have spurred me on as finalizing this manuscript has suffered from many distracting obligations.

Anders Linde-Laursen
Huntington, West Virginia,
June 23, 2009

Introduction

Bordering

I

We are constantly reminded that today's world is becoming increasingly globalized. Increases in transnational flows of people, goods, capital, and information ignore existing borders and call into question the geographical-political conception of the existing sovereign nation-states. The idea that existing borders might be undermined by such transnational flows has led some scholars to argue that both nation-states and their borders are becoming obsolete. Scholars making such claims join the ranks of those who have predicted, since nation-states came into existence, their downfall. Previously, nations were said to be doomed by military, cultural, and economic border penetrations; current doomsayers proffer globalization to be the end of the nation-state.

Surprisingly, given the amount of attention afforded to current and historical nation building processes, anthropological inquiries into the nature and effects of borders themselves have been largely absent (see, though, Wilson and Donnan 1998a, 1998b, Donnan and Wilson 1999, Haller and Donnan 2000). Anthropologists mostly stress that cultures, the objects of their science, cannot be clearly demarcated. However, anthropological findings are often communicated in ways that reify the existence of specific "cultures." Whether the cultures being investigated are geographically distinguishable (as in most "traditional" anthropological studies), are shared over great distances (as in studies of diasporic situations), or constitute flows and processes in a global context (a scientific endeavor, George Marcus has challenged anthropologists to pursue since 1998), most of these studies, often following in Barth's footsteps (1969), treat cultural processes as demarcating practices.

To complicate matters further, most anthropological studies take the very *place* for such processes for granted. This is especially true when it come to borders as they are often treated as lines that constitute "a national order of things" (Malkki 1995:5). Studies of the formation of and change in political borders have mostly been left to historians and political scientists, who generally have not paid much attention to the role such lines play in both reflecting and generating cultural processes. Our challenge then, is to rephrase Ernest Renan's more than century-old question, "What is a Nation?" (1882/1990) and ask instead, "What is a border?"

II

I offer the concept of *bordering* as a starting point for answering this question. The term bordering stresses that living with and thinking about borders constitute a situation of both barricading and facilitating transnational flows of people, goods, capital and information; thus, the border in this perspective is perceived both as a separator and a connector. This allows for a more meaningful examination of the continuous manipulation of self, other, resources, and power that occurs within and between adjoining groups. The concept of bordering furthermore facilitates the examination of flows within different frameworks of meaning: form of life, market, state, and (social) movements (Hannerz 1992:40ff). Approached in this way, borders have important implications for the concepts of national culture and identity as they exist as cultural practices; as will become evident, both have spatial and temporal, past and future, dimensions.

Applying the concept of bordering facilitates understandings of nation building processes by fostering a strong comparative perspective. This, in turn, allows us to examine hitherto mostly overlooked aspects of bordering processes. *Bordering* creates a framework for thinking about nations that while recognizing the importance of existing nations and states on both sides of a border, conceptualizes those nations and states as fluid manifestations of power and culture in both spatial and temporal dimensions. It is the very limit – the start/end (depending from which side it is seen) of the nation-state in time-space, as category, as event, as institutionalized form (Brubaker 1996:15ff) – that becomes the focus of analysis. From this perspective, I seek to theorize and understand the changing placements, forms, and practices of those limits. Needless to emphasize, such an understanding of borders also brings issues of power, especially its cultural forms and material manifestations, into the framework of exploration.

It must be recognized that interesting studies of borders do exist (see Wilson and Donnan 1998a:3-6, 1999:19-22). However, even where a border is an essential part of what is being studied, anthropologists tend to focus on other matters of culture and identity while the border remains "a backdrop to some other line of inquiry" (Donnan and Wilson 1999:26). A bordering approach facilitates our understanding of geographical distributions of culture across borders, not least in cases where we are discussing national minorities living "outside" the nation-state to which they refer – or are being referred – culturally.

László Kürti writes about the role of Transylvania (an area today inside Romania) in Hungarian nation building. He comments, "the borders set by the Treaty of Trianon…were largely an artificial, meaningless demarcation between the two states…What was important was what went on beyond the borders, not what they represented" (2001:123).[1] The point of the concept of bordering is the

---

1   One problem with such studies that I will not pursue further here is that while they often are very effective in regards to deconstructing the border, they take the existence of the "cultural" groups (minorities as well as majorities) for granted. Thereby they contribute

opposite – to approach what the border *means* to people, how they live with, understand, and manipulate it. Thus, the border is not a political fact disconnected from peoples' lives. When Kürti describes how an elderly patient travels from Transylvania to a hospital in Hungary, he describes this as an issue of belonging and the security of being able to communicate with physicians in one's own language (2001:73f). However, he fails to stress in his analysis that this is also an example of a specific cultural practice of border manipulation through which the patient gains access to resources paid by the Hungarian state that are not available to non-Hungarian Romanians.

That bordering can be a contradictory process with many seams of meaning is evident. In his interesting study of the US-Mexico border, Peter Andreas describes how the liberalization of cross-border movements, not least a result of NAFTA, goes hand-in-hand with a steady increase in policing and militarizing the same limit. He finds that the efforts to strengthen the border have not substantially decreased clandestine cross-border flows of people and drugs but have played a major symbolic role with the American public (2000). This conclusion points not only to the symbolic importance of the dividing lines but also, and more interestingly, to the paradoxical nature of border transformations. While it might be convenient to see the border as one line, it is in fact the location of very complex cultural phenomenon. Like Andreas, one might find that borders are simultaneously disappearing and expanding; in his case disappearance occurs through liberalization and expansion emerges through patrolling, fences, and so forth. Bordering, thus, must be regarded as a dynamic cultural process, always changing in response to historic developments and constantly being transformed by and transforming the social, cultural, and political contexts of the very nature of the limit.

Researchers discussing Europe's changing borders after 1989 often summarize their position as "the East disintegrated, the West integrated" (Kürti 2001:166). This position suggests that borders are becoming less important in tomorrow's EUrope. But to understand the effects of borders, we need to ask in more detail what is being disintegrated/integrated, by whom, why, and when. Furthermore, as we learn from Andreas and others, bordering developments are complex processes often saturated with contradictory yet interconnected components. It is too simplistic and conceptually limiting to understand developments in Europe as the direct result of, on the one hand, political and economic integration in the West (see for instance Guerrina 2002) and on the other hand, disintegration in what used to be the East. Processes of disintegration clearly are occurring also in the West (for instance, in Belgium, Spain, and the United Kingdom). And it is obvious that Russia is not prepared to accept continuous disintegration as its reactions to secessions on its southern periphery from the latter part of the 1990s

---

to essentializing one understanding of culture, at the same time they try to deconstruct another. I know of no study that manages to relativize simultaneously nation-state borders and ethnic (or otherwise defined cultural) categories.

have clearly testified. Furthermore, Russia is trying, at least to some extent, to reintegrate parts of the former Soviet Union through for instance, political and economic treaties with Belarus. Moreover, it is more problematic to regard what has happened or is happening today as a unilinear and causally related progression of events. The historic development we saw in Europe in the 1990s was swift; what became possible one day was not necessarily an option the next. For example, had Germany not been united "then," would such a move be possible today? It is equally challenging to consider what could have happened if the Baltic States had not been allowed independence in 1991 but instead had to fight for it today. And how would the West react to the kind of fighting that in November 1999 tore apart Grozny had it instead occurred in the suburbs of Tallinn or Riga?

It is also important to realize how interconnected are the contradictory processes of supposed integration and disintegration. It is, for instance, impossible to imagine Sweden applying to join the EU, holding a referendum and being admitted, prior to the dissolution of the Eastern bloc. Furthermore, while Swedish admission to the EU can be regarded as a clear sign of EUropean integration, the public referendum in Sweden preceding this integration clearly demonstrated national disintegrative tendencies. While the major cities, especially Stockholm and Malmö (with the southern county of Skåne) voted decisively for EU membership in November 1994, small towns and rural districts elsewhere were either split or voted against membership. In the northern, inland parts of the country, this "no" was overwhelming – in Jämtlands län the "no" vote accounted for 71.5 percent of the ballots cast.[2] It can be argued that this pattern represents a breakdown of the existing Swedish nation-state into distinctive pieces. However, the contradictory processes and locations of disintegration and integration must first be studied as bordering thoughts and practices before one can draw such a conclusion.

In many contexts, current transformations of culture, as well as of economy and politics, are being understood within a framework of "globalizations." This understanding conceptualizes cultural change as processes of border crossings and transnational flows. By explicitly negating the existence of previous as well as currently delineated "units" (see for instance Nederveen Pieterse 2004), proponents of this approach are forced to create convoluted arguments that should explain that these border transgressions form a critique of how the world has been and maybe is understood as a patchwork of separate nations (or other cultural manifestations). As Jan Nederveen Pieterse writes, "the current hybridity discussion seems superficial, for it is entirely dominated by the episodes of colonialism and nationalism of the last two hundred or so years" (2004:103). While I agree with the direction of his thought, I suggest a distinctively different course of argument. If people do reference the existence of such "cultures" as real, the cultural researcher's objective should not only explain that this is wrong. If we regard such ethnographic

---

2 Published results; www.riksdageen.se/arbetar/siffror/folkomro.htm, visited 27/12/2003. This split becomes even more evident in the subsequent Swedish referendum on joining the European Monetary Union in 2003, which failed (same reference).

observations only as distortions created out of specific historical conjunctures, a new discussion of "false consciousness" looms right around the corner. Instead our challenge is to understand how bordering, as *a cultural practice of making sense of and manipulating everyday lives and experiences*, is adapted and performed by individuals and groups. As such practices are performed in many different situations and contexts, such research must be furthermore what George Marcus refers to as "multi-sited" (1998).

III

To explore the concept of bordering, this book focuses on the Danish-Swedish border at *Øresund* (in Danish), *Öresund* (in Swedish), or as it is called in English, the Sound. Geographically, Øresund stretches north-south. In the north, it begins north of a line from Helsingborg to Helsingør (English; Elsinore – among English speakers probably best known as the home of Hamlet's Castle in Shakespeare's drama); it runs past København (English, Copenhagen) and Malmö, and ends further south where it widens and joins the Baltic Sea. Øresund forms a waterway that connects the Baltic, through Kattegat and Skagerrak, with the North Sea. The border itself follows this path. Thus, the Øresund border can be said to be located between southern Sweden and the Danish main island of Zealand, where the capital Copenhagen is situated. It is not least due to the proximity of the shores (at Helsingborg-Helsingør, the Danish and Swedish coasts are only a couple of miles apart, and at Copenhagen-Malmö about ten miles), that bordering processes at this particular line have been abundant.

Current discussions of nations, nationalism, and imagined communities often focus on and theorize divisions that were/are for different reasons contested and problematic. As Martin Stokes begins his article about the Turkish-Syrian border:

> Borders create problems for those whose lives they frame. [These problems include…] the efforts of the modern state to coerce or persuade local populations to accept its jurisdiction, political and economic decisions, notions of a unitary national culture and the dubious benefits of its military protection. These efforts are often acute in border regions. Here, the power of the nation-state and its symbolic apparatus do not fade out, but intensify […] This often places inordinate demands on minority populations, for whom a border often cuts across pre-existing and culturally more relevant ties with others excluded by these borders. Majority populations may apparently reap the benefits of such arrangements, but suffer the long-term consequences of living with problematic "others" in their midst. (Stokes 1998:263)

Studies of such problematic borders have multiplied recently, the result of added cases from Central and Eastern Europe after the Fall of the Wall, the breakup of the Iron Curtain, and the restructuring and multiplying of states. That conflicts have been in focus is no surprise given that many of these historic developments have left millions of individuals with citizenships and/or locations of living that do not correspond with their claimed or ascribed national identifications (Brubaker 1996, Gellner 1994).

Mathew Horsman and Andrew Marshall's statement, "If the principle fiction of the nation-state is ethnic, racial, linguistic and cultural homogeneity, then borders always give the lie to this construct" (quoted in Donnan and Wilson 1999:1), might well be true in a general sense. However, "always" and "lie" are very strong words when applied to discussions of bordering as political and cultural practices. Within this context Øresund is what one could term an "ideal" border. Here, ethnic, linguistic, and cultural distinctions correspond with the political divide. The border at Øresund has been undisputed for a long time. There are no continuous national minorities living in "the other country," none of the states claim territory on the other side of the water, and no policies of intimidation or involvement in the affairs of the neighboring country are practiced.[3] Thus, this particular border can be seen as a "cold," uncontested border; the kind of ideal divide that most people, who subscribe to nation-states as important parts of today's and tomorrow's world, probably envision surrounds most imagined communities (Anderson 1983). One question to consider, therefore, is how this ideal state came to be and through what political and cultural practices it is being sustained and constantly reinvented. As I will discuss in the following chapters, this "ideal" state is not a result of natural circumstances (for instance the water as a natural and true divide), but the result of hundreds of years of bordering carried out through state agencies, by international powers, and by people in their everyday lives.

IV

A discussion of the long and complicated history of the Øresund border with the broad perspective I intend to adopt would be voluminous. Accordingly, this book does not claim to cover all aspects of the border and its development over hundreds of years. It is rather a series of chapters linked together by concerns with what is a border, what people have thought and think of it, and how people have lived and live with it in their everyday lives.

---

3   This is true in general. However, now and then an issue appears that makes politicians comment upon the situation in the other country, and sometimes in a less than diplomatic fashion. For example, in the summer 2002, this was the case when the Social Democratic Prime Minister of Sweden expressed concern with the immigration policies of a new right wing Danish government. And for decades, Danes demanded the closure of a Swedish nuclear power plant at Barsebäck, just across Øresund from Copenhagen, as they feared an accident could pollute most of the eastern part of their country including the whole metropolitan area around the capital. In fact, Barsebäck was closed down in the late spring 2005.

The border in question has changed considerably over the years. The way people live with the border in 2009 is very different from how those living a hundred years earlier might have considered it. Consequently, many kinds of ethnographic sources can be used to discuss bordering at any given time. I do not pretend here to deplete or even catalogue all the possibilities, but I have deliberately tried to choose materials that encompass many different aspects of peoples' practices with and thinking about borders.

Obviously this approach to bordering has methodological and theoretical consequences, as ethnographic materials from primary and secondary historic sources from the 1600s, historic novels from the 1800s, and interviews conducted in the 1990s, must be treated differently. Where necessary, I have addressed those methodological concerns that have broader implications. Similarly, it is futile to start with developing a comprehensive theoretical argument that covers all the different times and ethnographic arenas that characterize this inquiry. Theoretical thoughts that are applicable to the lives and rationalities of peasants in the late 1600s do not explain much of what is stated in public debate in the 1980s. Consequently, I have chosen to develop the theoretical perspective for each chapter in relation to the ethnography presented and discussed.

The text is organized into five chapters. In each, I discuss a set of questions related to the border in Øresund employing different ethnographic materials and developing the theoretical tools that make these materials contribute to the understanding of bordering. In Chapter 1, I discuss how bordering processes between the Danish and Swedish kings' realms were affected by warfare in the 1600s. Central to the argument is how and why common people did or did not react when they suddenly found themselves on the other side of a border that was moved. This analysis is important not only because it situates the border in Øresund (which itself was a result of peace treaties in 1658/1660), but also because the history of what happened has in later periods been used as fuel for nationalist and regionalist agendas. Therefore, it is important to know something about what happened, how this affected both the geography of the border, and later logics of bordering. While this chapter creates a better foundation for understanding the historic events themselves, it also outlines how modern arguments (mis)understand the events and frame them in terms very different from those of the 1600s. Because this chapter is an investigation into the nature of the relationship between the state and community in the 1600s, it addresses the question of whether an imagined community could possibly exist at that time. The ethnographic material for the discussion in this chapter is mostly historical writings and published historical records.

In Chapter 2, I use material from a collection of historical novels written and published from the mid-1800s through the 1990s. These novels depict the wars and bordering practices of the 1600s. From this perspective, I discuss the emergence and dissemination of a hegemonic discourse of nationhood – the idea of a spatial organization of politics and people. This discourse joins together many different practices that both support and demand bordering as a central element.

Thus, this chapter discusses the solidification of national projects through what can be perceived as a period of continual crisis; it becomes evident that each crisis was probably not very different from the challenge posted to the nation-state today by globalization.

Chapter 3 pursues the idea of bordering through a variety of ethnographic materials that illuminate various understandings of what it meant to be Danish, what it meant to be Swedish, and what formed the Danish-Swedish border. The ethnographic materials include many kinds of bordering narratives gathered from schoolbooks, films, exhibitions, economy, and so forth. Together these materials demonstrate discursive continuities and discontinuities that lie at the heart of bordering. In this chapter, I also discuss how the discourse of nationhood in some time-spaces becomes interwoven with other discourses, for instance gender and generation. Through these ethnographic investigations it becomes clear how the discourse of nationhood during the 1900s constantly was created, recreated, and changed. The materials illuminate the role of bordering in such processes. In this chapter, I also review discussions in Scandinavian ethnology of conceptual frameworks for understanding nationhood.

In Chapter 4, I address a topic that is seriously understudied – the junctions among discourses of nationhood, bordering, and individuals' everyday life. I attempt to show how the modern Nordic nation-states have become materialized as practices in body movements, home constructions, and cleaning by ordinary citizens. From this perspective, I discuss bordering as an individual act and as a consequence of border penetration and narration. I also raise some questions about ethnographic practices concerning what people tell us, do not tell us, and what they might tell us but we are unable to hear.

In the final chapter, I return to the discursive framework established earlier in the book. However, instead of using the framework to scrutinize past events and practices, I turn the historic perspective around and look at how nationhood discourses are used and misused by different players in debates about the Øresund border after around 2000. Some participants in these debates interpret the discourse of nationhood in ways that previously have been marginal or non-existent, while others try to invoke a whole new discursive framework for bordering – and, consequently, also for nationhood. However, as I will demonstrate in this chapter, the existing discourses are difficult to escape and therefore such attempts result in bordering paradoxes. A methodological point here is that while it is perfectly possible in the past-time-space dimension to establish reasons for the continuity and discontinuity of bordering practices, it is impossible in the future-time-space dimension to predict how or why complex cultural dynamics result (or fail to result) in transformations of commonly held perceptions of bordering. Of course this ought to make historians and historical anthropologists more aware of those time-space alternatives that in the past have been subjugated and thus excluded from hegemonic narratives.

V

As will become obvious, especially in the fourth and fifth chapters and the postscript, I myself am very much located and in many capacities involved in the bordering between Danish and Swedish. Born in Copenhagen, where I also grew up and received my first terminal degree at Copenhagen University, I moved to Lund in southern Sweden where I received my second terminal degree. Parts of this book are translations from my Swedish doctoral thesis, published as *Det nationales natur. Studier i dansk-svenske relationer* (The Nature of the Nation. Studies in Danish-Swedish Relations; 1995). Other parts are newly written or completely revised for this book.

I recognize that what is presented here is not an objective selection and interpretation of bordering processes. Where the presented materials is directly associated with my person I write in the first person form. I chose this form for those passages to stress that I, in particular places in the text, play a role that might not be available to another researcher. Where this is not the case, I do not use the first person form. This does not indicate that these parts are more objective, but rather stresses that the ethnographic material used in these sections would be available to anyone else under similar circumstances.

While much of the scholarly material I cite when developing the theoretical arguments is available in English, most of the ethnographic materials as well as a lot of the scholarly debates about the particulars exist in other languages, of course mostly Danish or Swedish. As such, concerns about translations are obviously important. Where nothing else is mentioned, I have done the translations myself. Generally, I have tried to maintain the wording and structure of the Scandinavian texts to convey a feeling for how and what was expressed. This has, here and there, resulted in somewhat peculiar sentences. These oddities are not necessarily the results of the translation; some of the translated pieces also read outlandishly in the original languages. When previously published material exists in English translations, I have tried to find and cite them.

Chapter 1
A Border is Born

Via the officer on duty in Karlskrona [naval base in Sweden], contact was made with the Naval Command in Denmark. Permission to follow the fishing boats into Danish waters was requested. The Naval Command could not give an immediate answer but promised to call back. In about 15 minutes came the answer, which was negative. This with the motivation that the general permit from May 25, 1990, is valid only for passage and not for clampdowns in connection with monitoring of fishing. Consequently, the pursuit was cut off at the line of demarcation.

Contact was made to the police in Hilleröd [town in Denmark]. They were informed about what had happened and promised to make contact with the crews of the involved fishing boats when they came to harbor. This was done and the so-called interrogations were sent by telefax to the [Swedish] Naval Command in Karlskrona.

What is surprising is the Danish police's handling of the matter. Contrary to what they knew, that the crews on the trawlers were guilty of illegal trawling, the masters are questioned "to obtain information." They are not read their rights, though reasonable suspicion existed as they had been caught red-handed with their trawl in the water.

The question emerges. Is it due to incompetence that they [Danish police] act in this way? Or, what one of course does not hope, is it from pure aversion against helping Swedish authorities?

Trawl fishing in the Øresund is after all prohibited regardless of from which country one originates.

(The Swedish Coast Guard, management of region south, November 1, 1991)

There is a Line

The boundary between Denmark and Sweden is drawn down through Øresund, the channel that separates the main Danish island, Zealand, from southern Sweden. As far as this border, both states try to enforce their authority and maintain their monopoly on legitimate violence. As is indicated in the document quoted above, this is not always easy. People employ the border as a means to manipulate the authorities on both sides in attempts to negotiate the states as well as the border to their own advantage.

This happened for instance when Danes in the mid-1990s moved to Sweden but kept their jobs in Denmark. Due to differences in tax-laws, this move could make a difference in available income after taxation of up to around 20 percent.[1] Other groups also employ the border as an opaque wall between the countries. This is true for those who cross Øresund to buy cheaper or unattainable merchandise on the other side; for example, young Swedes go to Denmark to obtain alcohol while in the other direction go people who want fireworks, mostly around the New Year. This is also true for the fishermen who the Swedish Coast Guard complain about in the letter quoted above. The particular case from which this complaint originates can be summarized as follows:[2]

> Friday, October 22, 1991. The time is approaching 11:00 PM. The weather in the northern part of Øresund is fine; it is cloudless and bright moonlight. Half an hour earlier the Swedish Coast Guard had received a report that fishing with trawl, which is prohibited in all of Øresund, was carried out north of the small Swedish island Ven. From patrol-boat 286, in the harbor of Helsingborg, a dinghy with two men was launched. They are now approaching two boats carrying lights as merchant ships. They are, though, fishing boats jointly trawling: H 28 *Mona Nielsen* and H 220 *Marianne Hougaard* both of Gilleleje in northern Zealand. The men in the dinghy demand that the fishing ends and that the boats stop. The trawlers approach each other and all cables are transferred to H 220. H 28 then leaves at maximum speed on a northerly course towards Danish territorial waters. From H 220, the dinghy is warned about approaching while the gear is recovered. Immediately after the gear is on board, H 220 at maximum speed sails westwards towards Danish territorial waters. The men in the dinghy pursue the boat and demand through radio that the trawler stops. The commanding officer of patrol-boat 286 requests permission from the Danish Naval Command to continue the pursuit onto Danish waters. While this request is declined, the dinghy stops its chase at the limit of Swedish territorial waters. The time is not yet 11:30 PM.
>
> The Swedish Naval Command requests that police in Denmark take action. At 3:40 AM the same night, the masters of the fishing boats are questioned in Gilleleje. Both assure that they have not performed illegal fishing, that they have always carried correct lights, and that Swedish authorities have not tried to contact them.

---

1   Comparisons between taxation levels are always very questionable as systems in different countries work very differently. This is also very much the case between Denmark and Sweden; 20 percent is a reasonable estimate, not taking tax deductions of any kind into consideration. When the bilateral tax agreement was changed in 1997, many Danes moved back to Copenhagen.

2   The case is summarized on the basis of documents in Malmö Regional Court (Malmö Tingsrätt, Avd 4:3, Mål nr B 4252/92) and the High Court for Skåne and Blekinge (Hovrätten Avd 3, Mål nr B 329-93). This reference also applies to the letter quoted in the introduction from the Swedish Coast Guard, management of region south, November 1, 1991 (Kustbevakningen, Regionledning Syd, 1991-11-01, addendum 10).

The above is only one example of an encounter between Swedish authorities and fishermen from Gilleleje, who trawl in Øresund. Such encounters occur, however, quite frequently. The Swedish Naval Command knows how the "bandits," as they are called, are organized. To protect themselves the fishermen monitor the Swedish patrol-boats from deployed guard boats and from mobile patrols on shore. If Swedish authorities nevertheless approach, the trawlers take to Danish waters where they cannot be pursued. It has happened that trawlers trying to escape have hit patrol-boats. Danish fishermen, thus, employ the state boundary as a means against Swedish authorities who try to enforce bilateral Danish-Swedish fishing regulations.

Apparently, Swedish authorities feel a lack of support from their Danish counterparts for their efforts to stop the illegal activities. October 22, 1991, both bilateral regulations of fishing and the international laws of the sea were violated. H 28 and H 220 had performed illegal fishing, they had not carried correct lights, and they had not agreed with the authority's requests to stop. The authorities in the Danish Ministry of Fisheries responded by depriving the boats of their permission to catch herring from November 11, 1991, through January 15, 1992. They considered that this sanction "in itself is so severe that no other punishment, neither in Denmark nor in Sweden, should be enforced because of the illegal trawling." However, in the Danish Ministry of Fisheries it was recognized that the violation of the international laws of the sea *could* be the object of court actions in Sweden (letter to the Swedish Naval Command November 6, 1991).

In Sweden, the case was regarded quite differently. In Malmö Regional Court in November 1993 as well as in the High Court for Skåne and Blekinge in May 1994 both masters were sentenced to paying heavy fines for their offenses and to losing the value of both the employed trawl and the illegally caught fish, which totalled to around $20,000 (125,000 SEK). The verdict explicitly stated that since the fishermen had been punished by being deprived of their permissions to catch herring, and since a long time had elapsed since the crime, they were fined. Otherwise, they would have been sentenced to jail.

The Potential of a Political Border

In a study of one of Europe's most stable, undisputed, cold, and internationally recognized state borders, Peter Sahlins illustrates how a border is created and used in the Pyrenees (1989, 1998). In spite of much later warfare over the border between France and Spain, it remained where it was drawn in the Pyrenees in 1659-1660. Sahlins explores how the border after 1660 was perceived and utilized militarily, juridical, and economically. He argues that nations on both sides of the border in the valley of Cerdanya, which is the region he studies, were not constituted from their centers, Paris and Madrid, but in their mutual periphery. Here the populations learned as individuals and as groups to employ each their state as a means to manipulate struggles over limited ecological resources (1989:155ff,

229ff, 245f, 1998:42ff). In the long run, Sahlins argues, the inhabitants came to identify their opponents as either Spanish or French violators of rights, territory, and so forth despite both groups descending from the same linguistic, religious, and ethnic roots. He consequently finds that nationality is recognized at its limit, at the boundary, because populations can utilize the nation to manipulate their interests.

Sahlins' approach to bordering, however, falls short of explaining how and to what extent these manipulative and instrumental relations with the state were transformed into *recognition* of nationality and belonging. When he discusses such recognition, he references criteria such as language, habits, ways of cultivating the land, and clothing (1989:259ff) without exploring how people felt about or thought of the nation beyond that the state was a practical and powerful ally to be called upon in times of cross-border conflicts. Sahlins acknowledges, that "the two Cerdanyas were hardly distinguishable by their 'mores' or costumes or language" (1998:51). He also mentions that after the instrumental relation with the state was well established, many individuals continued to evade obligations towards the nation when they for instance, fled to the other side of the border to avoid the draft (1989:267ff).

Sahlins' thesis that nations are constituted along their demarcations through the manipulations of local populations and independent of the centers of power has been criticized. Responses argue that it neither takes into account local conditions in the valley of Cerdanya, nor sufficiently considers relations between center and periphery within the nations. William A. Douglass has pointed out that at the other end of the Pyrenees, in the Basque provinces on the Atlantic, there is a completely different evolution of nationalism and regionalism (1998, see also Ben-Amin 1991, Leizaola 2000). Through historic processes, he explains which different geographical, legal, and linguistic circumstances prevailed. Among other things, the Basque provinces maintained particular rights far longer than was the case at the Mediterranean shores. A historic problem, which Sahlins does not address, is that a border in the Pyrenees – although moved – was not a new phenomenon in 1660. The earlier border between the two states can likewise very well have been employed as a means of manipulation. An analysis of earlier periods and previous locations could reveal manipulative strategies both similar to and different from what Sahlins describes.

The present border in the Pyrenees was drawn in 1660 and has since remained stable; Douglass even holds that this border is "arguably the most stable in western Europe" (1998:88). The border in Øresund was drawn at the same time. Prior to 1657, Øresund was a waterway within the Danish kingdom. After 1660, it became a state border between Denmark and Sweden. Moreover, Øresund eventually became one of Europe's stable, uncontested, cold borders.

Like the boundary between Spain and France, the line drawn through Øresund can be used as a means. Manipulation, however, can backfire with double penalty the result as was the case when both Danish and Swedish authorities became aware of attempts to take advantage of the border; this is evident in this case when the fishermen were punished first by the Danish Ministry of Fisheries and subsequently by the Swedish courts. However, for most Danes and Swedes the border in Øresund plays no role in their everyday life. Due to Nordic treaties, they do not need a passport if they at all are interested in crossing the border.

Here however, I focus not on the possibilities of employing the border as a means of manipulation; similar possibilities have always been present at economic-political borders.[3] Instead, problems not systematically analyzed by Sahlins are in focus. After Øresund became a border between kingdoms, did the populations on both sides learn to recognize each other as belonging to different nations? If this was the case, was this process generated through manipulations by the locals along the border independently of or designed to appeal to the centers of power? In addition, did a continuous understanding of the border exist since 1660? To illuminate such questions it is necessary to follow developments from the time the border was born and to consider if it has always been recognized as a line dividing two nations. People's ideas about and practices at the border must be presented in order to conceptualize cultural processes of bordering at what we today recognize as a division between two nation-states. In this analysis, the possibility that different approaches to bordering are adopted at different times and by different groups must be kept in mind.

Between Øresund and the Mälaren Region

The border between Danes and Swedes has a complicated, and rather bloody, history. It has been estimated that a state of war between the two kingdoms has existed for no less than 134 years (Lindeberg 1985:7).[4]

After the decline of the Norwegian kingdom in the Middle Ages, and later after what was to become known as the Kalmar Union (between the kingdoms Denmark, Norway, and Sweden) was falling apart in complicated processes in the early 1500s, two centers of power existed in the North. Other powers, not least the Hansa city Lübeck and the Netherlands, intervened and strongly affected political developments in the region. Religious transformations were also important, as both centers during the break up of the Kalmar Union went through

---

3   This is not only true for state borders. It is also true, for instance, for the border between countryside and town. Until liberal reforms in the mid-nineteenth century in the Nordic countries, this was a line along which duties were paid and which separated zones in which different privileges to trade and craftsmanship existed.

4   The history of the border in the Øresund is here described from ordinary Danish and Swedish historical works and bibliographical encyclopedias. A colorful, but rather reliable overview can be found in Lindeberg (1985). A Skånian, regionalistic overview is Röndahl (1986, 1993). No discussion of relations between the Danish and Swedish kingdoms can be carried out not recognizing the substantial contribution by Knud Fabricius (1906a, 1906b, 1952, 1958). His work, thus, is also part of the foundation for this presentation.

Figure 1.1   *The Stockholm Bloodbath* **by Hjalmar Mörner (1794-1837)**

Notes: The Stockholm Bloodbath in November 1520 became a decisive episode in the final decline of the Kalmar Union. The Bloodbath was a drama in which two factions of nobility from the Mälar region clashed; one of which supported itself on the Union king from the Øresund area. The result was the execution of about ninety individuals. Subsequently, Gustav Vasa skillfully depicted the actions of Christian II as unChristian and thereby legitimized the powers he obtained for himself as a new Swedish king. Whether Christian II or archbishop Gustav Trolle was the main perpetrator has later been discussed intensively as summarized by Ingrid Markussen (i *Omstridte spørsmål i Nordens historie IV* 1973:31-37). Here the event is interpreted by Swedish artist Hjalmar Mörner (1794-1837).

Source: © Nationalmuseum, Stockholm.

Lutheran Reformations. One Northern power center mastered the agricultural regions around Øresund, controlled and imposed duty on trade into and out of the Baltic, and controlled the North Sea and North Atlantic coastlines. This center eventually emerged after internal wars and conflicts as the double monarchy of Denmark-Norway, ruled by a victorious fraction of the royal family, personified first by Fredrik I, and after new conflicts, not least the so-called *Grevens Fejde* (The Count's Feud), later by his son Christian III. The other emerging power center was in the region around Mälaren (the big lake where Stockholm is situated, at its outfall into the Baltic), had relations across the sea to the east and southeast, and possessed rich resources of strategic metals. This center eventually solidified as the kingdom Sweden (which included provinces in today's Finland) and was ruled by a new royal family, which was first headed by Gustav Vasa. Harald Gustafsson has convincingly argued that the disintegration of the Kalmar Union was only one possible outcome of political processes that, parallel to what happened elsewhere in Europe, strengthened the central authorities of the composite state and centered more of this power with the king (2000a).[5]

Through the subsequent two centuries, the two centers in the North were antagonistic poles in a struggle over influence in the Baltic sphere. Mostly the Swedish kingdom was on the offensive, while the Danish-Norwegian monarchy to the best of its ability tried to defend itself. Due to crucial economic and strategic interests in the Baltic, other European powers constantly involved themselves in these struggles as they supported soon the one and then the other center of power.

When the processes that dissolved the Kalmar Union ended, the two kingdoms for some time had partly complementary interests. Both new dynasties wanted to solidify their control of their lands in order to establish internal stability; both tried not to be embraced by Lübeck. In 1541, the kings concluded a strategic confederation for 50 years. Yet, one ought not to see this as a result of cordiality. Rather, the Confederation of Brömsebro must be regarded as an attempt to balance their mutual distrust. However, as the two kings recognized each other, the Kalmar Union can be said through the confederation to be definitively passed to history (H. Gustafsson 2000a).

Peace did not prevail. The political and geographical circumstances of the Baltic sphere doomed the two centers to permanent conflicts of interests, and thereby to war. Encircled by the possessions of the Danish-Norwegian king the lands of the Swedish crown were denied free access to the world market. Through his strong fortifications at Kronborg (in Helsingør) and Kärnan (on the opposite, eastern side in Helsingborg), the Danish king was in control of the only – and narrow – entrance to the Baltic.[6] To promote the trade of his countries the Swedish king was forced to try to break this containment. The Danish-Norwegian king, logically, was as determined to prevent the Swedish king from succeeding to maintain his weight in the Nordic balance of power. The result was a series of hostilities that kept the border between the kingdoms warm. In these wars, the border regions were often ravaged. Misery repeatedly was inflicted upon common people in towns and countryside. The Nordic Seven Years' War, 1563-1570, was followed by a period of relative tranquility. This was not least because the Swedish crown was preoccupied with expanding its influence in the eastern part of the Baltic sphere.

---

5   It is of course important to note that Harald Gustafsson (2000a) discusses alternatives in terms of territories and the number of state units, not in terms of alternative logics behind this directional development towards the centralized state, which is taken for granted. For the purpose of this text, it is the oppositional logic of the political-territorial development in the North that is in focus, not an understanding of the genesis of the modern state; see for instance Hechter and Brustein (1980).

6   The two alternative connections from the Baltic to the North Sea were not useable. The Great Belt was very difficult to navigate as it was filled with grounds and banks, and the Little Belt was blocked to north-south traffic due to its strong currents.

Bordering

Figure 1.2   Map of the Danish and Swedish kings' realms, 1563-1720

Notes to Figure 1.2: Map indicating changes to the borders between the lands of the Danish-Norwegian and Swedish kings from the dissolution of the Kalmar Union until the Great Nordic War (employing current Danish practices of naming the different conflicts):

• *Nordic Seven Years' War* (1563-70). No changes to the borders.

• *Kalmar War* (1611-1613). No lasting changes to the borders. However, Christian IV kept Älvsborg with Göteborg (Gothenburg), Nya and Gamle Lödöse as well as seven districts in Västergötland as a pawn until the payment of a one million *Riksdaler* war restitution was made in 1619. The young Gustav II Adolf was additionally forced to abandon any demands on his lands reaching the arctic North Atlantic.

• *Torstensson's War* (1643-1645). At the *Peace of Brömsebro* Danish-Norwegian king Christian IV ceded to the Swedish king for good the Baltic islands Gotland and Saaremaa, the two Norwegian landscapes Jämtland and Härjedalen, and additionally Halland as a pawn for a period of 30 years.

• *First Carl Gustav-War* (1657-1658). At the *Peace of Roskilde* Frederik III ceded to Carl X Gustav the Danish landscapes Skåne, Halland (this time for good), Blekinge, Bornholm and Ven as well as the Norwegian Bohuslän and Trondheim Len (except the northernmost parts of the country which was thus cut in two).

• *Second Carl Gustav-War* (1658-1660). At the *Peace of Copenhagen* the Swedish king returned: Trondheim Len and Bornholm to the Danish-Norwegian king.

• *Skånian War* (1675-1679). No change to the borders.

• *Great Nordic War* (1709-1720). No change to the borders.

Only one war involving the Danish king in the period 1563-1720 did not include hostilities with the Swedish king – the *War of the Emperor* (1625-1629). This war is excluded from the list, and it did not impart any changes to the borders. However, it meant a dramatic change in the balance of power in the Nordic realm as the Danish king, Christian IV, at the battle of Lutter am Barenberg (in what is today Germany) on August the 18th 1626, lost his army and consequently, his land military potential. The Swedish king, Gustav II Adolf, was much luckier with his involvement in the European *Thirty Years' War*, which is basically seen as a war between Protestants and Catholics. Gustav II Adolf secured Sweden's position as a major power in the Baltic until his successor, Carl XII, was defeated in the battle of Poltava (in what is today Ukraine) in 1709.

The seventeenth century became the century of wars around Øresund as the Swedish king demonstrated his aspiration to push forward to the West Coast and uncontrolled access to the world market. He founded a new merchant town, Göteborg (Gothenburg), on his only piece of land at Kattegat, where otherwise the Danish province Halland and the Norwegian Bohuslän blocked his access. At the same time, he tried to stretch his influence to the shores of the arctic North Atlantic. The war, which was the result of his efforts from 1611-1613, became nothing but an episode that could not obstruct the progress of the time in both kingdoms. New manufactories and trading companies were established, and new towns for trading and defense saw the light of day: Christianshavn (at Copenhagen), Glückstad (in Holstein), Göteborg, Kristiania (today's Oslo), Kristianstad (in northeastern Skåne), and Kristianopel (in eastern Blekinge).[7]

---

7   It is no coincidence that many of the towns are called Christian/Kristian. They are named after the Danish king of the time, the very much founding and building Christian IV.

It is one of the ironies of history that the Danish-Norwegian and the Swedish king were not at war with each other when the balance of power conclusively tilted to the advantage of the latter.[8] Both kings tried to extend their influence south of the Baltic. The Danish king, Christian IV, got his second son, Frederik, inserted as lord in the dioceses Bremen and Verden south of Holstein.[9] He thereby scrupulously stepped on the toes of the Duke of Holstein-Gottorp. The Swedish king, Gustav II Adolf, whose mother was the daughter of the duke, in the beginning had more interests nursing the eastern part of the Baltic sphere, not least in Poland. Of the Nordic kings, Christian IV entered first into the Thirty Years' War, 1618-1648. In 1626, his army was defeated south of Hanover and in 1627, the imperial army occupied all of Jutland. The peace treaty of 1629 forced Frederik to surrender Bremen and Verden. Gustav II Adolf in 1629 concluded a truce with Poland and entered in 1630 the wars over the German lands with considerable more success. When Gustav Adolf was killed on the battlefield at Lützen in 1632, he had subjugated more provinces and left a kingdom that was recognized as a European super power. His death, though, weakened the Swedish position. Consequently, Danish prince Frederik in 1634 was again inserted for a period as lord in Bremen.

The structural conflict of interests between the Danish-Norwegian and the Swedish kingdoms endured. Swedish advances along the southern shore of the Baltic, supported by an alliance between Sweden and the Netherlands from 1640, threatened to exclude Danish trade from the Baltic. Raising the duties paid by ships passing through the Danish kings waters at Helsingør turned out to be a very bad course of action taken on the Danish king's part. Warnings from the Danish king's envoy in Stockholm were ridiculed at the court in Copenhagen; in December 1643, one of the Swedish king's armies crossed the border to Holstein from the south and at the same time, another fell into Skåne. Under pressure from the Netherlands and promoted by French envoys, the Peace of Brömsebro was concluded 1645. The Swedish king's superiority was converted into extensive alterations of borders, which broke the Danish-Norwegian king's containment of the lands of the Swedish king. After the Thirty Years' War ended in 1648, it was the lands of the Danish king that were encircled. Dynastic connections between Stockholm and the Duke of Holstein-Gottorp were renewed and to the south, Sweden acquired several provinces: Bremen and Verden, Wismar, (Swedish) Pomerania, and Rügen.

It was obvious that interventions by external powers would be decisive in a forthcoming war. Under this impression, Copenhagen managed in 1649 to negotiate a defensive treaty with the Netherlands. In 1654, the Swedish queen

---

8   Relations between the two kingdoms were tense. To monitor his foe, the Swedish king for the first time in this period deployed a permanent representative in Denmark. From his reports it is clear how the relations between Denmark-Norway and Sweden were spun into the interests of other powers (see Tandrup 1971).

9   Frederik was later to become Danish-Norwegian king, Frederik III, as his older brother, Christian, died before he could succeed their father.

Figure 1.3   *Karl X* **by Nils Kreuger (1858-1930)**

Notes: Karl X directs his army to cross the ice on the Great Belt. This event resulted in the reconfiguration of the Nordic political geography and has been depicted by numerous artists, including the king's contemporaries Johann Philipp Lemke and Erik Dahlberg. Here is a less respectful representation by Swedish artist Nils Kreuger (1858-1930). Kreuger made a series of such royal portraits calling them "historical backsides."

Source: © Nationalmuseum, Stockholm.

Christina abdicated. Her throne was entrusted to Carl X Gustav, who immediately involved himself in hostilities in Poland. From Copenhagen, the situation seemed promising and in 1657, Frederik III attacked. The defensive treaty between Denmark and the Netherlands, therefore, was not activated. Frederik III's army soon conquered Bremen. However, the army of Carl Gustav, already well drilled, swiftly disengaged in Poland and marched towards Jutland. Soon the whole of Jutland was occupied. Beyond the battlefield, the fortune of war turned against the Swedish king. While he established a formal alliance against the Danish king with his father in law, the Duke of Gottorp, at the same time Poland, Austria, and Brandenburg concluded an offensive alliance against Sweden. From a strategic point of view, Carl Gustav was in a dangerous if not hopeless situation. From the south, he was threatened by the alliance; he could not escape to the north or east as the Danish-Norwegian navy blocked his path. Then, an atypical cold winter sided with the Swedish king. On January 29, 1658, his army set in motion across the frozen waters from Jutland to Funen and onward via the smaller islands Langeland and Lolland to Zealand, which they reached on February 11. Peace negotiations were immediately launched, mediated by envoys from England and France. At the Peace of Roskilde in February 1658, Carl Gustav's daring *blitzkrieg* was converted into a mutilation of the Danish-Norwegian kingdom.[10] Frederik III, who initially engaged in warfare to regain Bremen and Verden, had lost. A new border was born, and it followed Øresund.

However, Carl Gustav, soon regretted that he had not completely eliminated his opponent. Disagreements over the interpretation of details in the peace treaty became a reason, or at least a convenient excuse. The Swedish army landed once again on Zealand in August and advanced towards Copenhagen, which was besieged. While most of remaining Denmark soon fell to the Swedish army, Frederik sat well kept behind the fortifications of Copenhagen, which during the summer were reinforced and very much enhanced. Furthermore, as the Swedish king this time was the aggressor, the defensive treaty with the Netherlands was activated, and the besieged could therefore expect relief from the outside. In September, Trondheim, the main city of northern Norway, was liberated. At the same time a huge army from the allies of the Danish-Norwegian king, Poland and Brandenburg, was advancing north in Jutland. This did not result in relief for the sorely tried peasants and townsmen against whom one occupier behaved very much as the other. In October, a flotilla from the Netherlands forced its way through Øresund and relieved the besieged capital and in December, people on the island of Bornholm, located in the Baltic, succeeded in insurrection and turned the island back to Frederik. Increasingly isolated, Carl Gustav was forced to try an all out assault on Copenhagen. The attack, which came on the night between February 10 and 11, 1659, failed. In the town, there were a dozen casualties. Outside they

---

10 Carl Gustav Weibull offers a comprehensive picture of the negotiations that resulted in the Peace of Roskilde (1908). Askgaard treats the political and military developments 1654-1660, in particular the naval movements (1974).

were counted in thousands and included both soldiers from Carl Gustav's army as well as Zealand peasants who had been forced to walk in front of the attacking troops as shields. The Netherlands, England, and France were worried about the development and met in The Hague to agree on a compromise. Both warring kings rejected using the Peace of Roskilde as the basis for a new agreement. Frederik was promised the return of some of the provinces lost at the Roskilde peace accord as well as aid from the Netherlands to get rid of Carl Gustav; he accepted. In November 1659, the Swedish forces on Funen were crushed. Around Christmas, Carl Gustav went home to Sweden and immediately launched an attack on Norway. However, the Swedish king died and a path was opened for the mediation of the western powers. To the grief of Frederik III, the border in Øresund was confirmed with the Peace of Copenhagen in 1660.

The border was in the following decades not necessarily in accordance with the desires of the alternating rulers. Therefore, more hostilities followed in which the fortune of warfare often changed sides even if it was arguably the Danish-Norwegian king who generally fared the best. The Skånian War, 1675-1679,[11] and the Great Nordic War, 1700 and 1709-1720, did not cause any additional changes to the border between the southwestern and northeastern centers of power in the North; the border remained in Øresund. Eventually the kings abandoned warfare as a means to change this situation. The states with the power to decide had their own economic and strategic interests to consider. In the different conflicts, varying German states, the Netherlands, Poland, England, France, and Russia saw it as a manifest advantage that Denmark was found on one side and Sweden on the other side of the entrance to the Baltic. Thus, neither the hostilities in 1788 nor any in connection with the Napoleonic Wars had any consequence for the border. That the Øresund today is a cold border is thus not least caused by a prolonged chill first imposed by European powers. The determination of the international powers is evident. Due to the loss of territory and the war, the Danish-Norwegian king was seriously weakened in 1658. As support, the international powers agreed to let the Danish king continue to collect a toll from their merchants when they passed Helsingør even if the other coast of the waterway now belonged to a different kingdom. This Øresund toll on ships and cargo was not eliminated until the 1850s, when the US government declared that its merchants would no longer pay.[12] The international powers, thus, were seriously concerned with movement through Øresund. They regarded war between the Danish-Norwegian and Swedish kings and other kinds of cross-border hostilities as unwelcome attempts to keep the border warm. Those who had acted as midwives in 1658 and godparents in 1660 did not intend to accept any challenges to the border they had initiated, as such heating of the border held the potential to lead to a relocation of the established

---

11   The development after 1660 and until the Skånian War as well as the military history of this conflict and the political situation is treated in Askgaard and Stade (1983).

12   The Øresund toll was abolished from 1857 through an international treaty, and the Danish state was paid a once-and-for-all compensation by trading nations for its removal.

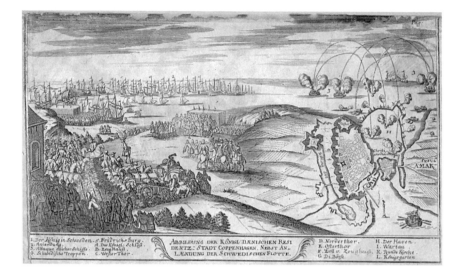

Figure 1.4   The Swedish king's forces attacking Copenhagen in August 1700

Notes: In the spring of 1700, there were skirmishes in Schleswig and Holstein, where the Duke of Gottorp, the brother in law of Carl XII, tried to gain an upper hand against Frederik IV. A fleet from Sweden, England and the Netherlands blocked Copenhagen and also bombarded Christianshavn (today part of central Copenhagen) and the Danish-Norwegian navy. When the Swedish king's army landed on Zealand and threatened Copenhagen in August, a peace was soon concluded under pressure from England, the Netherlands and Lüneburg. This is a contemporary graphic rendition of the events on Zealand.

After the Swedish defeat at Poltava 1709, Frederik IV invaded Skåne. However, after losing the battle of Helsingborg, on March 4, 1710, the Danish army for the last time left Skåne. The remainder of The Great Nordic War occurred outside the Øresund area. As seen from Copenhagen, the most important result of the war was that the great powers guaranteed that Gottorp no longer would be allowed to act in alliances with Sweden against Danish interests. Thus, a perpetual threat to the kingdom, and in particular to Jutland, had been removed.

Source: Reproduced with permission from the National Library of Sweden, Maps and prints, HP C XII A.26. Duplication: The National Library of Sweden, Stockholm.

dividing line. To maintain their own interests in movement in and out of the Baltic, they wanted this particular border to be kept cold, and if this chill was imposed by the international powers, it was to a large extent financed by the toll paid to the Danish king by their merchants.

It is too simplistic to regard the peacefulness of the Øresund border as only a result of pressure executed from abroad. One consequence of the wars around the Baltic in the 1600s and early 1700s had been that the two opposing kingdoms had worn out their offensive means, in part because they so persistently used them against each other. Furthermore, as time passed, the structural contrast between the two states gradually disappeared as they developed dissimilar roles in the international economy. Their resources and ways of utilizing them became quite different. This led to fewer feasible confrontations in the fields of economy and trade as well as to a corresponding reduction of possible ways of cooperation between the two countries.

Line or Ditch?

In general, natural borders do not exist beyond borders that insist on their own naturalness. Borders can only be understood in relation to the kind of states they separate; borders therefore must be understood as expressions of changing interpretations of history (Febvre 1973). Borders between states have not always and in all places been perceived of as *lines*. The demarcation of borders, and with that the acknowledgement of the rights of particular states to taxation and other benefits, seems at the earliest to have happened in regions where fixed settlements and cultivation were common. Prior to the making of a particular line, waste areas were not assigned to any power center. However, rather than understanding these as No Man's Land (Sahlins 1998:37) they ought to be understood as Every Man's Land in which centers of power mutually recognized, albeit not always with enthusiasm, each other's economic and strategic interests.

Thus, changing locations and understandings of border have been adopted through the times. The above short account of the birth and confirmation of the border in Øresund is necessary as background but it does not explain how individuals and groups conceived of the border. Nor does it explain how and why people and groups (re)acted (or not) to the different developments. To understand Øresund as a particular border, it is therefore necessary to consider how it was a border when it was born and confirmed, how it developed, how people reacted to it, and how people learned to live with it.

Even though the location of the border between Øresund and the Mälaren region was known and marked, and even if the populations were taxed by the king who happened to rule their local area, the states were not present at the line prior to 1658. Most fortifications were located behind the borders on the coasts, where trade routes and towns made the presence of the states desirable from other economic and strategic considerations.[13] The marked border that in the century before 1658 separated Halland, Skåne, and Blekinge from the adjoining provinces of the Swedish kingdom was not an impenetrable wall. Borderlines only existed for soldiers and princes – and for them, only in times of war (Febvre 1973:214).

---

13 Arguably this was changed when the Danish king founded two new fortified cities in the early 1600s, Kristianstad, in northeastern Skåne, and Kristianopel, in the far eastern corner of Blekinge.

Figure 1.5   *Differences over borders between Sweden, Denmark and Norway,*
solved by the kings Inge I Stenkilsson, Erik Ejegod and Magnus
Barfot in Kongelf, year 1100 AD by **Hugo Hamilton**

Notes: The border between the territories of the Danish and Swedish king was in the middle
of the 1000s marked with six rocks: four at the border of Halland, one at the border of Skåne
and one at the border of Blekinge, the latter in Brömsebro where meetings between the two
states also occurred. Elsewhere, borders were marked much later. Thus, it took centuries
to establish a line in the northernmost areas. A border between the interests of Sweden
(later Finland) and Novgorod (later Russia) was established as colonization and settlements
spread: in Karelia with the Treaty of Nöteborg 1323 and in Savolax with the Peace of
Teusina 1595. Agreements on linear divisions between Finland and Russia came in 1814,
and between Norway and Russia in 1826. Before the lines were established, Sami groups
for centuries were at times taxed by all sides: Sweden, Norway and Russia (see Gallén
1968, Gallén and Lind 1991, and articles "gräns-" and "rigsgrænse" in *Kulturhistoriskt
lexikon för nordisk medeltid*).

Source: Lithograph by Hugo Hamilton, from: Hamilton 1830; photo courtesy of Björn
Andersson, Historiska Media.

Until 1658, the inland was not cut off from the coasts just because the areas obeyed different kings. The river valleys through Halland and Blekinge were good and much used corridors of transportation and trade to the extent that it is reasonable to talk about the area across the border as economically well integrated before the border was moved (Bergman 2002:91-108). In this light, the moving of the border to Øresund can be perceived also as in agreement with the development of transportation technologies. The vast forests on both sides of the old border, which were difficult to travel through, became over the centuries penetrated by roads and later by railways. Water and mountainous areas, which earlier promoted contacts, were transformed into obstacles. Still, this development was only slowly taking off in the seventeenth century. What later can be seen as a sensible development cannot be seen as a reason for the events at that time.

The line between the kingdoms was for a long time hardly even unambiguous in political terms. It happened that authorities in one of the kingdoms summoned representatives for the population on the other side of the border when matters of mutual interest were determined (Åberg 1960:14, H. Gustafsson 2000a). The powerful political structures of the medieval church had not necessarily the same demarcations as the structures of the states. For instance, Gotland was a province obeying the Danish king until ceded to the Swedish in 1645, yet it was a part of Linköping diocese (in a town in central Sweden of the times). Similarly, powerful noble families could have possessions both in the Danish-Norwegian and in the Swedish kingdom and play leading roles in both states. When the Kalmar Union in 1450 was again in trouble and negotiations were necessary, two sons of the nobleman Aksel Pedersen Thott sat at the table – on opposite sides. Thus, the establishing of the Kalmar Union must also be understood in light of the existence of a Nordic high nobility.[14]

Approached from the point of view of the distribution of cultural artefacts, the picture of the border becomes even more complex. Surveying distributions of cultural elements was earlier the tool and to some extent the aim of the science of Ethnology.[15] The results of the work, as found in atlases, show many elements unequally distributed within Halland, Skåne, and Blekinge and thus, chopping the area into lesser regions. Moreover, many elements combined these regions with areas on the other sides of political divisions. The northeastern, forested region of Skåne shared features with the adjoining Swedish province of Småland

---

14   See for instance the map of Arvid Trolle's possessions around 1500 (Larsson 1974:105). One of the crucial prerequisites for the Kalmar Union was probably that it was stated in a Peace of 1343 that a nobleman could live in one Nordic state without loosing his right to control his possessions in the others (Rystad 1965:33); this situation prevailed also throughout the processes that ended the Kalmar Union (H. Gustafsson 2000a:100).

15   This science, known as *Volkskunde* in Germany, exists mostly in some northern and central European countries. While today often recognized as "anthropology at home," it earlier utilized methodologies very different from most mainstream anthropology. For a general overview see Rogers, Wilson and McDonogh (1996).

while the southwestern plain shared elements with Zealand on the other side of Øresund. This was true, for instance, for the thoroughly studied types of villages and houses including details as fireplaces and ovens.[16] The distribution of cultural elements and techniques reflected ecological conditions and not political divisions. Skåne in general was parted in three ecological regions with different methods of cultivation and fencing. Between the plains in the southwest and the forest in the northeast, a diffuse zone of woodlands running from the northwest to the southeast characterized the landscape (see Campbell 1928, 1936). Both at the border between the kingdoms prior to 1658 and at Øresund subsequent to 1660 it is impossible to point to any coincidence between a cultural criterion and the political border. Consequently, there is no foundation for perceiving Halland, Skåne, and Blekinge, or Skåne to itself, as separate cultural provinces. Moreover, it is impossible to establish cultural criteria by which the three political provinces with any significance were more or less associated with the rest of the Danish or with the Swedish kingdoms. As mentioned above, in the field of religion, both kingdoms were Lutheran; in 1658, due to nearly simultaneous reformations, each had established a state church.

In his discussion of "frontière," Febvre finds that borders between contemporary nations are understood as *ditches* separating people of different cultures. He finds that mores and feelings support this notion of difference (1973:214). By and large, historical sources do not contain evidences that people at Øresund in the 1600s reflected on how and when they belonged to a nation.[17] This silence made it very easy to claim later such sentiments as well as individuals' national motivations, which one often will find in writings and public debate attempts to establish a relation between the moving of the border to Øresund and a perceived imagination of a ditch between the Danish and Swedish nation. Without being distracted by the reflections of historical persons, the border provides a raw material from which national villains and heroes have been extracted; as Benedict Anderson emphasizes: "the silence of the dead was no obstacle to the exhumation of their deepest desires" (1991:198).

Following Febvre and the discussion above, when it can be demonstrated that historical persons and groups did act when the border between Øresund and the Mälaren region was moved, it must be explained from the conditions they were living under and the kind of state of which they were part. That their reactions

---

16   See the discussion in Sigfrid Svensson (1958). For summaries of the studies of houses, see Bringéus (1979) and Stoklund (1972). The method of mapping is illustrated very clearly through the works of Erixon (1957) and Zangenberg (1982).

17   Fabricius is discussing this in more places (1906b:105f, 191f, 250f). For instance, he finds that the "national" was aroused after 1670. But he also referrers to existing economic reason for the tensions between the people in Skåne and the Swedish authorities he discusses, not least the authorities' wish for an extra-tax for the military in 1673 (1906b:235ff). Bergman in more places discusses such notions; he also criticizes Fabricius for being blinded by his own time's understanding of nationhood (2002:263).

later have been explained otherwise should not obstruct an attempt to understand their actions in relation to their contemporary context. If no ditch existed between them and their neighbors, an alternative foundation for understanding what they did must be established. The following sections will analyze bordering actions taken by individuals and groups on different arenas in relation to the moving of the border in the seventeenth century. From their relations with politics on both local and state levels, and from the impacts their practices had on these levels, it will be unveiled how the structures and functions of the state were perceived at the time.

From this perspective, the objective of this entire book can be said also to discuss, in broad outlines as well as in ethnographical detail, how, when, and for whom the line between Øresund and the Mälaren region eventually was transformed into a ditch in Øresund. It is not any ditch but a ditch which became filled with difference, feelings, and morals through cultural processes associated with bordering practiced within different frameworks: everyday life, the market, the state, and social movements (Hannerz 1992:40ff).

Aliens, Drifters – and the Villain

One of the conditions of the Peace of Roskilde, 1658, was that two thousand horsemen immediately should be transferred from Frederik's to Carl Gustav's army. The 25 companies transferred were not complete – only 1650 men were delivered. A part of these had been recruited in Skåne while others were from Zealand or enlisted abroad (Johannesson 1981:222). Weibull claims that only 936 horsemen were received by the Swedish authorities since some deserted (1908:113f). This transfer was not unusual. In the seventeenth century, armies everywhere obtained enforcements through raising troops in occupied areas. During the wars 1657-1660 many of Frederik III's subjects were forcefully drafted for service in the Swedish armies south of the Baltic. Even if captives and peasants were enrolled in armies, the presence of such aliens was not always due to coercion. Many males traveled of their own free will abroad and enlisted (see for example O. Højrup 1963:414).

One of these traveling soldiers, who later became famous in Denmark, was Svend Poulsen, later nicknamed Gønge and today known as *Gøngehøvdingen* (The Göinge chief; Göinge consist of two districts in north-eastern Skåne). The stories about him are spun from myths and his biography is in doubt on many points.[18] Most certainly he started his military career with Christian IV's engagement in the Thirty Years' War and became a non-commissioned officer. As such, he subsequently served some years with the Prince of Orange before he

---

18 There are many uncertainties in and disputes between those who have studied the life and efforts of Svend Poulsen. See for instance Hansgaard (1956), Kjær (1992), and Wagner (2003). The picture drawn here should find support both from primary sources as well as from most writings about this historical figure.

returned to Denmark and became a "halberdier and royal servant." In the mid-1630s, he married a widow in the small town Knäred in Halland where he drilled the peasant militia of Laholm Len.[19] In the war 1643-1645, he was a recruiter and a non-commissioned officer in Skåne. After peace was re-established he stayed in Knäred, which was now an area controlled by the Swedish king, and continued functioning as a non-commissioned officer. From 1653, he was supposedly once again a servant of the Danish king. In 1657, he was described in a Swedish letter as "the deserted perjuring townsman and Hallandish rebel – Svend Poulsen" (Hansgaard 1956:55). At the outbreak of the war 1657 he was promoted to captain in the Danish king's army and enlisted dragoons. With alternating success his company fought in northern Skåne. During the winter 1657-1658, he was partly in Skåne, partly on Zealand. When hostilities resumed in August 1658, Frederik III ordered him to organize resistance in southern and eastern Zealand. It is from this irregular, but apparently quite successful warfare that the myths about his person stem (in the following chapter these myths will be discussed more). Later, Sven Poulsen appears again as a regular army officer during the reconquering of Funen. As payment for his services and outstanding debts, Frederik gave him after the wars two farms in Zealand. Svend also continued his military activities. He was both a spy in Skåne and a principal officer for the peasant militia in Zealand. In the Skånian War, he participated as a captain and from 1676, as a major and recruiting officer in Skåne. The time and place of his death are not known, but it is thought that he died around 1680.

A man such as Svend Poulsen could go a long way though a military career even if he apparently failed miserably in managing his farms. Real success though, required a person to be not only an officer but also a nobleman. The histories of the seventeenth century overflow with noble career-officers who drifted from army to army to achieve success. As long as they were loyal to their immediate master, there were no problems in changing and letting ambition steer the course.

The wars around Øresund contain illustrative examples of persons who succeeded. One of them was the organizer of the defense of Copenhagen while it was under siege by Carl Gustav's army. Hans Schack participated in Christian IV's campaign 1627-1629. Thereafter he took service in the Swedish king's army. From 1636 to 1650 (except 1647-1648 when he enlisted men in his native Saxony-Lauenburg), he was an officer in French regiments and fought with them in campaigns in today's Italy, Spain, and the Flanders. In 1650, he offered his services to Frederik III whom he knew from his time as a recruiting officer. However, this offer did not work out and he became a civil servant in Saxony-Lauenburg and from 1656 commandant of the free city of Hamburg. In the fall of 1657, he was offered enough compensation to join Frederik and towards the

---

19   Len (in Danish) and län (in Swedish) means "entailed estate." Historic len/län were still more or less recognizable as administrative units in Scandinavia until reforms in the early 2000s. The word is still in use in Swedish, while in Denmark it has been substituted by "amt." A reasonable current translation used here is: county.

end of the year, Schack arrived in Copenhagen where he, as commander-in-chief, became the driving force in the rearmament of the town's fortifications. Later, he was celebrated as the one who drove the Swedish army out of Funen. Even without being able in Danish, Schack became one of the pillars of Frederik's rule and after 1660, had great impact on the army and other matters. He became a Danish nobleman in 1657 and gradually gathered considerable properties.

Other individuals who became persons of consequence were the brothers von Arenstorff, natives of Mecklenburg. Carl had served different lords when in 1656, he took service in the Swedish army and became an adjutant. In January of 1658, he scouted the ice on Carl Gustav's path to Funen. Later when he felt passed over and lacking in promotion, he applied and received his discharge in 1661. Thereafter he took service with Frederik III and occupied several significant military posts. In 1674 he went to the Prince of Orange, but he later returned to the Danish king and died in 1676 from wounds received as commander-in-chief for Christian V in the Battle at Lund. His brother Frederik first took service with the Duke of Gottorp and subsequently, with Bishop Frederik of Bremen (the later Danish king Frederik III). In 1658, he was caught by the advancing forces of Carl Gustav and was trusted into the army as an officer. During the siege of Copenhagen, he was the adjutant for the Swedish king. After peace was re-established in 1660, he took service with Frederik III and occupied several high appointments. In the Skånian War, he refused promotions that would make him responsible for the outcome of the war. Despite this, he was blamed for the failure to relieve Kristianstad when a Danish force there was under siege by Carl XI. He escaped with fines and afterwards, became again assigned important military posts. He became a Danish nobleman and acquired several estates.

Numerous foreigners operated anonymous and unnoticed; the drifters later had the potential for becoming the raw materials from which the heroes of History were cast. This was only true though for those who succeeded. It could also end with an awful catastrophe. More than anyone else, Corfitz Ulfeldt has been assigned the role of the villain, who did not respect the ditch between nations and therefore committed treason. Ulfeldt was born into one of the foremost families of Danish nobility.[20] He received an excellent education, became a civil servant, and subsequently the chancellor for Christian IV. He also married one of the king's daughters. By virtue of his background and marriage, it was certain that he would play an important part in the political arena of the kingdom. In 1645, he negotiated the Peace of Brömsebro for Christian IV. Moreover, it was he who in 1649, returned from the Netherlands with the defensive pact that later proved so crucial. In 1651, Frederik III ordered an investigation into Ulfeldt's conduct as an administrator; consequently, Ulfeldt found himself compelled to flee. Thereafter he tried to counteract the Danish king

---

20 Heiberg has written an excellent biography of Ulfeldt (1993). However, his psychological approach to the life and circumstances of Ulfeldt and the complicated play around the sons-in-law of Christian IV (married to the half-sisters of Frederik III) falls short of explaining the historical context for the events.

Figure 1.6   Corfitz Ulfeldt and Leonora Christina, by I. Folkema, 1746

Notes to Figure 1.6: The nobleman Corfitz Ulfeldt married Leonora Christina, one of Christian IV daughters with his consort Kirsten Munk. With the other son-in-laws of the king, Ulfeldt became very powerful. However, his rapid ascent to power, association with both the state council and king (which often had diverging interests), and use of his position as an official to enrich himself immensely also gained him powerful adversaries. Despite Ulfeldt's success on behalf of the king as a diplomat, Frederik III, and in particular his queen, Sophie Amalie, loathed the ambitious and pushy Ulfeldt couple. These aversions became obvious when a woman (Dina) testified to non-existing plans of Ulfeldt and the king to murder one another. Due to these rumors and his dubious economic activities, Ulfeldt fled from Frederik III in 1651. He traveled to Swedish territory and took contact with Queen Christina whose abdication and travel to Rome he financed. In 1657, Ulfeldt took service with Carl X Gustav. In return for loans and services during the First Carl Gustav War, Ulfeldt was made Count of Sölvesborg in Blekinge and Herrevadskloster in Skåne. Soon he was accused of treason as he opposed the policies of Carl X Gustav and had been in contact with Bartholomæus Mikkelsen, the leader of a conspiracy in Malmö who helped Frederik III during the war. Ulfeldt was prosecuted in a Swedish court, but fled in 1660 to Copenhagen where Frederik III immediately had the troublesome nobleman and his wife, the king's half-sister, interned. They where first taken to the fortress Hammershus on the Baltic island Bornholm where Frederik III was already accepted as hereditary absolutist monarch and where the Ulfeldt couple, consequently, were beyond the powers of the state council. However, the couple's many close contacts with other royal houses made it difficult to keep them locked up and eventually they were allowed to settle on their estates on Funen. In 1662, they were permitted to go abroad as convalescents, but Ulfeldt immediately started again to conspire against Frederik III, who in 1663, sentenced him to death in absentia for treason against the king. Approximately at the same time, Leonora, while visiting London to recover old loans Ulfeldt had provided to the English king, ran into a trap and was taken to Copenhagen. Without any sentence, she was locked up for 22 years. Ulfeldt seemed to have vanished, so an effigy was symbolically executed in his place in Copenhagen. On the run, Ulfeldt died on a boat on the Rhine in 1664. Included in his death sentence was that Ulfeldt's property in Copenhagen be torn down and a monument to his infamy be erected on the grounds. The resulting square is today Gråbrødretorv in central Copenhagen. On this engraving from 1746, signed by I. Folkema, the traitor and his wife are seen with the square and monument to the left.

Source: Reproduced with permission from the Royal Library, Copenhagen.

from Swedish territory. When Frederik in 1657 attacked Sweden and therefore, the defensive pact with the Netherlands was not activated, Carl Gustav called upon Ulfeldt as both an adviser and lender to finance the war. Ulfeldt took service with the Swedish king and represented him during the negotiations towards the Peace of Roskilde. Subsequently, he was dissatisfied with the role he was assigned in the administration of Skåne, which he envisioned as a kind of republic of nobility (with himself as the *primo inter pares*) in a personal union with the kingdom of Sweden. He renounced the politics of Carl Gustav and was accused of treason. In 1660, Ulfeldt fled from Malmö to Copenhagen where he submitted his services to Frederik. However, the Danish king quickly got the troublesome nobleman and his wife (his own half-sister) interned. In 1662, Ulfeldt got royal permission to travel abroad as a convalescent but soon connected with foreign lords. His plans and schemes were reported to Frederik III who started proceedings in the Supreme Court. In 1663, Ulfeldt was sentenced to death for treason against the majesty. However, he never was caught but died a haunted man; his grave is unknown. His wife, the king's half-sister was appropriated by the authorities and spent the rest of her life in captivity. Ulfeldt's acts have since been understood as expressions of mental illness – perhaps caused by syphilis. There is no doubt that his schemes were overstated and that he consistently exaggerated his own importance and prestige. It is moreover probable that his conversion to Catholicism in his youth caused more of his seemingly desperate deeds. To be a papist, even in secrecy, was dangerous in Lutheran northern Europe in the 1600s. His conversion, which was concealed to his contemporaries, of course, might also be the cause for his close relation with the Swedish queen Christina who abdicated and also converted.

Because he was accused in court several times, Ulfeldt as opposed to the other drifters was forced to articulate reasons for his actions. In his defense, he consistently maintained his right to choose, as a *free* nobleman, which lord he offered his services. Likewise, he could also retract his service when he wanted without betraying the king (Heiberg 1993:142, 242). That personal loyalty between lord and subject existed was not an odd part of Ulfeldt's understanding of the world (see Österberg 1995:171-197). The pathetic end to Ulfeldt's life was caused neither by his overstepping of any ditch between nations nor as has later been claimed, his transferring of loyalty from one lord to another. Rather, he acted in ways that made him both hated and vulnerable. His many changes of allegiances basically rendered him untrustworthy (see also H. Gustafsson 2000a:186, 302) and when he subsequently surrendered to those he already had deserted, his life indeed took a turn for the worse.

That a person in the seventeenth century shifted his loyalty from one king to another was not unusual but it usually was combined with the geographical relocation of the person, irrespective if the move was voluntary or forced. The results of the wars 1657-1660, however, were the opposite. A multitude of people stayed where they were and had to be integrated into the power structures of another kingdom. Reactions towards this can bear witness to the interests of the people living at the borders. Their reactions reveal if they were acting independently of the centers of power and illustrate how the states of that time were structured.

Conspirators

Shortly after the hostilities resumed in August 1658, the tenant of the limestone quarry in Limhamn near Malmö, Bartholomeus Mikkelsen, became a key-figure in secret communications between Copenhagen and Skåne.[21] Mikkelsen was related

---

21   The conspiracy in Malmö is discussed by Fabricius (1906a:98-130) and Ebbe Gert Rasmussen (1990).

to leading circles in Copenhagen and managed not to swear allegiance to Carl Gustav when the area was handed over in the spring of 1658. A number of Danish noblemen used Mikkelsen as an intermediary between themselves and bailiffs running their estates east of the Øresund. One of his contacts in Copenhagen was the royal chancellor Joachim Gersdorff. Mikkelsen also became the link to attempts to raise the population of eastern Skåne against their new lord; he additionally composed a plan to take the fortified castle, Malmöhus, by surprise and hand it back to Frederik. Although the Swedish authorities had confiscated his tenancy, he expected to gain it back if the area came under the Danish crown again. He gathered a group of like-minded men that included a couple of priests and some tradesmen of Malmö. They later stated that they partook because they wanted to restore the trade of the town, which was harmed by new duties and trade regulation imposed by the Swedish authorities, and to secure the same privileges for Malmö that Copenhagen had obtained during the ongoing siege. Thus, a restoration of Danish rule would benefit their economic interests. From the beginning of November, Frederik III sent letters encouraging the conspirators. They also were connected with a circle around the nobleman Ove Thott who had fostered an identical plan. Many noblemen were dissatisfied with what had happened to Skåne and refused to swear their allegiance to Carl X Gustav. As punishment, they risked confiscation of both their estates and their movables. Furthermore, noblemen in Skåne had lost their prescriptive freedom from duty on trade.

A comprehensive plan to take Malmöhus was developed. It included 40 hands to be furnished by the nobleman Otto Lindenov and soldiers to be shipped over from Copenhagen. Towards Christmas, everybody was prepared. An attempt on the night after December 18 failed on account of wind and ice as the soldiers could not get across Øresund. A new attempt on December 26 failed as the ship carrying the troops ran aground the small island Saltholm outside Copenhagen. Interest in the plan now diminished in Copenhagen. The many informed parties in Malmö made the conspirators vulnerable and the Swedish authorities realized that something was going on. All through February and March 1659, Mikkelsen prompted for again having troops put at his disposal. Then the conspiracy cracked. A brother-in-law of Mikkelsen submitted, for unknown reasons, a written explanation to the authorities. On April 16, the first persons were arrested. Through summer and fall, both the conspiracy in Malmö and the attempts to raise the population in eastern Skåne were uncovered. More people were sentenced to death through proceedings in which Carl Gustaf himself interfered but only Mikkelsen and two young, prosperous traders were executed on December 22. Other conspirators were fined and their possessions were confiscated. Ove Thott, who claimed to be "a servant of the king of Denmark," was exchanged with a prisoner in Copenhagen. The court accepted a similar arrangement from a noble lady although she had to surrender her estates and was taken into custody. In practice, the court presupposed that a personal relation of loyalty ought to exist between king and subject (at least as

**Figure 1.7   Kronborg occupied by Carl X Gustav's army, from a drawing
by Erik Dahlberg for Samuel von Pufendorf's opulently
illustrated volume on the history of Carl X Gustav**

long as the subject in question was a free, noble person).[22] Some of the sentenced
priests and important townsmen later moved across Øresund and anew gained
prominent positions in the structures of the Danish state.

Thus, this whole project had a dismal ending. While the conspirators planned
and waited in Malmö, people on the island of Bornholm in the Baltic took action
in a different manner. The events on Bornholm clearly illustrate how and by whom
loyalty could be activated in conjunction with the moving of the border between
kingdoms. The works of Ebbe Gert Rasmussen have presented in detail both
the local actions as well as their relations with the political arena of the Danish
kingdom (1967, 1972, 1982, 1985, 1990). The severe winter, which had formed a
bridge across the Danish Belts and had propelled the Swedish king to victory, also
delayed the transfer of control on Bornholm in the spring of 1658. It was not until

22   This case was of importance to the Swedish court case against Corfitz Ulfeldt in
1660, as he had been in contact with Mikkelsen. The significance of serving and swearing
allegiance is discussed by Fabricius (1906a:63ff).

Notes to Figure 1.7: Kronborg, the royal castle in Helsingør and the guardian of the Øresund toll, came under siege immediately after hostilities were reopened with Carl X Gustav's landing at Korsør in western Zealand on August 8, 1658. The fortress was surrendered by Frederik III's not very courageous officers after an intense barrage during which, as it can be seen in this print, also a part of Helsingør was destroyed.

A conspiracy to retake Kronborg was soon established (Rockstroh 1907, 1908). As in Malmö and on Bornholm, this conspiracy was led from Copenhagen; the initiative seems to have been taken by Corfitz Trolle (Rockstroh 1907:68), overseer of stables, who together with the gamekeeper of Joachim Gersdorff, Ludvig Stats, was sent to Skåne to initiate a rising (E.G. Rasmussen 1990:58, 73f). Initiatives to fight for Frederik III were taken by a very narrow circle of powerful figures in Copenhagen.

The histories of the conspiracies in Malmö and at Kronborg are very similar. The leading persons were promised positive attention from the king and the plans depended on the participation of soldiers from Copenhagen. That relatively many individuals knew about the plans and that the plans were discovered more or less by chance by Swedish authorities are also parallels.

During the subsequent court proceedings, attention was, as in Malmö, paid to whether an individual could be held to be in the service of one or the other king. Thus, Oluf van Steenwinckel was executed; he was occupied with improving the fortifications and after the fortress fell into Swedish hands in September 1658, took service with Carl X Gustav as a master builder. Others convicted were forced to pay very significant fines, including the priest Henrik Gerner who was also sentenced to death but was not executed.

Source: Pufendorf 1915:634; photo courtesy of Björn Andersson, Historiska Media.

the end of April that a new administration and garrison had been installed in the medieval castle of Hammershus. On May 24, representatives for the ranks were in Malmö to swear allegiance to their new king, Carl X Gustav. Through the spring and summer relations between the new administration and the population were relaxed; soldiers were drafted and taxes paid. Even though the Swedish kingdom was in a state of war and claimed considerable contributions, these did not exceed what earlier had been demanded on behalf of the Danish king. However, from the middle of November, relations between administration and population suddenly and vehemently deteriorated.

At the same time as the conspirators in Malmö were encouraged, Frederik III from his besieged capital turned to key persons elsewhere, promising royal favors if they could advance uprisings which would relieve pressure on Copenhagen. On Bornholm, the key figure was Peder Olsen, the mayor of the tiny town Hasle. He organized a conspiracy and manufactured a plan: the commandant of the island, Colonel Johan Printzenskiöld, would be invited to a party and taken prisoner. Then, as a hostage, he would persuade the garrison of Hammershus to surrender. After a quarrel between Olsen and Printzenskiöld on the morning of December 8, the coup immediately went into motion even if it was not fully prepared. The abrupt start of events was maybe due to Olsen's fear that the colonel would not attend the planned party on account of the quarrel. No plan for the unforeseen situation existed. The incidents now followed in a haphazardous order until the course of events had reached a point where they were driven forward by their own logic with no possibility for retreat.

After the quarrel, the colonel rode towards Rønne, the major town on the island where the conspirators desperately tried to gather a posse. The pursuers were no more than a few persons, and while they rode into Rønne they did not seek Printzenskiöld. However, a Swedish bailiff had spotted the armed group. When they learned this, the group had to take action. They found the colonel with the mayor of Rønne. When they led him out into the street, Printzenskiöld was shot, maybe because he tried to escape. Regardless of what happened, the original plan was thereby spoiled. During the evening and night townsmen and peasants from the area were raised and Swedish officials outside Hammershus were apprehended. On the morning of December 9, the antiquated fortification was besieged, negotiations were inaugurated, and towards the evening, the garrison surrendered. Only a few people had been killed and the island was now in the hands of the conspirators. Though improvised, the insurrection had been successful not least due to deficiencies of the Swedish administration and garrison.

Subsequently, the Bornholmers gathered an assembly of the ranks and composed a petition to Frederik III, encouraging him to again take the island under his protection and to restore his laws and privileges. On December 20, a group led by Peder Olsen sailed to Copenhagen with the petition. On December 28 and 29, a number of the traveling companions received royal favors. Some were appointed royal officials. For example, Peder Olsen was appointed judge of the island while another assumed his earlier positions as customs officer and scriber for the island's court. Thus, ensuring the favor of the king, eleven members of the delegation, all known for their active participation in the rebellion, signed the later so-called "deed of gift," which established Frederik as hereditary and absolutist king of Bornholm. On January 7, the delegation was back on the island. In an assembly of the ranks, their achievements were approved. Many participants in the assembly, including Peder Olsen, were among the representatives of the ranks who swore allegiance to Carl X Gustav in Malmö only eight months earlier. What the islanders had asked for in their petition was not what had been offered with the "deed of gift" and later confirmed by the king. What the activists on the local arena had not realized from the start was that they were involved with the king's businesses on the political arena of the kingdom of which they had no overview.

The letters from Frederik III to key-persons in November are expressions of an energetic effort to improve the military and economic situation following the arrival in Copenhagen of the flotilla with aid from the Netherlands. That the king chose Peder Olsen as one of his agents was not strange. As an official, he had an extensive network on Bornholm. In addition, he had been the spokesman for the islanders in conflicts with their earlier royal representative Ebbe Ulfeldt, who was a half-cousin of Corfitz Ulfeldt and married to a sister of Corfitz' wife. The mayor of Hasle consequently had aided Frederik III with his decisive showdown with the husbands of the king's half-sisters (the "party of brother-in-laws") who had restricted his political maneuverability.[23] Olsen also had good relations with the successor of Ebbe Ulfeldt as royal representative on the island, Joachim Gersdorff, who as royal chancellor also played a role in the conspiracy in Malmö. The royal chancellor still had servants on the island who played decisive but obscured roles in the rebellion.

The letter from Frederik to Peder Olsen of November 8 provoked a series of events starting with the wrecking of relations between the locals and the Swedish king's administration and ending with the improvised but successful rebellion. The activists on Bornholm were locals but the initiative for their actions was cast among the highest circles in the besieged capital. The military reasons for the letter from the king must have been clear to the small group of local men who knew about it. However, they could not have known of the other roles Bornholm played in the king's plans. The besieged Copenhagen lacked all kinds of supplies to continue fighting. After the Swedish blockade had been broken with the arrival of the flotilla from the Netherlands, the king approached the city of Lübeck asking for food, clothes, ammunition, and weapons. In a letter to his envoy in the town from November 3 and thus before his address to Peder Olsen, Frederik suggested that the merchants provisionally could have Bornholm as security for payment for the supplies. He also offered to obtain and secure them this pledge. The merchants of Lübeck, who had been in possession of the island as a pledge from 1525-1576, were not interested in this arrangement, which openly would be hostile towards the Swedish king. They were even so tactless that they leaked the proposal to the Swedish envoy in town. However, the Lübeck merchants still meant business and as payment for their supplies to Frederik III got favors for their trade on Bergen in western Norway.

The king's solution to the financial aspect of his supply problems puts his subsequent address to the Bornholmers in another light. It was however, not only in this respect that the islanders were ignorant of their role in the games of the king. At least from the moment they arrived in the besieged capital with their petition, the islanders became tokens for the king's ambition of becoming absolute, hereditary king of the whole kingdom. By having the Bornholmers petition rewritten as a deed of gift, Frederik achieved what he passionately pursued even if only in a small part of the kingdom. Earlier on, he had accomplished what he had wanted in the royal parts of the duchies Schleswig and Holstein and possibly he had also achieved absolute, hereditary powers in Norway. This was, however, disputed by the Danish state council. Frederik completed his efforts during an assembly of the ranks in Copenhagen in 1660. As with the deed of gift obtained from the people of Bornholm, he succeeded in staging the event so it appeared as if the ranks requested that he become their absolute, hereditary king.

---

23 The so-called "party of brother-in-laws" dominated during a period the state council (*rigsrådet*). This body of high noblemen had to approve all the king's decisions to make them valid. The party of brother-in-laws had been created though the politic of the king's farther, the late Christian IV, who sought to control the high-council through getting his illegitimate daughters married into the high nobility.

It is impossible that this development was in the minds of the Bornholmers when they composed their petition. Nevertheless, when they were presented with the deed of gift, written by the king's advisors in Copenhagen, they could hardly refuse to sign. They had received royal favors, and they could not return the island to Carl Gustav who possessed it according to the international treaty, the Peace of Roskilde. If the Swedish king was reinstated, the conspirators would be punished for not having kept the allegiance to which they swore. In addition, the same persons had earlier experienced that Swedish troops could advance violently. In 1645, the island was surrendered to the Swedish king by local officials, among them Peder Olsen. Afterwards the king's army plundered and extorted contributions from the islanders under threats of burning down houses and farms.

State Patriots

The history of the conspiracies can be read as the king's rapacious manipulation of the ignorant commoners whose leaders he managed through personal favors. Or it can be read as the willingness of the people to risk all they had for their nation. Both these interpretations, however, offer no adequate explanation for why these individuals involved themselves in bordering practices.

The kings had interests in drawing the borderline between Øresund and the Mälaren region for the above-mentioned military and trade-strategic reasons. However, the kings, and therefore the states, had other grounds for being interested in the distribution of territories and their inhabitants. First, territory meant the right to levy taxes, which not least were used to strengthen the power of the king through paying armies and symbolic conspicuous consumption at for instance, royal weddings and coronations. Second, territory gave the right to draft men for the military. And third, territorial rights provided control of legislation that should secure stability and the possibility of maintaining the two first interests. This trinity of considerations is apparent, for instance, in the investigations the new Swedish authorities started on Bornholm just after the take over in April 1658 (E.G. Rasmussen 1972:18ff, 32ff). In turn, the people could demand that the lord keep taxes and drafts reasonable and additionally, protect them with his power against assaults from foreign armies and domestic thieves, extravagant noblemen, and other injustices to and disturbances of their peace and livelihood.

Thus, between lord and population in a society organized according to what might reasonably be termed state patriotic principles, mutual rights and obligations existed.[24] This reciprocity was build upon the subjects' loyalty towards their lord

---

24 I choose here the concept of state patriotic for this state. This choice accentuates the mutuality in the relations between lord and subjects (see H. Gustafsson 1994b and Österberg 1989, 1992), it also does not contain any reference to any more or less naturally demarcated territory. Other possible conceptualizations would be territorial state, early modern state, or feudal state.

and on his ability to attend to their interests. The mutual connections and obligations were on different levels managed by royal officials. If they did not consider the conditions of the local people, an individual or a group could complain to levels higher in the hierarchy or directly to their king; this often happened. Thus, people had the possibility to reject extravagant demands and act out of their own interests either through negotiations with the center of power or through disloyalty.[25]

On the local level, structures of power existed that reflected the composition of the state. This enabled, for instance, a nobleman such as Otto Lindenov to promise that 40 of his hands would participate in a conspiracy in Malmö even though they probably knew nothing or very little about it. Consequently, some subjects were more closely associated with the lord than others according to their position in the hierarchy, because the role of their family, or due to previous personal exchanges of favors upwards and downwards through the structures of the state. The hierarchy of such a state was purely political. A demand for any emotional engagement did not exist. Thus, this state was dependent neither on internal cultural homogeneity (as illustrated by the drifters and their inability to speak the language(s) of the kingdom) nor the need to establish ditches of cultural difference against other states. Those who lived in the state did not have anything necessarily in common – except their relations to their lord.[26]

It cannot be demonstrated that the burdens of taxes and conscriptions on Bornholm dramatically changed after the Swedish king took over in 1658 but the policies of duties and trade must have hurt. Early in the islanders' petition it is mentioned that the Swedish authorities had restricted trade. From the spring, export of among other things barley and oats was prohibited and that caused a lack of money on the island. It is uncertain how long this ban existed but it is certain that a measure of this kind harmed particular parts of the population more that others. The prominent townsmen and officials of Hasle and Rønne who led the rising not only fought because of vague promises of royal favors but also because of other, personal interests for which to fight – their positions in the hierarchical structures of the state and their private economy. Had they plunged headlong into this endeavor on behalf of the nation, the petition would clearly have demonstrated a ditch towards Swedes. A moral, emotional, or cultural involvement in being Danish as opposed to being Swedish does not surface in the text, however, beyond a complaint of being separated from family and friends, which in a general sense replicated an ordinary way within patriotism of depicting society as a family with a father (*pater*) as the lord. The whole petition addressed the relationship, the authors as subjects had to their lord and king. And of course, this relation was even more pronounced in the subsequent deed of gift.

---

25 For an illustration, see Linde-Laursen (1989b) on milling and milling privileges as well as locals' ways of complaining about their situation.

26 Harald Gustafsson (2000a) makes a similar argument about the structures of the centralized states, with an increased focus on the lord, that develop in Europe in the 1500s.

Figure 1.8   Title-page for Arent Berntsen's "The Fertile Splendor of
               Denmark and Norway" (*Danmarckis oc Norgis fructbar
               Herlighed*) from 1656

Notes to Figure 1.8: Arent Berntsen's description of "the Fertile Splendor of Denmark and Norway" (*Danmarckis oc Norgis fructbar Herlighed*) from 1656, which includes an overview of all taxes, made it clear that the ability of the people to pay their taxes originated with God and furthermore, that this was very important to the state. The images and vignettes on the title page clarify the message. Ceres, the goddess of grain growing, appears in the uppermost field: "Everything God created and maintains, though as the poet about me states; I give grain and land's growth, for the peasant's taxes, maintenance and food." To the left are Pan and Mercury. Pan is the god of shepherds and the fertility of cattle as well as the protector of paths and roads. Additionally, he is the god of hunters, nature and poets. Mercury is the god of trade, protecting the traveler, and the guardian of happiness and peace. The vignette states: "Animals and cattle, acorn and forest, we give the peasant to satisfy his taxes." To the right is the master of the sea, Neptune: "Cod and eel, flounder and herring, I give you for your taxes." In the lower field is the fortress Varberg in Halland, framed by a portrait of the author and his crest. The vignette here says: "With what God has provided we uphold our duty and taxes, the rest we enjoy."

Source: Photo courtesy of Björn Andersson, Historiska Media.

An imagination of personal and political loyalty towards the lord existed in the seventeenth century even though the nation or the people as a distinct group did not make itself noticed. On every level, an individual had to be loyal towards the lord as an embodiment of the state, and in Denmark-Norway as well as in Sweden also as the leader of the Lutheran state-church. This background reveals the graveness of Corfitz Ulfeldt's violation of both these bonds. The events on Bornholm demonstrated how true state patriots behaved towards their lord and king. Therefore, they were described in propagandist prints appearing as early as during the siege of Copenhagen and later during the Skånian War (E.G. Rasmussen 1985:10, 13). In a reader for schools by Ove Malling from 1777, which was used for a long time, the events were mentioned under the heading "Love for the fatherland" (Malling 1992:115f) although they could as well had been placed in the section on "Allegiance towards the king."

When peace was negotiated in Copenhagen in 1660, Frederik III was eager to maintain Bornholm. His status there, as absolute, hereditary king, could be employed in the *coup d'état* he accomplished the following year, not least with support from prominent townsmen of Copenhagen. Through the coup, the high nobility was stripped of the formal (though less and less executed) power it possessed in the state council. Even the Swedish crown was interested in this change of the Peace of Roskilde. The Bornholmers had clearly demonstrated that they were not to be relied upon and also, the Swedish crown could demand something in return from the Danish king. And that exactly was what happened after additional negotiations. The Danish king had to hand over to his Swedish counterpart estates in Skåne corresponding to around ten percent of the total tax base of this rich province as substitution for Bornholm. While the Danish king's own possessions in Skåne already had been confiscated with the Peace of Roskilde, Frederik had to exchange land with owners of estates in Skåne who received compensatory estates elsewhere, especially in

Jutland. Through this deal, the Swedish crown received comprehensive estates in the captured province, which it could distribute among officers and officials whom it could trust to be loyal. At the same time it got rid of people it might never have been able to trust such as, for instance, the Danish royal chancellor, Joachim Gersdorff, who turned over all his Skånian possession corresponding to nearly one third of the total demand (Fabricius 1906a:157ff). The establishment of new landowners in Skåne who were loyal towards the Swedish crown was likewise promoted in other ways. Estates earlier belonging to the Danish king and the estates belonging to the chapter in Lund were also available. When redistributing the estates, Swedish nobility often received extensive Skånian (Danish) noble privileges while lover ranking officers and officials had to be content with corresponding, lesser Swedish ones. The Swedish crown in this way distributed huge amounts of estates to new owners as support and payment for expenses. These new owners subsequently remained loyal to the Swedish crown, which they owed both their positions as officers and officials and their noble privileges. Turning the political orientation of the Skånian nobility away from Copenhagen and towards Stockholm was also actively promoted in other ways. Noblemen with estates on both sides of Øresund and who wanted to keep their possessions in Denmark were forced, after the wars, to sell their Skånian estates. The logic, contrary to a century earlier when the Kalmar Union was dissolving, was that a person could not remain loyal towards both the Danish and the Swedish crown. In addition more Skånian estates were made available for the new, immigrating nobility through sale and bankruptcies (Fabricius 1906b:68-108).

The desire of the Skånian nobility from the very start to pledge allegiance to the Swedish crown had been very limited. However, when it became known how the Danish nobility was deprived of its political influence when absolutism was introduced 1661, more recognized it as an advantage to make the leap. Instead, they fought to maintain as extensive noble privileges as possible within the structures of the Swedish state. On this point, they had support in the peace treaties from Roskilde and Copenhagen, which stated that all old freedoms should endure. This new sworn Swedish nobility continued henceforward in most cases to maintain their loyalty to their new lord. Regardless, they also continued to travel often to Copenhagen and to other places in the Danish kingdom, as did the newly arrived Swedish officials.[27] Their travels were often family visits but probably also just trips to Copenhagen, which was the nearest town of European appearance and with aristocratic circles. Thus, the new borderline was not an impenetrable wall.

Fabricius found that in the tax protocols from 1673, 60 to 70 percent of estates in Skåne were in "Swedish possession," indicating that the owner was faithful to the Swedish crown (1906b:109-119). This result, obtained in less than a decade, is a dramatic development. To this came, as indicated above, that most of the nobility continuing to live in the province had sworn a new allegiance – and that the majority kept this oath.

---

27   See Fabricius (1906a:153ff and 1906b:14-51, 65ff).

The relation between the common people of Skåne and their king was usually administered through this nobility and officials and thus, the orientation of these groups affected the political orientation of the general population. To secure the newly acquired provinces for the Swedish kingdom, bailiffs and officials in towns and countryside naturally were substituted, as soon as possible, with persons the crown trusted (Bergman 2002:233-247; Fabricius 1906b:165-176, 221-223). By substituting nobility and officials with persons from the Swedish kingdom or elsewhere, the crown more or less ensured that the secular authorities would remain loyal in the transferred provinces. As a consequence of these moves and through a series of political maneuvers by Copenhagen and Stockholm that secured both centers of power their respective goals, Bornholm came to play an important part in the process by which the political loyalty of Skåne was transferred to Stockholm; at the same time the island remained an enclave in the Baltic, faithful to the Danish crown. Obviously kings, as well as subjects, were very active in employing bordering practices that were calibrated towards the structures of the state patriotic society and aimed to strategically affect their situation.

In 1666, the Swedish crown established Lund University, which opened in 1668, thereby creating an organization that over the course of time would provide allegiant clergymen – the persons who usually were the most local royal officials in the Lutheran kingdom. Moreover, the crown could expect this clergy to have a great impact on the loyalties of their congregations.[28] While today it is possible to distinguish between the secular and the ecclesiastical loyalty that the lord could expect from his subjects, these two sides of reality were integrated during that period's Lutheran state. In his secular reader, Ove Malling emphasizes the religious foundation for the power of the lord:

> It is likewise impossible to distinguish between faithfulness to the fatherland and faithfulness to the king [...] The good and wise commoner realizes this, and he loves his king, as a well-behaved child loves his father. My king, my father, he says and he feels it. It is natural and easy for him to obey his king, to execute his will accurately, to see to his advantage, even to venture property and life for him and in his service; for the king's person is sacred to him, and in his wellbeing he sees also the father-land's and his own.

---

28   On the establishing of the University and the importance of priests as instruments to secure the allegiance of the common people, see Fabricius (1906b:151-164, 137). Until the Skånian War, most priests installed in vacant benefices were from Skåne, some came from the north, some from west of Øresund and a few from German states (Bergman 2002:260-264; Fabricius 1906b:135). The founding of the university was in line with the Swedish crown's attempts to establish loyalty among people in other parts of its kingdom. Lund University was thus, the fifth in the Swedish realm, after Uppsala (in mid Sweden, 1477), Tartu (in today's Estonia, 1632), Åbo (Turku, in today's Finland, 1640), and Greifswald (in today's Germany, obtained with a university at the Westphalian Pease in 1648, university from 1456).

> God, who erects states and consolidates thrones, is also the one who placed kings, these kings to whom we swear obedience, allegiance, and manhood. To obey them is thus to obey God, and to be faithful to them is to be faithful to oneself, to one's oath. (1992:146)

Martin Luther made similar observations about the secular rights and power of the lord as reflected in his *The Small Catechism*, which was published in many languages and many printings over centuries:

> What constitutes the subject's obligation towards his authority? Answer: It is threefold:
>
> 1. Thou must thank and pray to God for Kings and for all in authority; that we may lead a quiet and peaceful life in godliness and honesty.
> 2. Thou must obey them and submit yourselves to them, as they have the power to punish.
> 3. Thou must give them tribute, as that is your duty.[29]

Here the lord's interests in territory are stated clearly. In a speech prepared for the event of his execution, the priest Henrik Gerner clearly advances harmony between the ecclesiastical and the secular. Gerner was involved in a conspiracy to capture the castle Kronborg (in Helsingør) from the Swedish army and restore it as a stronghold for the Danish king in 1659:

> Solomon says, that a three-stranded thread does not break easily. Likewise with an honest patriot there is between him and his king a three-stranded thread, by which he is tied. 1) The law of God: Honor the king. 2) The natural love for the fatherland, as phrased by Ovid:
>> I cannot substantiate,
>> Why most men
>> Both now and always
>> Love their own country.
>
> 3) The humble oath, which they owe their king to keep. Who break this thread is an idolater, worse than the beasts, yes, even a perjurer, and consequently a traitor. [...] Therefore it says: *Honor your king* with thoughts, words, actions,

---

29   Luther's *The Small Catechism* exists in countless editions. With a glance to the wording in an American edition from the early nineteenth century this is a translation from a Swedish version from 1667: *Catechesis eller Summa af then helga Skrift/ innehollandes then Christeliga Lärones Hufwudstycker/ både huru wij här som Guds Folck Christelige skole lefwa/ och sedan hoos honom ewinnerliga salige warda.* För ungt och annat enfaldigt Folk uthi kårta Frågor och Spörssmåhl författat aff D. Martino Luthero. Stockholm/ tryckt aff Georg Hantsch. The English editions reviewed are structured more as quotes from the Bible while this and other Swedish versions are organized as questions and answers.

with estate and fortune, yes, even with life. Never take to despising thoughts about your king, never curse the king in your thoughts... (from Møller 1903-1905:152f)

In addition, Gerner expressed his confidence that Frederik III would reward his surviving family for the faithfulness he had demonstrated. However as Gerner was not executed, he had the opportunity himself to enjoy the fruits of his loyalty to the state patriotic structure, and he later became bishop in Viborg, Jutland.

The provisions taken in the acquired provinces, the founding of the university, and the conscious politic of substituting sworn, faithful noblemen and officials are the most important premises for understanding why the population in Skåne, Halland, and Blekinge so relatively soon became loyal towards the Swedish crown. For state patriots it could be possible, if all other matters were equal, to act the same no matter which king they obeyed.[30] When people, as on Bornholm, locally acted to affect decisions that were outlined in internationally recognized peace treaties and developed within the political arena of the state, they had several objectives; they balanced loyalties towards present and previous lords, they had expectations of future favors from the lord/state, they created movements, and acted to protect as well as promote their own interests. However, these different objectives were not created independently from the states' centers of power as Sahlins' interpretation suggests. Actions by authorities as well as by individuals, which were caused by the moving border, were constantly and scrupulously considered in light of how they associated with the state patriotic relations that constituted the states at the time.

Snapcocks

The politics adopted by the Swedish crown in the acquired provinces east of Øresund have later been depicted as attempts to annihilate the Danishness or maybe Skånianess of the population. The resistance the crown met from people in the provinces has in the same way been perceived as peoples' natural aversion against switching nationality – in this case, becoming Swedish. This explanation has been used to explain forms of armed resistance from locals in Skåne and Blekinge. These irregular fighters, so-called snapcocks, were on several occasions involved in the wars between the Danish and Swedish king in the seventeenth century.[31]

---

30 Bergman (2002) convincingly demonstrates that local populations in Blekinge addressed central authorities before and after the moving of the border with their livelihood and not identities in mind. Consequently, there are profound similarities in what they requested and how they acted towards the authorities of their earlier Danish king and their later Swedish king.

31 Snapcock is a literal translation of the Scandinavian term: *snaphane*. There are several explanations to the origin of the word. One hold that it refers to a fast trigger mechanism that was often used on guns fabricated in a part of Skåne, the Göinge districts.

In writings about snapcocks, a single person or event instead of general surveys and explanations has continually been in focus. A systematic investigation has not been carried out, nor has any comprehensive explanation to the phenomenon been offered. However, while the activities of snapcocks are less analyzed, they are at the very center of the history of the border as these activities are bordering practices formed by the society in which these individuals lived.[32]

To explain the activities of the snapcocks by referring to the natural inclination of a people to remain Danish ascribes to them interests in the preservation of national unity. As discussed above, such a sentiment is not a reasonable explanation for events in the seventeenth century. However, snapcocks did involve themselves in the conflicts of kings and states. Their actions must be explained from the goals they tried to reach, as was similarly the case with the drifters and the conspirators noted above. The snapcocks' deeds must likewise be explained from how people in that time's society could revolt against their lord and against the rights and obligations imbedded in the state-patriotic structures of society; one must consider what kinds of political practices of rebellion commoners had at their disposal.

Others besides lords maintained institutions that could be employed for their own interests as illustrated through the so-called border peace treaties.[33] At least from the fourteenth century, separate arrangements between secular and clerical magnates are known in which parts of the kingdoms established peace in times of war. As Eva Österberg has pointed out, one of the warring kings often could be interested in or even initiate such agreements (1971:117ff). In connection with the dissolution of the Kalmar Union and the subsequent wars over the drawing of a border between Øresund and the Mälaren region, such arrangements often were established across the existing border. The accords could include that both parties

---

32   Much has been published about single persons or events when it comes to the activities of the snapcocks (see the overview in Wiggers-Jeppesen and Boisen Schmidt 1981). Here I generally refer to publications in which snapcocks are discussed more or less in detail: Blom and Moen (1987), Hansgaard (1956), Liljenberg (1963), Osvalds (1993), Rosén (1943), Sundin (1986), Sörensson (1916), Tomenius (1984), Tuneld (1960), Åberg (1958, 1975, 1981, 1995). In addition, I refer to published materials from courts and other instances in: Cronholm (1976), Edvardsson (1960, 1974, 1975, 1977), Johnsson (1909 and 1949:73-82). In most texts, a set of elements emerges and therefore, is critical to the image of snapcocks disseminated. These elements include the Göinge-chief (has been illustrated above and will be discussed further below), the coup at Loshult (will be explained below), the battle at the entrenchment at Hönjarum (a local fortification which was captured by the snapcocks from a Swedish force), and the seven brothers, or *Uggleherrarna* (literally: "lords of the owls"), who were all involved in snapcock-activities. Add to this that every author describes the violent punishments that could be imposed on snapcocks (will also be discussed below).

33   For discussions about and case studies of border peace treaties, see Larsson (1974:102ff), Lindberg (1928), and Rystad (1965). It must be stressed that border peace treaties not only embraced commoners. In most cases townsmen, nobility, and/or clergymen were actively involved. The earlier often used term: peasant peace treaties, is therefore misleading.

continued to pay taxes to the lord to whom they had been paying, that they would notify each other of approaching armies, and that they would not participate in warfare across the border. Therefore local people did obstruct armies by refusing to support their advances beyond the border. The agreement could also contain an assurance of continued free trade across the border and that the parties together would fight lawless gangs. While the border peace treaties were not official, as both kings did not simultaneously approve of them, they were at the very least public acts. The agreement on participating in a peace was established as a legal, collective action at the district thing (*herredsting*). Although the arrangements were defensive, they could also contain stipulations of more active performances. For instance, both sides should arm themselves and together fight to maintain security and peace. The same institutions, the district thing, which could declare peace across the border, could also call upon a local or regional active movement of resistance. When such a call took the form of open uprising against the lord, this was associated with the social relations of the time. A basis for rebellion existed if the lords tried to increase their rights and demanded more than their subjects regarded as reasonable or if the subjects did not receive the protection to which they were entitled in the state patriotic structure. This of course is not to say that rebellion would occur; when it did happen, it was associated with weaknesses with the central power(s). A central demand was always the reestablishment of earlier existing conditions as for instance, to insert bailiffs and distribute fiefs on earlier terms, return to previous taxations, and so forth. Thus, reasons and institutions, which could be enacted and could take the form of a popular rebellion, existed (see Bøgh 1985, Würtz Sørensen 1985, Österberg 1979).

Thus, political institutions that could attend to local or regional interests across the border existed. It was important for both sides to keep the war from their door and to let social life proceed. People married across the border (Strid 1970), and it was crucial to keep trading routes open between the towns in the Danish king's territories at the shore and the inland areas, which obeyed the Swedish crown.

The last widespread popular rebellions in both Denmark and Sweden occurred in association with the Lutheran reformation but had more than religious backgrounds. If approached as expressions of local or regional interests a more complex picture emerges. *Grevens Fejde* (The Count's Feud) in Denmark is most often described as warfare between different wooers for the throne and fractions of noblemen. The last popular rising in Denmark, *Skipper Clementfejden* (the Master Clement feud) in Jutland in 1534, is understood as an offshoot of this conflict. In today's Sweden, the last serious popular attempt was *Dackefejden* (the Dacke feud, see Åberg 1960). This occurred in 1542-1543 and had its center in Småland. It has been explained as a response to increasing taxation and cases of assaults by bailiffs.

Similarities between these two feuds are numerous. In both rebellious regions, ecological environments were under pressure due to enhanced cultivation and fighting over land as a scarce resource. In northern Jutland the increasing and profitable trade with oxen, which the nobility tried to monopolize for themselves at the expense of peasants and townsmen, was a central element in the uprising

(Tvede-Jensen 1985). The right to trade also played a role in Småland where the new Swedish crown tried to force the inhabitants to trade through its town Kalmar instead of what had customarily been down through the Danish king's town of Ronneby. The Swedish king, Gustav Vasa, wanted both to tax the trade and to secure provisions of food for the raw-material producing areas around Mälaren; here, oxen also played a role (Myrdal 1988). Increased taxation on smallholders in the forests and attempts from the center at Mälaren to curtail local autonomy provided additional reasons for rebellion (Larsson 1964, 1974, 1992).[34] Participants in both uprisings included not only commoners but also some priests and at least in northern Jutland, several noblemen. The general participants were treated similarly and sentenced collectively while the leaders were punished individually. Furthermore, both rebellions had connections with the Hansa town Lübeck, which had obvious trading interest to protect in the North, and both used a restoration of the Catholic Church as legitimization for their actions. Thus, popular uprisings must be regarded as resistance to changes of what was perceived as customary levels of rights and duties embedded in the structures of state patriotic societies; taxation and increased pressure on ecological resources under the right conditions could cause widespread social unrest. The rebellions also became associated with protests against the intervention of the central power (also with regard to the establishing of a new Lutheran church) in what used to be local affairs and the rights of commoners to trade. These practices thusly follow general trends of rebellion as a way to restore earlier conditions – very similar to what the people of Bornholm requested in their petition to Frederik III. Rebellions therefore, contrary to modern revolutions, are reactionary movements that try to recreate earlier forms of mutual dependency and it is evident that, in the 1500s, commoners were able to actively express their interests and oppose processes that potentially would dilute their living (Bergman 2002:196-198).

There are similarities between this praxis of rebelling, which from the reformations can be traced back through the Middle Ages, and the activities that can be summarized as snapcockery. One similarity is the summoning of the population in an area for action. This happened on Bornholm and spectacularly in connection with "the coup at Loshult" in northern Skåne. There, a pack of summoned peasants, supported by a few horsemen upon whom they had called from the Danish king's army, seized the Swedish king's convoy with his (the state's) money in 1676. Locals on the other side of the old border looted what money the escort succeeded in rescuing.[35] Shortly afterward, the parishes on both sides of the old border, which had cooperated at the coup, contracted the last known border peace treaty in the North.

---

34   On the importance of the export of oxen and the development of this trade in Denmark, see Frandsen (1994) and Møllgaard (1988), and in Sweden: Bjurling (1940), Linge (1969), and Myrdal (1988).

35   Sweden lacked silver in this period. Therefore Swedish coins were made of copper. Because they had to contain equal value, the coins became very bulky and thus, it was literally tons of money the peasants stole and hid away in the forests.

Yet, there are more reasons to point to interesting differences between this treaty and the older border peace treaties. A decision on popular rebellion or border peace earlier belonged on the district thing, thus the collective punishment of participating regions after feuds was completely logical. However on Bornholm, in Malmö, and around Loshult the risings were developed secretly. Consequently, any organizational similarity with earlier rebellions is not traceable. The absence of the district thing in the processes indicates that relations between the center of power and the local community had been strengthened in ways that made this local political arena inaccessible for such radical denunciations of loyalty (see H. Gustafsson 1994b, Österberg 1989). The events still must be conceptualized as expressions of the interests of the local participants. However, the centralized, Lutheran state in the 1600s had extended its control of local and regional arenas to a level that made earlier known practices of local social and political unrest inaccessible.[36] This extension from the center must be understood as a crucial aspect of how the central state developed and expanded through professionalizing its institutions (Bergman 2002:233-247, 263-275). The process furthermore maintained that locals had possibilities for actively shaping decisions that affected their livelihood. Obviously, armed rebellion was an option when, for whatever reason, such dialogue was no longer feasible (Bergman 2002).

Snapcockery is the name for a variety of activities including the summoning of local commoners, the enlisting of local companies of irregular soldiers in the king's service, so-called free riflemen (*friskytter*) under the command of officers (at least some times elected by the soldiers), and entirely individual deeds. Snapcockery could be a single action but in more cases, snapcocks continued their activities long after the termination of the state of war between the kings. Thus, there was a scale of actions from the well-regulated and coordinate to individual acts. The same individual, of course, often could be found as a participant in changing circumstances spread along this scale as the conflict developed.

In connection with the wars 1657-1660, people with close relations to the center of power in Copenhagen organized the most noticeable activities of resistance. The conspiracies in Malmö, at Kronborg, and on Bornholm were developed by traders and people with contacts in the Danish royal administration. Attempts to raise

---

36 Harald Gustafsson (2000a) in his discussion of the dissolution of the Kalmar Union in the early 1500s provides material that would allow for a very interesting comparison of the different organizations of state patriotic structures in Sweden, Denmark and Norway, what he calls "political culture." He would be able to relate differences with regards to environmental and historical contexts (not least the roles of nobilities) and to provide an in-depth discussion of the different roles of the district things in the very differently organized kingdoms. However, he chooses to include instead a discussion on culture and shared identities, referencing Anthony Smith's discussion (1988) on "ethnies" that leads him into contradicting himself. This is the case, for instance, when he debates the role of languages. In some places, he finds that a common language meant a lot to people in the early 1500s; contrary to this, he finds it unimportant that many Swedes spoke Finnish (see H. Gustafsson 2000a:300, 303, 330).

local commoners, as it was discovered during the trial against the conspirators in Malmö, were performed by bailiffs and deployed officers – people with relations to leading circles. In some instances, such officers of the state patriotic structure also summoned commoners of a particular area to participate in local acts of war, as for instance during the defense of the island Langeland against the Swedish king's army in 1658-1659 (Skaarup 1994). Free riflemen were also enlisted, as for instance the company of dragoons that Svend Poulsen in 1657 gathered in southern Halland and northern Skåne. In the fall of 1658, Frederik III sent Svend Poulsen to southern Zealand to try to raise the commoners. However he did not manage to gather more than a casual horde. In all these cases, the organizers were people who were hurt by the politics of customs and trade adopted by the new Swedish administration or were already previously involved with the Danish center of power on an individual basis. What they have in common is that the transfer of the provinces threatened their position in the state patriotic hierarchy.

Unrest in the countryside of Skåne and Blekinge occurred 1658-1660, presumably mostly due to imposed quartering, drafts, and deserting soldiers who hid in the forests. One person in the countryside who rose against moving the border was the powerful, rich peasant Olof Persson in Holje, Jämshög parish in Blekinge. The Danish king, Christian IV, borrowed money from him, but paid him back with estate; he held 54½ tax-units as pledge, exported timber, and sold oxen. He was accused of arranging an ambush on a Swedish troop and other evil deeds, and from 1658 until 1661, Olof walked in and out of prison when finally he was released. Then, with his connections with Danish royalty, Olof was no typical peasant.[37] The irregular soldiers recruited by the Danish king's officers had no easy life as they were pursued persistently. In June 1660, the priest in Sørbymagle in Zealand buried again Peder Jørgensen: "Prior to Christmas of 1659 he was found by the occupying Swedish, with around 9 others in Skovhastrup, where they stayed as snapcocks. He was shot, others died differently, some even escaped. Until now he has been hid in the ground, now moved here, and buried in the churchyard" (O. Højrup 1963:250). However in general, no activities that are not connected directly with Copenhagen occurred in the countryside during 1657-1660 (Fabricius 1906a:79ff, 131ff). How the incorporation with the Swedish kingdom would develop and how it would affect the peasantry was not yet intelligible on a local level.

The subsequent Skånian War, 1675-1679, framed widespread snapcockery. Calm prevailed in the Skånian towns where the population apparently supported the king who at the moment, was in control locally. After 1660, the Swedish authorities worked on getting trustworthy officials installed and in more towns, big garrisons were accumulated. Thus in the towns, power could be put behind the state patriotic relations with the crown in Stockholm (Fabricius 1906b:165-176). In the old towns of Blekinge on the other hand, there were many reactions.

---

37   On Olof Persson in Holje, see Cronholm (1976:138f, 269f), Fabricius (1906a:138, 144), and Hallenberg (1940:183).

These small towns were not only under pressure from the enforced policies of duties and trade, as were the towns in Skåne. They were threatened additionally and directly with abolishment. Or, they were already degraded in connection with the establishment of new towns, first Karlshamn and then later, the new naval center Karlskrona, which already during the war was on the drawing board. The townsmen of nearby Sölvesborg thus complained that privileges to trade and craftsmanship for their town had been reduced (Petersson 1940-1943, IV:5-24).

Nobility, as depicted above, mostly had been substituted with people the Swedish crown could trust. Noblemen whose loyalty Carl XI doubted were pressured to permit themselves be interned in the neighboring landscape of Småland. During the war a few noblemen were sentenced for having helped Christian V – their estates were confiscated and in several cases, they were given a death penalty (Fabricius 1952:83-87, 121-146).

Priests in the countryside actively tried to keep a delicate balance between fighting lords; often both lords would approach the individual priest as his local representative. Some priests, though, expressed sympathy for one side or the other, which brought a few in conflict with their congregations who were of the opposite mind.[38] In some places in the countryside strong reactions and widespread snapcockery occurred. Who became snapcocks has been a matter of debate. Some authors have asserted that they were the anti-social elements of the forests[39] while others maintain that they were settled peasants and craftsmen who rioted because they wanted to remain Danish.[40] While the result is decisive for how to evaluate their actions in the twentieth century, the ditch between the different views has been dug deep. If they were anti-social elements, their practices were immoral and a plague for the population. If on the other hand, they stood up for their nationality, they were real members of a resistance movement, fighting a just war for their compatriots. More authors mention both possibilities and distinguish between the nationally minded free riflemen and the unorganized snapcocks who were anti-social.[41] The free riflemen themselves seem also to have been eager to dissociate themselves from the unorganized snapcocks (see Sörensson 1916).

It is possible that some snapcocks seized arms due to a personal recollection of having served with and sworn allegiance to the Danish crown and thus, acted from the loyalty embedded in state patriotism. This, for instance, could be claimed for Olof Persson in Holje mentioned above. Nevertheless, this does not mean that they fought for a nation. To juxtapose the settled, nationally disposed against the anti-

---

38   See Blom and Moen (1987), Fabricius (1952:46f, 80-83, 186-195), and Johnsson (1908).

39   Alf Åberg (1958:98-107, 1981:61ff, 78, 88, and 1995:111ff, 122f) and Jan Sundin (1986:163f).

40   Lauring (1961:120-131), C. Sørensen (1880:267f), and Sörensson (1916:62). Hansgaard depicts snapcocks as that time's partisans and saboteur, equalizing them with members of the resistance movements of the Second World War (1956).

41   See Blom and Moen (1987), Fabricius (1952:101-106), and Liljenberg (1963).

social thereby affords a far too simple depiction of snapcockery. There were many settled peasants who only once or on a few occasions performed snapcockery; afterwards they returned to their everyday doings. The many participants in the coup at Loshult illustrate this. The Swedish authorities' quite frequent announcements of pardon must be understood as their acknowledgment of this fact even if it could be supposed a reluctant recognition.[42] In addition, there is clearly a development of snapcockery during the war. To be more accurate, rebellious practices had already started before the war and they continued after the kings had agreed on ending their hostilities. The period of snapcockery, thus, was longer than that of warfare.[43]

From the very start snapcockery seems to be what Hobsbawm termed "social banditry" (1969). Snapcocks were men who felt exposed to injustice, for instance increased taxes, quartering, conveyance and road building, or drafts, which made them react and defend a real or expected freedom or right.[44] The insecurities, which the war caused, promoted the demonstration of persons' or regions' own interests; it did not create them. Individuals practicing social banditry shared the surrounding community's values and worldview, which they demonstrated through their efforts that they were ready to defend. They also continued to be in touch with their community. From records, it is perfectly clear that snapcocks were closely associated with people around them. They not only robbed and murdered; they also resided, cultivated the fields, traded, and fathered children. Thus, snapcocks were known by and most often related to the locals or to the priest or other persons with local authority. This is part of the background for the dubious position many priests held in the eyes of the Swedish authorities. Thereby, the authorities could employ a priest's knowledge of the locals, as a priest was the local representative of the Lutheran state. They, for instance, were ordered by the Swedish authorities to submit lists of all persons from their congregations who had sided with snapcocks (Åberg 1981:99). Their knowledge of locals, however, could also be used in favor of individuals. When Nils Ubbesen, an officer of a company of free riflemen, in 1679 was prosecuted at a district thing in Småland, a certificate from a priest was produced, which described the accused as "a merciful snapcock" (Edvardsson 1975:139). Social banditry was also an element in the local cash economy. The riflemen's pay as well as what they seized of value was transformed into food and equipment; and often snapcocks immediately asked the individuals they had robbed if they wanted to buy their property back for cash (see Edvardsson 1975).

---

42   The coup at Loshult is thoroughly illuminated by Vigo Edvardsson (1960). He includes an announcement of pardon from December 3, 1677 (1975:37f) and a couple of personal letters of pardon (1960:150ff).

43   See Brockdorff and Tägil (1985:99ff) and Åberg (1975:44, 66ff, 119).

44   Just after his army landed in Skåne 1676, Christian V distributed letters to "his subjects." In them, he argued that the freedoms and rights the surrendered provinces had been guaranteed in the Peace of Roskilde and Copenhagen had been liquidated by the Swedish authorities (Sörensson 1916:8). This claim is both an appeal to already unsatisfied persons and a promise of the return to earlier conditions so central to popular rebellions.

In the early phases of the war the Danish king, Christian V, summoned in more cases with success commoners in Skåne and Blekinge. After the battle at Lund, in which Christian's army was defeated, it became increasingly clear that any return of land to the Danish king was not topical – at least not this time. Consequently, the population became more reluctant to openly support the Danish king's efforts. Thus instead, Christian V started to organize companies of free riflemen and his officials started heavy-handedly to collect taxes and provisions for the army (Fredriksson 1983:355). As the free riflemen, through the war, became less and less able to support their operations on the retreating Danish army, local communities became exposed to pressure from both parties in the war. In the spring of 1677, the Swedish crown sent the officer Johan Gyllenstierna through Blekinge and the northern parts of Skåne. In each congregation, he demanded that the parishioners swear allegiance anew to the Swedish crown. In some places, even more pressure was put on the locals. Their oath stated that if anyone in the parish henceforward had contacts with snapcocks, heavy collective fines would be demanded or a proportion of grown up men would be hanged.[45] The most brutal expression of this strategy was the treatment of Örkened parish in Göinge district. In April 1678, Swedish troops from several sides entered the parish with orders to burn all buildings (except of course, churches) and to execute all males capable of bearing arms. There were not many casualties, as the population had taken to the forests, but all farms except three were burned (Tomenius 1961, 1984). From the Danish side the pressure on the communities came from the army that from the start had been kept disciplined; later in the war, the tactic increasingly changed to purging the land (Fabricius 1952:52-59, 158-180). While at least one of the regular soldiers participating in the coup at Loshult in 1676 was punished by the Danish army for plundering against orders, the troops later were let loose. Possibly, this was a deliberate strategy to ruin the future tax base of the Swedish king. In any case, it made both regular and irregular troops, the latter often commanded to do the dirty work that was extremely unpopular with local communities. Consequently, the irregular soldiers increasingly were marginalized partly due to the pressure imposed locally by the state and partly due to their own activities. It became more and more difficult for them to act in agreement with the norms and interests of other locals and thus, they were reduced to thieves and highwaymen who took what they needed by looting. The surrounding local community therefore, no longer had any interests in protecting them. Had they from the very start acted as

---

45 It ought to be mentioned that the Swedish king, Carl XI, also found it necessary to achieve a similar renewed oath from the Sunnerbo district of Småland (Edvardsson 1975:126) and that similar oaths were gathered by other officials in other parts of Skåne (Fabricius 1952:92-98, Åberg 1981:78-83). Demanding renewed oath was earlier employed in the second Carl Gustav War, 1658-1660 and was targeted at snapcockery (Fabricius 1906a:149). This reassurance of loyalty from the population was repeated in 1710 in connection with the Great Nordic War (Strid 1973:72).

anti-social elements the commoners would have known how to handle them – as was illustrated from the border peace treaties, the settled population knew how to keep itself armed and fight gangs of robbers. That snapcockery later became such a tremendous problem was partly due to the Danish king's efforts to recruit free riflemen. The commoners ability to fight snapcocks was also reduced by the pressure they were subjected to from both sides of the war, and as arms in some areas were confiscated by the Swedish authorities.

A development of snapcockery through the war, its legitimacy, and its organization can be discerned: it started as summons of peasants, continued as free riflemen, and ended as pure robbery. At least in the beginning of the war both the Danish army and the locals all had reasons to distinguish between the more or less well-organized companies of free riflemen under command and the armed individuals of questionable character who drifted through the parishes. However, the development of the war meant that as the first became fewer, the number of the latter kept growing. It is evident that the Swedish authorities vehemently chased all groups – they had armed themselves against their king, their lord appointed by God, whom they owed absolute fidelity.

When the State Punished

In September 1679, some snapcocks were interrogated at the thing of eastern Göinge district. Among them was Tue Trulsen. He had been drafted as a soldier to the Swedish king's army but had deserted together with others who had been drafted. Three weeks before Christmas 1678, he joined the irregular company of riflemen led by Har-Åken. Tue stated in court that he was forced to join but he also admitted that he received ten *rigsdaler* and a horse for enrolling. During the winter the company was stationed at the fortified town Landskrona, which was held by the Danish king's army. Tue though, had stayed behind in Göinge, participating in the tasks Carl XI's authorities had ordered the peasants to perform. In addition, he conveyed information to hiding comrades about Swedish troops who were in pursuit. Once he accompanied other snapcocks, but he turned back home. Tue confessed that he was with Har-Åken's company when it plundered the priest's farm in Hallaryd in June 1678 (the chronology of the court document is not linear). Later the clergyman was released and his cattle returned for 160 *daler* silver from which Tue received three. Departing, the snapcocks had been shot at from a farm that they then looted and burned. Together with two others, Tue had also killed a Swedish soldier. Tue Trulsen was sentenced for desertion and thus "rebellion" for homicide, arson, and "other more unseemly achievements." From his confession and "above-mentioned unchristian achievements" he should be "punished on his life, the head off with an axe, put on a pole and a gallows above, then the body on four wheels, as an example and warning to other similar vicious" (from Edvardsson 1975:42ff).

Figure 1.9   Snapcocks attacking in the forest, by Johann Philipp Lemke (1631-1711)

Notes: Snapcocks kept an eye on supply routes and other paths between the army and its staging areas behind the front. By attacking singley horsemen or smaller units, who took supplies or mail to the front, snapcocks significantly disturbed the Swedish war efforts. Ambush in a forest; these highwaymen could be snapchocks. Contemporary sketch by Johann Philipp Lemke (1631-1711), who from 1684 lived in Stockholm and for the Swedish king painted scenes from the Skånian War.

Source: © Nationalmuseum, Stockholm.

It is unknown how many snapcocks were punished. Except that Tue did not claim that he had been given a letter of pardon, both his acts and his sentence appear typical. On some occasions, he had been with smaller and bigger groups but also in periods stayed at home and fulfilled his obligations towards the (Swedish) state. In many cases, it is impossible to determine how snapcocks can be differentiated from thieves and murderers with whom they certainly associated in some contexts.

Some ordinary criminal acts, thus, were probably judged as snapcockery by the courts. They could even be presented as such by the delinquent. If he could plead later to have been given a letter of pardon, he could perhaps be pardoned. Thus, snapcockery did not necessarily stand out from ordinary patterns of violating the laws although the insecurity of times of war in general promoted violent crimes. Both the actions and sentences of snapcocks are parts of the history of the transference of Skåne, Halland, and Blekinge from the Danish to the Swedish crown. There is good reason to believe that persons actually living this history regarded the wars and the reprisals as violent. However, the means employed are no more violent than in other places and not in themselves unusual.

When the state punished snapcocks, it used the ordinary measures of the time – hanging, beheading with an axe, the head on a pole, and the body on wheels. In some cases, breaking on the wheel was also used. Executions, however, could be given a much more macabre form. Even if it was unusually violent, few texts on snapcockery fail to mention the punishment of Hans Severin, captain of a company of riflemen. He was sentenced: "to be pierced alive, not internally but between the back and skin [from his rear] though his neck, afterwards the pole to be raised, the feet nailed and hands tied behind his back, under a new gallows with the rope around the neck, not pulled. His name on the gallows" (from Åberg 1958:107). Some writers append, that some individuals were hanging like this for days before succumbing.

The rigorous punishments ought not to surprise. Convicted snapcocks had committed the worst conceivable crime – they had rebelled against their king who was their lord by grace of God. Moreover, Severin had been the leader of this. The penalties employed were expressions of the state patriotic society and in general punishments chosen to deprive the convict of his honor (see Egardt 1962 and Matthiessen 1910, 1919). Symbolically the punished was expelled from the society of both living and dead; he could not be buried therefore in consecrated ground. Thus, it could be requisite that already buried snapcocks had to be dug up and put on wheels (Edvardsson 1977:196). On the other hand, captured enemy soldiers were not handled with kid gloves by the snapcocks: an eye for an eye… (Fabricius 1952:155).

In some cases, where the accused had committed obvious snapcockery and had confessed, the courts decided for immediate punishment. Thus, the execution could be carried out before the next case was on trial. However, this is not the same as if summary courts were employed.[46] The courts moreover took into consideration the

---

46   Snapcocks caught in connection with military operations could be executed immediately. When the fortified town, Kristianopel, in Blekinge, surrendered in February 1677, captured enlisted German troops were put into the Swedish army, drafted soldiers from Denmark were forced to march all the way to the border of Norway, and 70 snapcocks (those born in the lands of the Swedish king including Skåne, Halland, and Blekinge) were executed. Contrary, while the fortified town, Kristianstad in Skåne, surrendered in August 1678, free riflemen were secured departure.

pressure the commoners had been subjected to from the troops and administration of the Danish king.[47] From published documents, it is even evident that courts sometimes acquitted the accused. This was the case if the accused proved self-defense when killing a soldier or, if he was no doubt a snapcock but later had obtained a letter of pardon (Åberg 1975:74, 92, 119).

In 1684, Sven from Boarp was on trial at the court in Båstad, Skåne. He was sentenced to death because he had been around on more than one occasion when Swedish officers had been killed and robbed as well as because he had at other times been with snapcocks. Many gave evidence on his participation in the different actions. Sven also admitted that he had been with snapcocks but explained that he had been forced to join them as they otherwise would burn his farm. Because it could not be proven that Sven personally had killed anyone or had received any part of the booty, and while Sven was covered by an announcement of pardon, he was later acquitted in the high court.[48] Thus, convicts could hope to have their punishments mitigated. Capital punishments were in some cases converted into huge fines and confiscation of poverty; it was the case for more of the conspirators in Malmö and at Kronborg. Consequently, as it was common practice at the time, the punishing state could also be merciful.[49]

The Interests of Common People

What then made individuals engage in activities that later could, and eventually in many cases would, endanger their honor and life? Who were the snapcocks and from where did they come? Moreover, why did some of them continue fighting a lost battle long after announcements of pardon had provided them a feasible path out of the whole wretched state of affairs?

The exact numbers of free riflemen and snapcocks is unclear. Liljenberg finds that they made up around five percent of males in Skåne, Halland, and Blekinge capable of bearing arms – something like two thousand five hundred men (1963:173). Even if his calculation is not more than a qualified guess, it is clear that they formed much more than a handful. The five hundred free riflemen who were demobilized on Zealand after the end of the Skånian War alone show this, even if some of them probably did not come from the transferred provinces

---

47   See Edvardsson (1974:43ff and 1975:87ff).

48   This particular case is discussed by Edvardsson (1977:130-142) and Åberg (1975:70ff).

49   An investigation by Rudolf Thunander of 1,342 death sentences from Småland from the period between 1635 and 1699 shows that only 280 were confirmed by the higher court. In more than 70 percent of the cases (982), the higher court changed the death sentence to another penalty – fines and/or different kinds of penalties on the body, mostly forms of whipping (see Österberg 1995:154).

(Sörensson 1916:61).[50] It is agreed by most authors that snapcockery was most common in the western part of Blekinge and the northern part of Skåne, and in particular in the two northeastern Göinge districts.[51] From this, it has even been suggested that people in this forested region traditionally were less involved with authorities and more rebellious.[52] However, a more reliable explanation of snapcockery might be reached if the activity is considered as an expression of the population's own interests. Such an explanation makes it comprehensible why Gyllenstierna's expedition to collect renewed assurances of allegiance in 1677 focused on this particular region, why the biggest companies of free riflemen were recruited from there (Hansgaard 1956:168-184), and why priests on the Skånian plain in the southwest corner of the landscape reported that snapcocks were not to be found among their parishioners (Åberg 1981:99). The forested region situated between the war zone in Skåne and the Swedish king's deployment area in Småland made the efforts to control the area by the authorities more rigorous; relatively more cases of snapcockery could have resulted in trials. On the other hand, this location also meant that the population was bothered more by marching troops passing through, requisitions to forage, and other un-pleasantries; thus, there might have been more reasons to take action. More authors directly point to this connection. This does not explain, however, why even prior to the beginning of the Skånian War there seems to have been alarm among the population in this region where instances of snapcockery had also occurred during the war 1658-1660.[53] It is not unlikely that the uncertain calculation that five percent of males, capable inhabitants of the provinces, were snapcocks during the Skånian War, is an exaggeration in general. However in the forested region, which was sparsely populated and which population constituted a relatively small part of the total, this number of participants in snapcockery must be understated, probably even immensely underestimated. Thus, to explain this regional concentration seems to be one possible approach to understanding the whole phenomenon.

The peasant economy of the forested region depended on a number of advanced, marketed commodities such as oxen, horses, iron, wood and timber

---

50   Fabricius finds that there must have been between one and two thousand free riflemen and fewer unorganized snapcocks. He mentions a number not very much different from Sörensson's (Fabricius 1952:206).

51   Christopher Cronholm holds this opinion as early as the 1750s when he wrote his chronicle (1976:56f).

52   For an example of this explanation, see E.G. Rasmussen (1990:55). That the peasants actually had some liberties not recognized in most other areas is evident. They were for instance not subjected only to use mills owned by the nobility or crown. But in this matter, the region was no different from other less populated districts in Denmark and places where the distance to such central facilities was recognized as being too long or difficult (Linde-Laursen 1989b).

53   See E.G. Rasmussen (1990:55-61), Fabricius (1906a:131-151), Wägner (1886:2f). See Tuneld (1960:52ff) on the report on raids for snapcocks in connection with the trials of Olof Persson mentioned above.

products, tar, saltpeter, and hops. The region, thus, was very much involved in a money economy. Arent Berntsen described the economic situation of Skåne and Blekinge in 1656, the year before the war that eventually resulted in the secession of the area from the Danish to the Swedish king:

> [On Skåne:...] Because of such an abundance of meadow and pasture the population of this land breed much of all kinds of splendid cattle/ thus their oxen are recognized as the biggest and best in the kingdom [...] So is also the forested region convenient for orchards, for flax and hemp/ and for hop gardens/ which the inhabitants in the same place also know how to cultivate/ and they are unusually hard-working with this in the Giönge districts/ from where immense quantities are brought to the market towns and sold. Moreover the people in the forest know how to operate and take advantage of their area/ with burning charcoal/ chopping firewood/ timber for houses and wagons to cut and bringing all to the market towns and the plain to sell/ both the wood itself/ as well as finished products as wheels/ whole wagons and their accessories/ as grass rope/ besoms/ shovels/ troughs/ and more/ wooden as well as iron articles/ because the same people in the forest in general are inclined to more crafts/ and practice these [...] The fruitfulness of Blekinge/ this is nearly of the same kind/ as is Skåne's only that there is in general even more available of oak forest [for pig raising and fattening]/ of pine/ spruce/ birch/ hazel/ so that the occupations of the population to a large extent is chopping timber and wood/ of which they sell large quantities to foreigners/ and a part they themselves on their own crafts/ bring to other countries/ so that many/ through this/ establish great fortune. (1,1:70, 73f, 88)

In this area, there was thus a relatively well-developed economy well prior to the Skånian War. However, it was an economy that from external causes was already impaired by poor conjunctures when the provisions of the Swedish authorities hit this region fiercely after 1658. In Skåne, and especially in the forested region even prior to 1657, there are sustained signs of an economic crisis. For example, lots of taxes remained unpaid (Jeppsson 1983:98f, 104). After being transferred to the Swedish crown, the economy was hit further by the politics of customs and trade adopted on some areas and increased competition on others. After 1660, the production of iron was for instance, in crisis as it now had to perform in an integrated market with the iron works in Småland and the Mälaren region. And, now the Swedish authorities often prohibited the earlier and extensive export of horses.[54] The sale of wood and timber was limited partly by increased rates of duty and partly because foreign buyers were shut out from their earlier trading places along the coast.[55] The authorities hoped that closing such peasant harbors would reduce "all sorts of trickery" going on there (Bonde 1658:169). The peasants of

---

54   See Bjurling (1945:70, 95, 110, 151) and Weibull (1923:155f).
55   See Wägner (1886:102) and Petersson (1940-1943, II:57, III:8ff).

Jämshög complained to the Skånian Commission in 1669 that the custom had ruined their export of horses and wooden articles (Petersson 1940-1943, IV:125f).

A century and a half earlier, the export of oxen to the continental market had drawn southern Scandinavia into the world economy and a regional division of labor (Wallerstein 1974).[56] This transformation had led to popular risings and reprisals in association with the Lutheran reformations, the Master Clement and Dacke feuds. After having culminated around 1620, this earlier lucrative trade fairly rapidly disappeared from southern Scandinavia (Frandsen 1994:diagram 4 and 63). Eventually it became isolated in parts of Jutland where it continued far into the nineteenth century (Damgaard 1989, Møllgaard 1988). Even if fattening oxen for sale to the domestic market continued in Skåne until the end of the eighteenth century, its significance for the economy soon was much smaller than before 1658.[57]

The peasants in the forested region, therefore, were not any more free or rebellious from tradition. However, the positioning of their region in an economy and division of labor that reached far beyond the borders of the kingdom gave them specific interest for which to fight. That being the case, they indeed conducted this fight in a most unfortunate situation. On the one hand, their economy was seriously harmed and continuously threatened, and on the other hand, they were forced by the Swedish authorities to contribute continuously more in taxes and other fees. Furthermore, these taxes were converted from being paid in goods to being paid in money – and for the taxpayers, a disadvantageous rate made it even more important to obtain cash. In addition, a new tax per capita was introduced and the previously not taxed smallholdings, which were plenty in the forests, were surveyed for taxation (Jeppsson 1983:100f). Entire villages as well as individual peasants complained about the oppressive taxes to the Skånian Commission 1669.[58]

---

56   The production and export of oxen has been discussed by Björnsson (1946), Bjurling (1940), Frandsen (1994), Møllgaard (1988), and Weibull (1923). It is also mentioned by Bonde (1658:170-173) and Dahl (1942:plate 11). The forested region must have functioned as a breeding region in the same way as Møllgaard describes that the districts Thy, Mors, Himmerland, and Fjends sold steers for final fattening in Salling (1988:81, 83). This corresponds with the structure of property in Skåne. The woodlands (between the plain and forest) were dominated by nobility who had the privileges to export the oxen after final fattening and thereby gather the profit. The area from where they bought the breeding on the free market was to a lesser extent controlled by the nobility, as it was the case in the forest. Fabricius discusses the economically risky trade with oxen and the falling conjuncture as a reason for the bankruptcy of the Skånian nobleman Ove Thott in the 1660s (1906b:69-73).

57   See Bjurling (1945:50f), von Linné (1959:24, 304, 394f), and Weibull (1923:188-197, 213f).

58   See Bergman (2002) about this commission, the complaints, and its importance as a tool for the central authorities and locals to negotiate with each other. Bergman argues that the commission, which was a normal administrative tool of the time, allowed locals and central authorities to negotiate outside the normal structures of the developing central state,

During the Skånian War, this heavier taxation of the forested region accelerated (Fredriksson 1983:352f). In some cases, bailiffs demanded even higher taxes, keeping the money to themselves (Petersson 1940-1943, Wägner 1886).

In the period between the transfer in 1658 and the Skånian War, money became scarce in the transferred provinces. Prohibitions against export of grain hit the economy of the plain, while the peasants' principal commodity was redirected from the foreign to the Swedish domestic market, which paid less. Still they could find a market for their commodity – the Skånian plain was about to become the granary of Sweden.[59] The people in the forested region seem in every way to suffer more. Their regional clientele on the plain lacked spending power, and they themselves were cut off from foreign buyers who had previously bought their products. Scarcity of money undoubtedly hit relatively hard where the economy depended on the cash sale of advanced articles on the domestic and foreign market.

While opportunities for selling articles from the forests and obtaining money were diminishing, the taxation of common people in the region increased. If they had earlier been fairly prosperous, they gradually became if not penniless, at least relatively poorer.[60] On top of all this, the forested region was from 1660 until the outbreak of the Skånian War disproportionately burdened with quartering cavalry. The cavalry was quartered on lands owned by the crown and freeholds, and these lands dominated the forest.[61] The rights of the horseman to claim one-fourth of a farm owned by the crown (instead of receiving quarters/housing and food provisions), the ability to claim one fourth more if the peasant died, and the right to possess all of the farm if the widow died or married again put the region under extreme pressure.[62] Furthermore, quartering was drastically increased just prior to the Skånian War as Carl XI gathered his available troops in Skåne where they had to find their livelihood from the peasants (Fabricius 1952:27f). Additionally, this concentration of troops heavily burdened people in the forest as

---

which could be part of the problem. Thus, the state needed to amend to appear as just to the locals. Consequently, the commission itself was a tool for the integration of the obtained provinces into the Swedish kingdom.

59   See Bjurling (1945:96-102) and Weibull (1923:112f).

60   The Skånian Commission estimated at the end of the seventeenth century, for instance, that Jämshög parish in the heart of the forested region was "firm" (Björnsson 1946:47). In Danish historians' writings, it is often held that forested regions and the moors of Jutland were less prosperous that the plains. This is because grain growing is regarded as the true meaning of agriculture. But it is not always in accordance with historical circumstances where regions with more alternative productions could be prosperous, while grain growing plains could be poorer, and especially more sensitive to conjunctures (see Berntsen 1656, 1,1:74f).

61   Fabricius discusses questions about conscription, quartering (mostly among the peasants on lands of the crown and freeholders in the forest), and the tax burden (1906b:192-205). Quartering is a continual complaint in the works of the Skånian Commission 1669-1670 (Bergman 2002:177-187; Petersson 1940-1943, Wägner 1886:81ff).

62   See Björnsson (1946:37-40), Jeppsson (1983:105f), and Åberg (1981:48ff).

**Figure 1.10   Peasants commandeered to haul many heavy loads for the
Swedish army, by Johann Philipp Lemke (1631-1711)**

Notes: One of the burdens put on commoners in times of war was to transport the army
and its supplies. Peasants in the forested areas, which were between the Swedish army and
the staging areas for its supplies, were called out to pull many heavy loads. Contemporary
sketch by Johann Philipp Lemke.

Source: © Nationalmuseum, Stockholm.

they were obligated to serve by transporting the military and building new roads;
these roads not only joined the kingdom with its obtained provinces but also held
military significance for the control of snapcockery (Fabricius 1906b:224-227).

It can be concluded that the crown's demands to its subjects considerably
increased after the transfer of the provinces east of Øresund in 1658. The drafting
of soldiers also belonged to the list of burdens. While persons from the transferred
provinces were considered unreliable, they were often sent to the armies in the
Swedish possessions south of the Baltic. Some deserted before departure and

took to the forests where they became certain participants in snapcockery.[63] Many snapcocks defended their actions by explaining that they were forced to join the rebellious activities by other snapcocks or to secure food (Åberg 1975:111). This was probably true. Yet, the people in the forest could recollect that their interests were looked after better during the times they were subjects to the Danish king. Perhaps the people realized the end of the trade with oxen was a structural phenomenon, but the Swedish politics of high rates of duty on this and other articles did not make the adjustment any easier (Bjurling 1945:104f). Moreover, the argument could be employed that the duties and all these other burdens that were forced upon the common people mobilized the locals, and made them rebel for reactionary objectives in a time of possible transition.

It is understandable from this context that early in the Skånian War it was possible to summon the peasants in the forest in all of northern Skåne and Blekinge to support the Danish king's army – and their resistance did not appear like a bolt from out of the blue. In 1675, governor of Skåne Leyonsköld wrote Carl XI, "One cannot trust the peasant at all, especially as he is now so depressed by the manifold burdens laid upon him, which are becoming more, so he from desperation and the memory of *the good old days* easily can be seduced to some blunder" (from Fabricius 1952:24). Thus, even prior to the Skånian War snapcocks were what Hobsbawm termed social bandits (1969). The snapcocks articulated the accumulated frustration, which was partly a result of the structural decline of the forested region's economy; in their eyes, it was partially a result of unfairly extended burdens to which they were subjugated since their homes and lives had come under the authority of the Swedish crown.

A consequence of the military development of the war and a result of the politics implemented by the Swedish *and* the Danish authorities, the relatively few, persistent active snapcocks were eventually reduced to thieves. Consequently, the communities for whom they, from the start, armed themselves to defend gradually marginalized them, and they had to perish or flee to Zealand or Bornholm. Still, the many who had participated in a summoning or had accompanied a party of snapcocks on a single occasion found their way back to the community for which they had been willing to fight. On their way back, they found protection under one of the Swedish authorities' proclamations of pardon and maybe they paid a fine or returned to the proper owner what they had stolen. Thus, the few eventually were separated from the many. The more active a person had been, the more difficult it was to pursue the path back, although it is still unclear exactly what acts a person might have committed to be denied a return. At the end, circumstances decided on which side the single person would find himself – who had been drafted as a soldier, which households had been loaded with heavy quartering (by any of the warring parties), who had fired what happened to be a fatal bullet, and so forth.

To regard the snapcocks as either freedom fighters or anti-social elements is both to understand their actions from a conception of the relation between

---

63   See Fabricius (1906b:193ff) and Åberg (1958:81f, 1995:86f).

individuals and state that was not present in the seventeenth century and to neglect the complex reasons for and dynamic development of their involvement. Economic conjunctures, changes in the ways of calculating taxes, limitations to earlier rights to export from their own harbors, quartering of troops, and much more played together as the background for the fights of the snapcocks. Since the bulk of these developments hit harder in the forested region than elsewhere, it is understandable why the snapcocks were concentrated in western Blekinge and northeastern Skåne. Most of the population in all the transferred provinces tried, often deliberately, not to take a stance, as an expressed identity could very well become troubling if the political and military winds did not blow their way. Thus, contrary to the idea that identities were more pronounced and more stable in earlier eras, this situation shows that identities by many were seen as something highly undesirable. Those, however, who made a stance utilized the means available to them as they performed bordering to protect or enhance their own situation in relation to the state patriotic structures within which they lived their everyday lives.

Swedification …?

Christian V's army left a ravaged land after the Peace of Lund, which after French intervention ended the Skånian War. Many farms were deserted and the population decimated – perhaps as much as one out of three individuals had vanished (if compared to times prior to 1657) (Bjurling 1945:diagram XXII). Many, especially children and old people, had died of starvation and from sicknesses; fewer were killed in acts of war. During the war, people from Skåne had also been encouraged and ordered temporarily to move to Zealand, a policy Copenhagen continued to adopt even after the war. With promises of help and reduced taxes, a little more than ten thousand Skånians moved westwards. Priests, officials, and especially noblemen were helped to new positions. However, for commoners there was little aid available. Most of them came from the most war-torn regions and were thus, penniless; they ended up among groups of poor people without access to land in the eastern part of the island (Fabricius 1952:185-224).

After the war, Skåne surprisingly was restored quickly. Many deserted farms were again tilled (Åberg 1958:132f) and the level of violent crimes from the time of the war was normalized (Sundin 1986:152ff). As early as 1690 the population was back at the level of 1657, due to a great excess of births over deaths and immigration, including more than five thousand horsemen and infantrymen (and their families) who were moved to Skåne by the Swedish crown. The conjunctures just after the Skånian War were not good. A short upturn caused by a Swedish-Danish-Norwegian treaty of neutrality from 1691 and a few good years for agriculture after 1700 meant a temporary relief. However, largely there was an insignificant expansion of economic life in Skåne, Halland, and Blekinge in times up to the Great Nordic War.

From the Peace in Roskilde 1658, it was uncertain how the transferred provinces would be regarded in relation to the Swedish kingdom. Were they conquered lands that could maintain some autonomy, as was the case with Pomerania or Livonia? Or, were they to be regarded as a part of the Swedish kingdom and become incorporated? That the composite states embraced lands with various legislation, churches, and other differences can today be regarded as odd, because the modern state, as an ideal, has the same relation to all its citizens.[64] However, the European states of the seventeenth century were only at different paces in motion towards this unifying, homogenizing, and centralizing state. Sweden was somewhat ahead of Denmark-Norway in this development. As the Danish-Norwegian composite kingdom, the Swedish state, though, administered different relations with various groups of subjects (see H. Gustafsson 1994a, 2000a, Gustafsson and Sanders 2006).

The treaty from Roskilde was open both for incorporation of the transferred provinces and for maintaining their exceptional status. In paragraph nine, it says among other things:

> All ranks, nobles and commoners, clerics and non-clerics, townsmen and peasants, will also maintain their usual justice, laws, and old privileges and freedoms unlimited and unhindered as far as they do not run against or resist fundamental laws of the Swedish crown with which these ceded lands and counties for ever henceforth will be incorporated. (Weibull 1908:134, see also 146)

This paragraph maintains existing laws and rights but only as far as they do not run contrary to fundamental Swedish laws. While these are not specified, periods up until the Great Nordic War embraced different interpretations. Just after the Peace of Roskilde, Carl X Gustav worked for swift incorporation but it was stopped by the resumption of hostilities and his death. At the Peace of Copenhagen, the same provisions were repeated.

In the same way as Frederik III let the ranks delegate to him absolute power, Swedish authorities and officials tried in the subsequent period to make the population in the acquired provinces express its desire to become incorporated (Bergman 2002). In this way, the alterations could not be perceived as a violation of the treaties. Even if some results were achieved, it was not until the end of the Skånian War and thereafter that real power was infused into the efforts to accomplish uniformity between the transferred provinces and Sweden. This development was advanced when Carl XI, by the ranks in 1682, was assigned absolute authority. Swedish courts and laws were implemented in towns and counties; estates were subjected to the same reduction (the crown claimed back lands earlier sold or donated) as in the rest of Sweden and other provinces. In combination with bad

---

64   See the interesting discussion of the meaning of "land" (Olwig 1995) as the unit of what here is called the composite state.

conjunctures in the 1680s, this caused an end to the old Skånian nobility, which took Swedish service, married Swedish nobles, or sold its estates.[65]

The most controversial direction of the incorporation was the shift from what is most often perceived as Danish to Swedish church government and the related shift to using Swedish language during services. As language played such a decisive role for the formation of nations in the nineteenth and twentieth century (Fabricius 1958:294) it is not surprising that these questions have been regarded as essential for the understanding of cultural processes associated with the relocation of the border. At the same time, however, it is difficult to reach a thorough understanding of what happened. Research clearly demonstrates that Swedish authorities put pressure on remaining priests to learn to write Swedish and employ Swedish in their sermons. Even if some clergymen did not immediately yield, there was a clear transition towards the requested (see Fabricius 1958). However, it is also evident that in some cases, and maybe in most where the accounts refer to "Danish," the authorities were not as concerned with the language actually spoken during sermons but demanded uniformity in the performance of services.[66] This uniformity was not only a problem in Skåne, Halland, and Blekinge; the new Swedish church legislation of 1686 demanded uniformity in all parts of the Swedish realm where different dioceses earlier had established themselves with their own variants of the (Lutheran) church government (P. Nilsson 1993). It is significant that commoners' protests in the period were directed against changes in the church government, which in some cases were regarded as attempts to change their religious beliefs rather than against the language of the priest (Fabricius 1958:178, 180, 185). However, the spoken language could be a problem as congregations complained about clergymen they could not understand but this was regardless if they were from north of the old border or from west of the new (Fabricius 1906b:133, 146).

On April 10, 1678, Carl XI wrote the Swedish bishops, who had received transferred Danish and Norwegian landscapes to inform them that he found it important,

---

65   Fabricius depicts the development in detail (1958). The account here is only on the situation of Skåne and Blekinge. In Bohuslän, the same arrangements were made (Arcadius 1886, Lönnroth 1963) while a similar development in Halland, as Fabricius depicts it, had started already after the Peace of Brömsebro in 1645. The implementation of Swedish laws is discussed by Jerker Rosén (1977) and Inger Dübeck (1987).The reduction considerably strengthened the economy of the Swedish king/state (see S. A. Nilsson 1977). In the transferred provinces, the reduction was extended also to apply to donations made by the Danish crown prior to 1658 (Fabricius 1958:217-238).

66   Bergman argues that most local groups rarely used labels as "Danish" and "Swedish" and when it happened, they employed such words very pragmatically and not associated with any understanding of nationality. It could for instance be used to designate "old" times versus "now" (2002:199, 249-251, 281).

that there [be] introduced between our Swedish and the mentioned lands such uniformity and similarity, that the inhabitants and subjects through that *in the course of time* must be won for the Swedish from the Danish customs and usages and, thus, eventually *lose love for Denmark*, which otherwise undoubtedly will persist between them, because agreement of ceremonies, constitutions, customs, and language unfailingly keep the tempers together. (from Fabricius 1958:21f)

This wish can in the nineteenth and twentieth centuries be read as an order to strive for all kinds of cultural similarities between the obtained provinces and the Danish-Norwegian kingdom. But in agreement with the perception of society in the seventeenth century, it rather ought to be understood as a directive to work with and strengthen all forms of relations that bound the king (and state) and his new subjects together.

After the Skånian War it was demanded and met with some success that the population learned to recite *The Small Catechism* in an edition printed in Sweden. As a logical consequence of the change in the church government, the churches' books and bibles of course also had to be exchanged for Swedish editions. Nevertheless, this was not followed by attempts to change the speech of the new subjects. Prior to 1660, the vernaculars of Skåne shared similarities with both dialects in Zealand and Småland. The speech of townsmen was closer to their western neighbors while the people of the countryside spoke more like their northeastern acquaintances. There is no reason to believe that this picture immediately changed (S.Ö. Ohlsson 1993). For the minority who could write, investigations by Stig Örjan Ohlsson (1978-1979) point foremost to the complexity of the processes of change after the moving of the border. Any real orthography for the spelling of each word did not exist before much later. Prior to 1660, people in Skåne, Halland and Blekinge used Danish-like spellings; especially after 1680, and earliest in the towns, many officials seem to employ instead Swedish-like spellings. This can be seen both as a consequence of the immigration of officials from the north, who probably first settled in the towns, and as an illustration of the pressure for uniformity that was exerted on all writers. Around 1700, the written language was a means for the center of power to strengthen its communication with its local representatives and through these scribes, the commoners. As expressed by Bishop Hahn in 1681, the educational effort such as the teaching of *The Small Catechism* was aimed "at connecting the tempers of the common peoples in fidelity against the authorities and in good mutual comprehension" (from Fabricius 1958:65). Education was a political and institutional tool; it was not enforced either to encourage that the common people would feel unified within a cultural community or with the center of power or for that matter with each other.

The Swedish crown did not try to enforce similar uniformity in the other provinces under the Swedish crown. Finnish survived hundreds of years of Swedish government and likewise, there was no attempt to enforce Swedish language in the provinces south and east of the Baltic. In the disputed Bremen and Verden, which in 1648 again came under the Swedish crown, many efforts were made to strengthen the inhabitants' loyalty to the king. However, an idea of the kingdom as a whole or any perception of Swedish nationality was not part of this propaganda.[67]

In Europe, the attempts to implement uniformity in conquered territories are exceptions of the time (H. Gustafsson 1994a:62). That such an extraordinary procedure was invented must be understood in the light of the Swedish crown's reasonable concern that more wars might be fought over the provinces transferred from Denmark-Norway. All measures had to be taken to securely anchor the population's political relations to Stockholm. The common "ceremonies, constitutions, customs, and language" requested by Carl XI promoted this process. And because the Swedish kingdom, not least because of the lessons learned during the Skånian War, had to conceive of the Danish-Norwegian kingdom as an unpleasantly present threat to its inner security, it was principal for it to secure the loyalty of the population of the transferred provinces. While similar drastic measures were not employed elsewhere, in the territories obtained from the Danish king, Swedish authorities experimented with an idea, which resonated with Machiavellian ideals, that loyalty and integration were promoted by commonality in language, customs, and institutions (see Harald Gustafsson 2000a:307, Machiavelli 2004:14f). As the processes of dissolution of the Kalmar Union two centuries earlier (H. Gustafsson 2000a), it is fitting to regard the wars between the Danish-Norwegian and Swedish kings in the latter half of the 1600s and early 1700s as processes that propelled these states towards the modern, centralized state. The wars' contribution to establishing absolutist rule in both kingdoms is one such important step, while the concern with implementing homogeneity not in cultural but in political and institutional terms, is another. From this perspective, the measures implemented in the Swedish areas during and after the Skånian War make perfect sense. It was the structural connectors in state patriotic society – the officials, nobility, and clergymen[68] that were at the center of attention. The university in Lund (revitalized in 1682), the legal system and laws, and the freedoms and rights of the nobility were standardized. That the state patriotic structures were of primary concern is obvious when the Swedish crown at times employed earlier institutions, which existed in the Danish kingdom, but not in the Swedish, to promote the political loyalties in society even if the same measures did not contribute to uniformity. Thus, it is obvious that the Swedish state improvised and tried the capacity of alternative institutions to promote the ultimate aim – securing the loyalty of the subjects to their king. Several times, the administrative structures of the earlier Danish areas were reorganized within the Swedish state (G. Larsson 1997). When considered useful, even older Danish institutions were employed. For instance, in 1683 Bishop Hahn in Lund revived the former Danish *landemode*, a gathering

---

67   See Rainer Brüning: *Herrschaft und Öffentlichkeit in den Herzogtümern Bremen und Verden 1697-1712* (Stade 1992) summarized by Harald Gustafsson (1994a:53).

68   On the role of the Lutheranism and the clergy in the Nordic states, see Ingesman (2000), Gustafsson (2000b), and Thorkildsen (1997).

of the region's clergy, which was an institution unknown in Sweden, as a tool to promote uniformity and the change from Danish to Swedish liturgy (Fabricius 1958:124f). The Swedish state created institutions as vehicles for the promotion of the state's political geography by institutionalizing the break-up of previous connections and by hampering movements across Øresund. This was the case, for instance, when scholars from Lund University were prohibited from having contact with colleagues in Copenhagen.

Around 1700, the policy of uniformity had resulted in widespread ability to read and a common written language; these in turn, became crucial conditions for the processes by which a nation-state, Sweden, was constituted out of what remained of the former Swedish Baltic realm. In his analysis of the general background of nations, Benedict Anderson finds that nations became possible when the change from one written language (Latin) to many different languages coincided with "print-capitalism," or a new way of thinking economy (capitalism) combined with a new printing technology (1983:49). This merger of economy, technology, and language became the framework for changing conceptualizations. In novels and newspapers, contemporary events and peoples who were independent of and maybe unknown to each other, became connected. A universe of simultaneity and fellowship was created (1983:28-40). Anderson's analysis stresses the connection between nations and fields of communication, and thus, a common language is given a prominent place.

> What the eye is to the lover – that particular, ordinary eye he or she is born with – language – whatever language history has made his or her mother tongue – is to the patriot. Through that language, encountered at mother's knee and parted with only at the grave, pasts are restored, fellowships are imagined, and futures dreamed. (1983:140)

Anderson's aim is to examine how such fields are transformed to feelings for and experiences of fellowship – *imagined communities* (1983:122f, 129ff). He thus shifts the analysis towards the psychosocial in an effort to understand what Febvre called ditches, without regarding nations as effects of the individual's psychological needs.

> To understand them [nations] properly we need to consider carefully how they have come into historical being, in what ways their meanings have changed over time, and why, today, they command such profound emotional legitimacy. (B. Anderson 1983:13f)

From this, Anderson defines nations in what are undoubtedly the lines most quoted by researchers of nation building.

> ...it is an imagined political community – and imagined as both inherently limited and sovereign.

> It is *imagined* because the members of even the smallest nation will never
> know most of their fellow-members, meet them, or even hear of them, yet in the
> minds of each lives the image of their communion. (1983:15)

However, Anderson stresses that all other kinds of communities – with the
exception of the smallest face-to-face ones – are imagined in a similar way. Even
if his work is on nations and nationalism, he does not reserve the term "imagined
communities" for only this particular sense of solidarity.

Around 1700, the policy of uniformity had laid a solid foundation for the
making of a Swedish nation with its unique written language. However, the
realization of this nation was a later development – as history evolves forwards
but is comprehended backwards, it is easy to confuse this later effect with original
intentions. This backward view, however, conceals rather than clarifies the
processes. Among other things, the attraction of nations is that they edit history
and transform coincidences into faith; each community becomes equipped with its
infinite past and boundless future (B. Anderson 1983:19).

The forested region, where snapcockery had been most common during the
Skånian War, became the area in which the learning of the catechism progressed
most in the years following the war, and where additionally, many people learned
to write. However, this ought not to be understood as if people there adapted to the
"complete consequences" of the outcome of the war (Fabricius 1958:169). Their
eagerness must instead be seen in the light of their own interests and as a defense
for their outward economic activities that could be supported by such knowledge.
At a visitation in Osby parish in 1708, most of the nearly one hundred absentees
were out travelling to Skånian markets (Sundin 1986:151f). Also in centuries
following the Skånian War, the people from Göinge were known as manufacturers
and merchants of many various articles (Hanssen 1977:175-195).

After a defeat of Denmark by Prussia and Austria in 1864, it was written on a
mass grave for soldiers at the Dybbøl burial ground:

> From that, we have recognized the love,
> that You lost your lives for us.

This *"us," the nation as an imagined community*, cannot be traced in the events at
the end of the seventeenth and the beginning of the eighteenth century. Then, states
embraced groups of many different vernaculars, and the social and cultural (though
not necessarily the physical; see Hanssen 1973b, 1977) distance between master
and servant was pronounced and legitimized through the structures and functions
of the state. Two of the central officials for the implementation of uniformity in
Skåne, Halland, and Blekinge, Rutger von Ascheberg, governor general 1680-
1693, and Chr. Papke, bishop 1688-1694, were born respectively in Kurland and
Swedish Pomerania. Even if there is no reason to doubt their loyalty towards the
Swedish crown, which also rewarded them generously for their services, there is

equally no doubt that they did not belong to any Swedish cultural community.[69] No feeling of nationhood is concealed beneath state patriotism and loyalty to the king, as Fabricius suggested (1952:225). The establishment and dissemination of such a feeling must be sought in later historical conjunctures. Thus, the population of Skåne, Halland, and Blekinge were not Danes prior to 1657 in more than a strict political sense as loyal subjects to the Danish-Norwegian king. Moreover, they were not re-nationalized to become Swedes. If any Swedification happened, it meant that they were intensively instructed about and reminded of their continued obligation to be loyal subjects within the state patriotic structures, albeit to a new lord. And if they chose to disobey, this was a bordering practice that was finely adjusted to the society in which they lived.

The intensive efforts by the Swedish king's officials to strengthen state patriotic relations between people and king proved successful at the outbreak of the Great Nordic War in 1709. The generation subjected to the Danish king prior to 1657 had died, and nobody held any longer a personal recollection of any other relation of loyalty than the one to the Swedish crown. Even if times were harsh, the restructuring of the economy in the earlier centers of unrest had progressed. In addition, the tax burden had been regulated and reduced. Moreover, the earlier problematic system of quartering changed with the effect that the hosting peasants' power, in relation to the soldiers, was strengthened, and the peasants had economic advantages of the system (Åberg 1981:115ff). While the Danish king's army advanced and then retreated through Skåne in 1709-1710, the local population remained passive. There were no incidence of snapcockery and no attempts to establish companies of free riflemen. Thus, no attempts to re-establish earlier relations of loyalty and achieve advantages from this were made – not from the population and not from the Danish king. Or as Christopher Cronholm expressed it 40 years later: "The people of Blekinge and Skåne proved, during all this, their honest fidelity towards their king and fathers-land and ameliorated with this what they had committed the previous time" (1976:61). Possibly, the people in the earlier transferred provinces also had a quite realistic notion of the military resources of the Danish king as well as the interests of the decisive foreign powers in perhaps moving the border again. *Not* acting in the Great Nordic War and taking to arms during the Skånian War were both expressions that proved common people related to the border and its movements through practices that supported their own interests.

The story of the border supports that people practiced bordering within the frameworks of their everyday lives, the market, the state, and social movements in efforts to pragmatically manipulate and affect the border as they knew it – as a division between two similarly Lutheran state patriotic structures. From the above, practices within the frameworks of everyday life, the market, and the state are obvious. It is debatable whether a framework for a social movement existed in the 1600s. If limited to such a movement in a truly modern sense, it did not. However,

---

69   See Fabricius (1958:260f) on the immigration of quite comprehensive groups of German speakers and others to Skåne following 1679.

it reasonably can be argued that there existed a framework for a social movement as a possibility for acting collectively in accordance with social mores, as in "social banditry." This was the case for the earlier rebellions, the border peace treaties, the snapcocks during the earlier period of the Skånian War, and "non-activity" as a collective practice during the Great Nordic War. That the public arena for such a social movement was closing during the period discussed here was also illustrated through the central authorities' increased control over the district things. Furthermore, it is significant that people did not involve themselves in such bordering processes to obtain or maintain identity, which in relation to the state patriotic structures of power made no sense. Identity as allegiance to one of the possible structures was seen as potentially destructive and dangerous rather than desirable, as the fate of the snapcocks during the later phase of the Skånian War testified.

The Great Nordic War hit hard the population in Skåne, Halland, and Blekinge with drafts, taxes, contributions, and a subsequent pestilence. Nevertheless, it was not taken as an opportunity for the people in the transferred provinces to manifest dissatisfaction with their conditions and situation. The commoners, thus, seem not to have had interests which differed from the state's/king's in ways that called on them to take up arms and take advantage of the insecurity and weakening of local state authorities which were results of the war. The heroes, this war provided for the (later) national History, indicate the shift that had happened. The Swedish hero became Skånian governor, Magnus Stenbock, who with very limited resources managed to defend the border in the Øresund. The Danish hero had nothing to do with the transferred provinces. He was a naval carrier officer born to bourgeois parents in Bergen and named Peter Wessel. In 1716 while the war was still ravaging, but long after the last of the Danish king's soldiers had retreated from Skånian soil he was ennobled with the name Tordenskjold (literally: Shield of Thunder).

Skåne, Halland, and Blekinge were later to be Swedified also in a cultural sense. In Swedification, the widespread abilities to read and write played a significant role as the policy of the late seventeenth century carried the long-term side effect that the linguistic continuum later naturally broke down into an "east" and a "west" of the Øresund. But that implementation of uniformity, that Swedification, and digging of an emotional ditch all around the country was evolving through the same means and simultaneoeusly within the same progression of processes elsewhere in Sweden – in Stockholm, Dalarna, Västergötland...

Alternatives to the Course of Times

Prior to the Skånian War it had been debated on numerous occasions and on both sides of Øresund whether it was possible to move Skåne back to the Danish crown (Fabricius 1906a:249f). However, thereafter the Great Nordic War the border remained stable in Øresund. This, on the other hand, is not to say that alternative drawings of the border and alternative formations of states were not later topical possibilities especially concerning dynastic relations across the Øresund, which sporadically received acute attention in the following century. Two decades after the Peace of Lund, which ended the Great Nordic War, the question was raised in relation to the election of a Swedish heir to the throne after the childless Fredrik I. The Swedish government, which came to power in 1739, had on its program an attempt to recapture the former Swedish possessions south and east of the Baltic. Fearing an attack from the rear, it wanted to make sure that the Danish king, Christian VI, would not take advantage of the situation. During negotiations, Copenhagen preferred Bohuslän for Skåne as repayment for its passivity, something it seems the Swedish government made a show of negotiating (Lönnroth 1963:277ff). While the Swedish-Russian war from 1741 developed poorly for Swedish arms, negotiations were resumed. Christian, though, wisely kept from entering any alliance that would irritate the Russian emperor. Instead, he promoted more energetically, through bribes among other things, his son, crown prince Frederik (V), as a possible heir to the Swedish crown; this was done not least to prevent a person from the house of Gottorp from ascending the throne. Among the European powers that engaged themselves in the election (see O. Nilsson 1874-1905), St Petersburg had the best cards. As a precondition for an acceptable peace, the Russian empress Katharine II got her nephew Adolf Frederik of Gottorp elected as the heir to the Swedish throne in June 1743. Even if the rank of commoners and some noblemen from Skåne had supported Danish crown prince Frederik's candidacy, this was not an attempt to shift the political border. Those involved were anxious to protect the prevailing circumstances and the Swedish borders (Fabricius 1958:288ff). These negotiations on dynastic association were the last time a complete Nordic union, mimicking the Kalmar Union, was a realistic alternative. The next time such discourse was heard, Finland had been created and separated from the other countries.

Tensions between the Danish-Norwegian dynasty and the persons of the house of Gottorp on the Swedish throne continued after 1743. To promote peace, Gustav III, the son of Adolf Frederik, was married to princess Sofia Magdalena, the sister of Christian VII. Nevertheless, quarrels surfaced all the time, for instance, on the scene of the theatre. In 1779, Johannes Ewald wrote a "Romance" to his ballad opera *Fiskerne* (The Fishermen), which later became one of the two Danish national anthems. The efforts by the naval heroes in the verses are directed against Sweden:

> King Christian stood by tow'ring mast,
> In mist and smoke.
> His sword was hammering so fast,
> Through Gothic helm and brain it passed
> Then sank each hostile stern and mast
> In mist and smoke
> "Fly!" shouted they, "fly, he who can,
> ¦: Who stands 'gainst Denmark's Christian:¦
> in fray?"[70]

---

70   http://www.ambwashington.um.dk/, accessed September 9, 2006.

In Stockholm, Gustav III himself participated in working out a countermove, the opera *Gustav Vasa*, which was staged for the first time in 1786.[71] The intrigues within the courts and on the stage were replaced however, by open hostility when during the Swedish-Russian War in 1788, a Danish-Russian alliance implied that an army was sent from the Norwegian border into Bohuslän. A theatrical battle was fought and the province was captured. However, English negotiators intervened and peace was soon restored (Björlin 1890). In connection with this conflict, no attempts were made to move the border in the Øresund.

During the wars following the French revolution the two kingdoms for a long time stood side by side and cooperated among other things with convoying their merchant ships. Nevertheless, during the Napoleonic wars they came again to stand against each other. French and Spanish troops under the command of the French carrier-soldier, Marshal Jean Baptiste Bernadotte, Prince of Ponte Corvo, were sent to the North to participate in a planned but never realized invasion of Sweden. Swedish troops were with no success send against Norway. The leader of the resistance in Norway, Prince Christian (Carl) August of Augustenborg, was then in 1809 elected as the heir to the Swedish crown but he died the next year. In spite of Frederik VI's attempts to be considered, the choice in the second round fell

Notes to Figure 1.11: In 1808, the vicegerent of Norway, Prince Christian August of Augustenborg (in Schleswig), completely outmaneuvered Swedish attacks. Then he, according to an agreement, kept his forces on the Norwegian side of the border while officers in Stockholm deposed Gustav IV and instituted a more liberal constitution (Regeringsformen 1809). In return, Christian August was then elected Crown-Prince to the childless Carl XIII and became tremendously popular in Sweden. Danish King Frederik VI, who had his eyes set on the Swedish throne, got another chance; the following year, during military maneuvers in Skåne, Carl (Christian) August had an apoplectic stroke and dying, fell off his horse. Other contenders for the Swedish throne were Frederik VI's brother in law, Duke Frederik Christian (the brother of deceased Carl August), and the kings cousin, Prince Christian Frederik (later, for a short summer in 1814, the king of Norway, and from 1839, as Christian VIII king of Denmark). However, in the end, Marshal Bernadotte, one of Napoleon's generals, was elected as the successor to Carl XIII with the name Carl XIV Johan. Engraving signed J.F. Martin.

Source: Reproduced with permission from the National Library of Sweden, Maps and prints, HP C XIII. A.23.

---

71   In the original version of "Romance" it said "the Swede" in stead of "the Goth"; but this was changed under the impression of Scandinavianist tendencies in the 1830s (Østergård 1991b:146, note 32). The naval heroes of the song are: Christian IV (the battle at Kolberger Heide July 1, 1644), Niels Juel (the battle in Køge Bay July 1, 1677), and Tordenskjold. Thus, it is not completely logical, that the song is often talked about as the Danish "song of the king." On *Gustav Vasa* and an additional Danish countermove, see Lindeberg (1985:262f, 270f).

Figure 1.11 *Svea mourns the loss of Crown-Prince Carl August on May 28, 1810. The Pillar of the Throne, the Hope of the People* by J.F. Martin

on Marshal Bernadotte. With this, the latest (until present times) and fully realistic chance to establish a Scandinavian monarchy had failed.[72]

Even if the Napoleonic Wars had no effects on the border in the Øresund, they greatly influenced the two states. As a consequence of the policies of Christian VII and Frederik VI the Danish-Norwegian double-monarchy was dissolved at the Peace of Kiel and the Congress in Vienna in 1814-1815. At the same time, the new Swedish king, Carl XIV Johan (Bernadotte), had to surrender Swedish Pomerania and Rügen, the last of the Swedish possessions south of the Baltic. As a reward for his alliances with the enemies of Napoleon during the wars, Carl Johan obtained a royal union between Sweden and Norway. This was a substitution for the Finnish part of Sweden, which became a Russian grand duchy after being conquered in 1809. As a substitution for the loss of Norway, Frederik VI ended up with the small duchy Lauenburg, south of Holstein.

Many alternatives have been possible in the course of the historical developments towards the firm borderline in Øresund. Different results of single events could have changed the course of history. However, it is not possible to guess what would have been the outcome in the long run – another border between Øresund and the Mälaren region, or no border at all? Speculating in what could have happened if, for instance, the winter had not been so dreadfully cold in the end of January and the beginning of February of 1658 is as uncertain as the result of the frost is known.

Despite huge human sacrifices, the political border in Øresund had proven to be a fact that was hard to change during the one and a half century that followed the Peace of Roskilde. From the 1800s, the Øresund proceeded also to become a ditch between nations and to be recognized as a cultural division. A century later Claes Krantz could truthfully write: "Even it Öresund is not wider than one can see from the one shore to the other, one finds soon after going ashore [in Copenhagen] that one has come far, far away from Swedish circumstances" (1951:44). While digging this border ditch, the history of the birth of the border came to play its role as a source from which manifestations of different kinds of processes of bordering could be extracted and performed.

---

72 For a concentrated review of later ideas of a united Scandinavian, see Stråth (2005).

Chapter 2
An Idea of a Space

Why is Holland a nation, when Hanover, or the Grand Duchy of Parma, are not? How is it that France continues to be a nation, when the principle, which created it, has disappeared? How is it that Switzerland, which has three languages, two religions, and three or four races, is a nation, when Tuscany, which is so homogeneous, is not one? Why is Austria a state and not a nation?

(Ernest Renan 1882/1990:12)

Political and Cultural Communities

In 1827, the Danish poet Christian Wilster wrote a poem that celebrated the Danish national playwright Ludvig Holberg and his contributions to developing theater in the Danish language. In the poem, Wilster pointed to a community of language as something imperative for the nation and he described the situation of the Danish language in Copenhagen when Holberg arrived there from his native town Bergen (in western Norway) in 1703:

> Every man, who with wisdom worked on learning
> Latin on paper only scribbled,
> with ladies French, and German with his dog
> and Danish with his servant he spoke.
>
> He [Holberg] taught the Danes, that the Dane is born
> to talk with ancestral tongue,
> for home-brewed was indeed the magnificent mead,
> which strengthened heart and lung.[1]

Being Danish was to Wilster closely associated with the writing and speaking of a particular language. From his comprehension of this cultural criterion, he perceived the situation in 1703 as problematic, as Danes did not necessarily speak Danish. However, before the middle of the eighteenth century, there existed a public sphere with newspapers and literature in Danish. Wilster regarded Holberg

---

1   Christian Wilster: Ludvig Holberg, verse 2 and 11 (1827, in *Digtninger*: "Poems"). This poem can be found in "*Folkehøjskolens sangbog*" (The Songbook of the Danish Folk High Schools).

to be the paramount champion of this sphere and consequently, he saw him as a savior of the nation (Feldbæk 1991a:118ff).

However, what Wilster understood as problematic in 1827 was not perceived necessarily in the same way in 1703. In 1694, a collection of epistles and gospels in three languages was published in Stockholm for "The studying Swedish youth."

> You find here three languages for triple benefit. The Swedish, your mother tongue; the Latin, which must be used by the learned; and finally the German language. The first allows you in your younger years to look into the secrets of God's truth; the second is a means for you to get to know all secular scholarship; the third makes you capable of associating with that nation – of which a lot since many years have lived in our dear fatherland; and with which – as more beautiful provinces have had the luck, to be counted among the loyal subjects to our most gracious king – you cannot evade to [be] associated with every day.[2]

The two quotations, while separated by more than a century, tangibly illustrate that state patriotic loyalties were being dismissed as the only foundation for society in favor of what Benedict Anderson describes as the *imagined community* (1983). In 1694, the studying youth had to make themselves proficient in their mother tongue to access truth (which was religious and Lutheran), Latin to be scholars, and German to communicate with other faithful subjects in the king's lands. Obtaining these abilities made them more useful as state patriots to the majesty and of course, increased their possibilities for attracting his favor. In 1827, a new perception was materializing: it was now crucial to be included and practicing in an abstract, national community with one language. Of course Anderson is not the only scholar who has discussed how such profound alterations of peoples' way of perceiving society can and ought to be analyzed (see reviews by Alonso 1994, Brubaker 1996, Calhoun 1993, Foster 1991, Kellas 1991, Löfgren 1989, Østergård 1991a, 1992c).

Ernest Gellner has argued that the transformation must be explained as a complete rupture between earlier agrarian societies, which were divided into a food producing peasantry and a separate elite (1983:8-18), and later nations. In his conception, the nation was not the answer to a psychological requirement but a structural, functional demand from modern industrial society (1983:34f):

---

2   *Epistler och Evangelia. Oppå Sön- och Helge-Dagar/ öfwer hela åhret; Den Swenska studerande Ungdomen til nytta och gagn/ På Theras egit Moders- Latiniske och Tyske Språken utgifne/ Och uti thetta beqwäme Format/ nu nyligen af trycket utgångne. Sampt Herr Joh. Heermans Exercitium Pietatis tillagdt.* Tryckt i Stockholm. Uti thet af Hans Kongl. Maytt. Alernådigst privilegerade Burchardi tryckeri Åhr 1694. Quotation from the introduction p. 1. An interesting aspect is that the king's subjects are not asked to learn Finnish, which was indeed spoken by many of them. This reflects that the Finnish speaking parts of the king's lands were governed and administrated by Swedish speakers.

Nationalism is *not* the awakening of an old, latent, dormant force, though that is how it does indeed present itself. It is in reality the consequence of a new form of social organization, based on deeply internalized, education-dependent high cultures, each protected by its own state. It uses some of the pre-existent cultures, generally transforming them in the process, but it cannot possibly use them all. (Gellner 1983:48)

According to Gellner, only the modern state's standardized and standardizing educational system could provide the necessary mobile, generally educated, cultural homogenous workforce with one common language that was a requirement for (economic) growth (1983:19-38).[3] Nations, in Gellner's conception, are a consequence of nationalism (1983:55) because they create spaces for the swift and simple communication that was a necessity for the proliferation of industrialized societies (1983:65f, 74f, 114).[4]

Contrary to Gellner, Anthony D. Smith argues forms of historic continuity exist from modern nations and (far) back in time. He distinguishes between *modernists*, who find that nations are a product of modernity, and *primordialists*, who hold that nations are always changing appellations that have existed as natural categories (Smith 1988:6-18). Smith's division corresponds with Uffe Østergård's distinction between functional and historical explanations for nationality (1991a:168). Smith tries to establish a theoretical understanding of the nation between the two points of view. He asserts that a nation is always a continuation of a pre-modern grouping, an *ethnie*, as he terms it, with a common name, mythology, history, and culture. Within the ethnie, people feel solidarity and belonging to a particular territory. Importantly, Smith mentions that he cannot prove conclusively continuity from any specific ethnie to a particular nation (1988:32). A general problem with Smith's analysis is that he constructs very loose criteria for the demarcation of an ethnie, and therefore sees "community" concealed, for instance, behind similarities in style and architecture; he does not discuss how imaginations and practices of community are established through these particular artifacts (1988:44-46). Thus, it can be argued that Smith suffers from the same problem as Peter Sahlins. Smith does not explain what is needed as a minimum or what is sufficient for something to constitute an ethnie or a nation. Consequently, ethnies need not be the predecessors of nations. Nevertheless, an ethnie of course can very well play a part of the History that a nation employs as an argument for its own authenticity.

---

3 Gellner indicates that classes only constitute political potentials if they are associated with national groups in modern society (1983:121). He reiterates here Michael Hechter's theories about a cultural division of labor (1975). Gellner furthermore contradicts a basic Marxist assumption as he argues that agrarian societies were horizontally socially divided (had classes) while industrialized societies are not characterized by classes (1983:12).

4 John Edwards has discussed relations between languages and identities. He argues that there is no absolute tie between the two even if they often coincidence (1985). His argument is partially contrary to Gellner's findings.

Smith finds three processes crucial for nations to become the most suitable among alternative ways of organizing society in the modern era.[5] These processes are the three West European revolutions (often associated with three nations – England, France and Germany); the dissemination of capitalism and a regional division of labor; the establishment of the rational, centralized state; and the capability of this state through museums, educational systems, and other media to establish what he calls cultural standardization (1988:131ff). His thoughts on modern nations basically follow the same lines as Gellner although Smith has a completely different conception of the historical dimension of the possible pre-modern pasts of particular states.

Eric Hobsbawm also has discussed the historical dimension of nations. He finds that it is a fact that cultural differences existed in pre-modern societies. Criteria that demarcate nations can be established. However, he argues that every principle of difference – language, ethnicity, symbols, and consciousness of political affiliation is both too extensive and too confining to be understood as the basis for nations in a modern sense. These differences though, could be employed easily to legitimize nations-to-be (1990:59-77, 92). Consequently, the authenticity that nations usually wear does not exist (Hobsbawm and Ranger 1992).

Thus, scholarly arguments differ concerning the historic preconditions for nations. Nevertheless, most researchers agree that the requirements for the modern (European) nations were established in the late eighteenth century through the encounter of an economic reorganization (capitalism), the Enlightenment, and the political consequences of the French Revolution (A. Knudsen 1991). With the execution of the French king, the Revolution not only eradicated a person but also broke a then authoritative organizational principle for society where the king was appointed by God to rule the state – state patriotism. The Revolution, consequently, had to establish a new legitimization for both the organization of the state and its *demarcation*; borders needed new, trustworthy and practicable explanations. This legitimization was borrowed from the Enlightenment's conception of the natural and was connected with forms of representation of the people (democracy). While the state earlier was where the king ruled, this transformation meant that the right coincidence between territories and the distribution of peoples had to be established: Swedes, Danes, French, and so forth were to be understood as different by nature.[6] Therefore, each lived in and ruled their sovereign state as each constituted a different nation. Here, there are obvious risks that single events are ascribed far too much significance for these developments. Events such as

---

5 Smith does not distinguish clearly between nations and states (see for example 1988:129f, 214ff). He has generally repeated his argumentation in a later book (1991).

6 The thought about natural demarcation was not least influenced by the work of Johann Gottfried von Herder. He was, as Uffe Østergård has pointed out: "a thinker of the Enlightenment, who used the philosophy of the Enlightenment to erode the universalistic pretensions and hopes of the Enlightenment to substitute it with a thinking in difference, which of course if studied more closely also is universalistic, only in a different manner" (1992a:87).

the French Revolution must be understood as *expressions of* this new perception rather than as its principal cause. The transfer of power from the sovereign king, inaugurated in accordance with God's will, to the nation and the simultaneous reinterpretation of France from a land for taxation into a naturally demarcated crowd with a common past and future is, thus, a realization of the idea of the imagined community as political practices.

Economic and political conjunctures and events in the late eighteenth century promoted the idea of the nation as an adequate cultural practice that answered particular concerns in regards to what constituted a proper and sovereign state. These conjunctures, however, did not determine which practices ought and ought not to be associated with the nation, who ought to practice them or for that matter, which territories were or were not to become sovereign nation-states. As Gellner states from his functionalist perspective, not all of the possible units could become nation-states but the minimum size would be determined by the educational system: the schools where teachers taught, the colleges where teachers were taught by professors, and a minimum of one university that educated professors (Gellner 1983:34). Thus, in Gellner's understanding, the minimum size of a nation-state is decided by the minimum size of a rational educational system, which is carried by at least one university that teaches in the language of the nation. An earlier suggestion that "a language is a dialect with an army," often attributed to Yiddish linguist Max Weinreich, can be rightly rephrased as "a national language is a dialect with a university," even if an army might be necessary to defend that learning institution. Following this idea, it is obvious that today's world system not only contains realized national projects, but also a lot of never tried, tried and jettisoned, and frustrated, fighting, and frightening ones.[7] Throughout history, alternative outcomes have always been possible. Yet, history is not accidental (Fink 1992:29). The number of possibilities is always limited, indicating that in principle, it is feasible to survey all the turns and distractions after-the-fact in order to explain why events unfolded as they did. Therefore, it is possible to explain why some projects were successful while others failed in their attempts to create themselves as imagined communities with not-so-imagined political and policed borders around them.

Benedict Anderson (1983) and Eric Hobsbawm (1990) have contributed significant ideas to the problem of studying nations. While others focused on the nation as a result of political and economic conjunctures, Anderson and Hobsbawm investigated how the idea of "the nation" affected history including which forces in societies have employed, in different eras, the idea as an argument. They demonstrate that not until modern times did the nation achieve its significance as an idea about the coincidence between a state, a territory, and a people unified through (the invention of) a common language, an ethnicity, a religion, and a history.[8]

---

7   See for instance Steen Bo Frandsen's discussion of a possible Jutlandish imagined community in the first half of the 1800s (1996).

8   Benedict Anderson includes in his study non-European conditions; interestingly, he finds an American nationalism with republican institutions, citizenship, and national

According to Hobsbawm, the nation from the late eighteenth through the late nineteenth century was regarded as a stage in the evolution towards a whole, undivided world. The idea was progressive, turned against conservatism and traditions, and it promoted easily transgressed borders (1990:37ff). This radical conception of the character of nations went hand-in-hand with the development of the modern state in which subjects were transformed into citizens. The people was made aware of this change through paying taxes, being drafted, being in contact with the administration, the police, schools, the railroads, the telegraph, and through filling out census forms (Hobsbawm 1990:80-100). In the process of forming this *political community*, the political center (the elite) closely observed the citizens' relations to the state, as loyalty from the citizen was balanced by the political rights assigned to or acquired by the people.[9]

As a reaction to this radical employment of the idea, dynasties let themselves be nationalized from the mid-nineteenth century and implemented an official nationalization policy of their lands (B. Anderson 1983:80ff). The nation, consequently, was utilized in new ways resulting in political mobilization based on ethnicity and language (Hobsbawm 1990:43), and turning nationalism into a defense for and generator of history and traditions. The conception of the nation became cultural and conservative; consequently border maintenance became essential to practices associated with this idea (Hobsbawm 1990:102).[10] This new *cultural community* was not carried forward by the elite or by workers. It was promoted by the educated middle class who was in a socially vulnerable situation and through their alliances with peasants. Through these processes, languages became politicized and essentialized not least through schools because citizen and state used the media for communication with the other (Hobsbawm 1990:93ff). Each state-carrying language became the vehicle for its particular nationalism because it was the only proper means for expressing experiences and feelings.[11]

---

representations prior to the French Revolution (1983:78). As this study is concerned with Europe, this phase is not discussed in details here. However, in general the Eurocentric bias of the discussion of nation-state building and nationalism is a problem of my concern.

9   It is obvious that the rights and duties did not encompass all previous subjects but a certain restricted portion hereof. Restrictions were made according to changing criteria: gender, age, economic conditions, and civil status. Kirstine Damsholt has in her thesis analyzed how the question of defense and understandings of the state, the people, and the nation were intertwined in Denmark in the late eighteenth century and the mid-nineteenth century. She argues that democratization and the will to participate in the defense must be understood as parts of the same perception of the organization of the state (see summary in Damsholt 1995).

10   Fabricius also discusses this development in the introduction to his work (1906a:5ff).

11   This in itself, as Benedict Anderson argues (1983), should have called attention to the Americas where nationalisms continued to employ the languages of the former colonizers; even if these languages later developed some national characteristics. For

In summary, Benedict Anderson and Eric Hobsbawm both distinguish between two conceptions of the nation, one succeeding the other. Each formulated a hegemonic idea of nationhood that was the constitutive idea for the relations between nation and state in a particular era. Both understandings can be traced back to the times of the French Revolution (Hobsbawm 1990:21ff). Uffe Østergård, similarly, finds that two conceptions of the national – one political, subjective (you are what you are because you choose to be) and the other cultural, objective (you are what you are because you are) – were presented by Johan Gottfried von Herder from the 1770s; he introduced the national as an idea about difference and a notion of belonging.[12] Smith and Gellner also discuss two forms of nationalism but reference them (based on Hans Kohn's works) within a geographical distribution in European space. In the west, there was a principally territorial, bureaucratic, and inclusive nationalism, and in the east, there was an ethnic, genealogical, and separatist nationalism.[13] These different thoughts about spatial and chronological differences can be combined however. The *political community* (western, bureaucratic, inclusive) dominated the time from the French Revolution through approximately 1880, and the *cultural community* (eastern, popular, exclusive, and possibly separatist) materialized in the period from 1870 until 1918 (according to the chronology of Hobsbawm and slightly earlier, according to B. Anderson).[14]

While it is not uncommon to distinguish between an earlier centralist state and a later nation-state, from the above it is clear that such a distinction does not separate two distinct stages in a historic development. The development from one to the other covers at least two cultural transformations. In the first, an earlier identification by subjects with a political-religious structure that privileges the prince becomes substituted by an identification of citizens as political subjects (*citoyen*) with a corresponding political geography that is most often a territorially defined state. In the second transformation, the established geographical-political object of identification is overwritten, as citizens identify instead with a cultural geography that also comprises a territorial entity. How the transformational processes evolve in each case tremendously influences understandings of bordering and the history of each nation-state.

One important consequence of different conceptions of the nation is that one community might not always understand the other. Orvar Löfgren has pointed out that in general, here is the root to the evolutionary view that northwest European

---

instance, the version of Microsoft Word in which this text is written, contains 19 different versions of Spanish, each identified with a particular nation-state.

12   See Østergård's discussion (1990 and 1992a:84-113).

13   See Smith (1988:129-152, 1991:80ff) and Gellner (1983:94ff).

14   East and west ought not to be taken literally. It can be argued that the cultural community was the foundation for the dissolution of the Swedish-Norwegian personal union in 1905 as well as for the simultaneous wishes among people in North-Schleswig to be affiliated with Denmark and Icelanders to achieve independence from Denmark in the latter part of the 1800s and the early part of the 1900s (Vollertsen 1992).

nations foremost employ when they (with irony and worries) look down on mainly central and east European nations, not least the Balkan, and are unable or unwilling to understand "them" and their overt nationalisms (Ehn, Frykman and Löfgren 1993:28ff).[15] Similarly, here lies the basis for distinguishing between the good political communities, which are peaceful and well balanced, and the bad, outwardly aggressive and inwardly exclusive cultural communities (Balibar and Wallerstein 1991). Both "good" political and "bad" cultural conceptions of the imagined communities are present in every national project; however, in certain periods and areas one asserts hegemony over the other. Furthermore, one has to recognize that the "good" conception of the imagined community is only good in contexts that celebrate specific kinds of political processes, namely democracy and elected representatives as the best (or the least deficient) form of government. On the one hand, this stresses the importance of acknowledging that the "good" nationalism in some contexts needs to be recognized as distinctively "western" (see Chatterjee 1993). On the other hand, this equally stresses that the difference between good and bad nationalisms are not due to their "essence," but due to their different radical potentials. While the early, political conception of the imagined community questions the political structures and institutions of existing state-patriotic lands and paves the way for democratic development and representation, the later cultural conception questions the geography of the existing nation-states, threatening to disrupt borders where there is not a "proper" coincidence among state, culture, and language.[16]

Michael Harbsmeier has in an article (1986) discussed relations between nations and has contributed to the analytical potential of the concept of imagined communities (Østergård 1992c:35). Through a review of Benedict Anderson's works, Harbsmeier offers an important perspective that adds to the understanding of nations and national identities. According to Harbsmeier's discussion, nations are different from empires and religions. While the latter imagineered identities contain asymmetrical relations (*they* are not as good as *we* are), the relation between nations in modern time is constructed as symmetric, schismogenetic identities that presuppose the existence of other similar nations from which a nation can separate itself. Nations, thus, obtain identity through bordering processes in which ideas and practices of *mutual recognitions and acknowledgments of differences* are crucial; *they* exist, therefore *we* exist (Mead 1970).

---

15   Michael Billig coins the term "banal nationalism" for toned down or unnoticed nationalism, pointing out the example of a flag hanging unnoticed from a public building as opposed to a flag waved with passion (1995:8). However, using representations of nations (languages, flags, coins, news) as examples, he fails to stretch his social psychological perspective to cover the truly banal such as the unnoticed aspects of everyday being, which will be discussed later.

16   It can be argued that political discourse in the US and parts of Europe in the early 2000s (the George W. Bush government) embraces the first of these qualifications (as in the War on Terrorism and the idea that the War in Iraq will propel democracy there and in other Arab nations) while their actions, though, demonstrate that they are oblivious to the other.

This perspective has the potential to improve significantly the conceptual tools available for studying imagined communities. At the same time, it shifts the research interest from a nation to the relational process between imagined communities and thus, to what is here conceptualized as bordering. This shift in itself will be discussed further in the following chapter. What is important to recognize is that if there are two supplementary understandings of imagined communities – political and cultural communities – then there will be two equally and supplementary kinds of bordering processes. *Political bordering* focuses on the line between two nations as a mutually recognized divide between two political subjects, while *cultural bordering* develops the perspective that such a ditch must reflect real, objective cultural differences between the people on either side of the border. This argument follows Febvre's idea (1973) and indicates that there exists an intimate connection between the kind of nation-state and the kind of bordering processes that can be documented.

One-State or a Danish Nation

Distinguishing between political and cultural conceptions of imagined communities and bordering is clarifying on a general European level. However, it ought not lead to the conclusion that the formations of nations appear as definitive processes. If a nation is to last, it must continually be recreated or transformed as well as simultaneously re-legitimized.[17] Consequently, the study of nations must be elevated above the classification of general types and the establishment of chronologies. It must be contextualized instead within historical circumstances that explain nations' power of attraction and their ability to survive though many political transformations.[18] A short description of developments focusing mainly on the question of language within the Danish nation will shed some light on situations in which political and cultural perceptions of borders play together and succeed each other.

Questions about Danish-German bordering processes and their associated discussions about the relation between the kingdom Denmark and the duchies Holstein and Schleswig (where in parts the Danish king was the duke), advanced and illustrated ideas about the nation, as these questions came to significantly

---

17   See discussions in Balibar and Wallerstein (1991:93-105, 228) and in Löfgren (1989:22f).

18   This perspective furthermore should facilitate an analysis of how nationhood has been adopted and supported by very different political and economic structures and state systems. In this context it is interesting to note that Benedict Anderson's curiosity towards nations and nationhood originated in his interest in how similar, or not so similar, communist states in southeast Asia could end up fighting wars against each other (1991:1f).

inform debates about Danishness in the early and mid-1800s.[19] One particular problem with Danish-German relations was that people in the Danish public sphere disapproved of German, which at the same time was the language used by parts of the royal state administration.[20] The German language was perceived as a symbol for the regime's induction of foreign born civil servants, and at the same time, it was the common language of parts of the state (the duchy Holstein and the southern part of Schleswig). Thus, in debates, political loyalty towards the state and king collided with cultural bordering processes over the question of Danish-German relations within the One-State as well as between this state and other German areas.[21]

To argue from the vantage point of the relation between language and state became, through the eighteenth century, an exclusive cultural argument for a part of the Danish bourgeoisie who sought to reserve for themselves possibilities and careers within the administration of the absolutist state. Ole Feldbæk has argued that this constitutes an original and early demonstration of Danish national identity in the European context (1991a). As Hobsbawm estimates the start of processes related to cultural communities to be around 1870 and as Anderson suggests it to be in the middle of the nineteenth century, this could be true. However, the result from the discussions on these issues in Denmark, the so-called Law on Naturalization from 1776, is built upon an explicit political comprehension of the state. Without regard to cultural (linguistic) criteria, the offices of the state became reserved for persons born in the king's lands. Although culture was given some political significance, this relation was not (yet) associated with the demarcation of the state. This process and its legislative result, if anything, ought to be seen as an example of the co-existence of cultural and political conceptions. It furthermore highlights that both possible understandings of bordering, at that time, conformed to the hegemony of state patriotic ideas and practices, which promoted patriotism and loyalty to the king as integrated political elements.

Ove Malling, quoted above for his plea for the divine source of the Crown, wrote in 1777 about "love to the fatherland."

---

19   Questions about Danish and German bordering processes mark the work *Dansk identitetshistorie* (Danish Identity-History, Feldbæk 1991-1992). See for instance, Feldbæk (1991a), Feldbæk and Winge (1991), Winge (1991a, 1991b), and Rerup (1991, 1992). The thorough work by Inger Adriansen on belonging in southern Jutland is also most illustrative (1990).

20   Among other parts foreign affairs, administration of the duchies, and until the 1790s the army was managed in German while the navy was commanded in Danish (and Norwegian; written Norwegian and Danish being identical until the end of the nineteenth century).

21   One-State is a translation of the commonly used Danish term: *Helstat*, which literally means "Whole-State." This state encompasses Denmark and the duchies as one multilingual European state. In these debates, other lands under the Danish king at the time played a minuscule role and included Iceland, the Faeroes, Greenland, the Danish West-Indies and colonial enterprises in India and West Africa.

Nature itself has called us to love our fatherland. Even in the younger age, we start to feel more than usual for the house, the area, the air, in which we usually live; for the persons among which we are risen. We grow up and around us, we find a crowd of people with the same language, same customs, same church, same authorities – our feelings themselves make them dear to us because they are like us. (1992:89)

This assertion resonates with the earlier quoted wish from the Swedish king in 1678 – uniformity in costumes under the same authority would create harmony between the Swedish kingdom and the territories ceded to it by the Danish king. Carl XI's wish was built upon the state patriotic foundation for society, even if today it can be understood as a step towards the nation. Malling though, refers in the spirit of the Enlightenment, to nature as the creator of feelings that in turn become the foundation for unity and generosity towards the fatherland. In a sense, the imagined community as well as its history, which his book is about, existed for Malling. However, it was a natural community of a particular political kind under a divine authority; the community possessed no political sovereignty or exclusive cultural quality of its own. Thus, it very much resembled earlier understandings of a state as a family led by a father as discussed above. That the power of the king in late eighteenth century was perceived as integrated with the fatherland and did not stand above it was, however, a significant discursive difference from the situation in the seventeenth century (Feldbæk, 1991a:126). State patriotic society was in no way lacking dynamics and therefore, it changed significantly over time as the establishment of absolutist kingdoms demonstrates.[22] That these dynamics eventually led to establishing imagined communities can easily be proved through an analysis of what later happened, but this is not the same as to say that it was the aim of practices in the past, or for that matter, the aim of Carl XI. Even if radical thoughts sometimes were aired in the Danish public sphere (for instance, in 1767 on the liberation of peasants from adscription, and in 1772 on establishment an assembly of the ranks), they continued to be under the complete hegemony of arguments for absolutism and the dynastic One-State throughout the eighteenth century. The history of Malling's book itself illustrates the complex situation in 1777. Malling looked upon language as something that tied people together; he deliberately wrote his book to teach school children, "to master Danish as the language of the father-land, speak and write the same correctly" (Malling 1992:567). The efforts of Malling must thus be seen as a part of efforts to establish a correct Danish spelling and to protect the Danish language. From the 1770s this (among other things) implicated a deliberate effort against mixing Danish and German linguistic elements (Winge 1991a). However, the Danes, Norwegians, Schleswig-Holsteiners, and so forth of the One-State were exactly

---

22 Eva Österberg has interestingly argued that the development of political institutions in the history of early modern Sweden can be analyzed as a dynamic interplay between communalism and patriarchism (1995:190ff).

not talking or writing the same language. Thus, it lasted no more than two years before Malling's book appeared in two different German translations, both faithful to the original.[23]

Even if Danish-German bordering processes endured, gradually new battles were fought in this arena. From the end of the eighteenth century and approaching the wars of 1848-1850 and 1863-1864 (over the association between the Danish kingdom and the duchies), new groups with new arguments invaded this field. It was, thus, not the same absolutist state that implemented the Law on Naturalization in 1776 and approved the first democratic, liberal constitution in Denmark in 1849. Nor was it for the same inhabitants that these laws were created. While from the Danish side the battle over Danish-German relations at the end of the eighteenth century was directed against foreign (German speaking and writing) nobility and officials, it was at least from the 1820s that German-oriented Holsteiners were the opponents targeted by bordering practices. A political perception of the outer demarcation of the state was increasingly being connected with exclusion on cultural grounds of groups within the One-State. Thus, when university students from Copenhagen associated with their Scandinavian colleagues (Nilsson 1997), their peers from the (Danish king and) duke's university in Kiel (in Holstein) attended similar German student meetings.[24]

In agreement with the chronology discussed above, the practices at the time, of a political community in Denmark, thus developed a radical argument, which became a forceful propellant in driving the transformation of the framework through which the nation would come to exist – the state. The critical presence of Danish-German relations in this process illustrates the complexity of the situation where cultural arguments were continually made possible. Still, only small groups in the capital and provincial towns engaged themselves in these discussions. While the pattern of the state today can appear obvious, it was in the mid-nineteenth century evolving through an uncertain and hesitant process driven forward not least by bordering concerns. In the discussion about the political organization of the state, advocates for absolutism (state patriotism) and parliamentary rule (political, later cultural community) alternately held the momentum. In Denmark, the first was supported actively by foreign interest, not least the Tsarist regime in Russia. When the Constitution of 1849 reached its fifth anniversary, the Danish government

---

23   The book of Malling was also used to teach Danish in the so-called "mulatto-schools" at the Danish fortresses on the West African coast (in today's Ghana). The governor, though, applied for a special permission to exchange it, as its stories (history of the Danish kingdom) were difficult to explain to the children (Malling 1992:571ff). Feldbæk also discusses the book of Malling (1991a:190ff).

24   The Nordic students through these meetings became proponents for a movement usually identified as "Scandinavianism" that argued that a unification of the three Scandinavian kingdoms would be a progressive political development. Fredrik Nilsson discusses the political technologies (rallies, moustaches, steamboats) and radical arguments of the Scandinavianists (1997).

opposed a committee's wish to celebrate democracy through a commemoration at the Hermitage (a small palace on the royal hunting grounds north of Copenhagen). The Danish king, Frederik VII, allowed it because he was to be presented with a medal in honor of the occasion, yet speeches in the open were prohibited. At the same time, an organization for the protection of the Constitution was founded in Copenhagen (Rerup 1992). Many things had to be thought, practiced, and developed into routines before parliamentarism (nearly) was a fact in Denmark from 1901.[25]

From the exercises of a few, practices that imagineered the nation solidified and gained importance through the nineteenth century as the idea became meaningful to and practiced by more and more individuals and groups. The idea could explain what was happening around them, and it could be employed as an argument for interests and causes they wanted to advance. Establishing and maintaining the state as an organizer and field of communication became important to the development of the nation. But the cultivation of this field happened as much through all the movements and organizations started by the citizens; many of which were soon again abandoned. Constantly, new arenas were created for cultural practices. For instance, popular meetings or public associations were founded and such practices were transformed later into routines (Nilsson 1997, Rerup 1994, Østergård 1994). In Denmark, the late 1800s development of organizations for gymnastics and voluntary home guards, of which the latter were distinctly anti-German, illustrates such processes. It was through participation and practice that new generations subsequently learned to think in ways that were not possible previously.[26]

With alternating intensity, bordering discussions concerning Danish-German relations continued throughout the nineteenth century. Between 1848 and 1864, cultural bordering was employed as an argument for a new border ditch that excluded from the Danish kingdom at least the duchies Holstein and Lauenburg, and perhaps even the German-speaking part of Schleswig. Such an argument consequently also advocated the abandonment of the One-State. After the military defeat of the Danish army in 1864, which resulted in the secession of Lauenburg,

---

25 The complications in the processes can be illustrated by N.F.S. Grundtvig. Though often assigned a principal part in the construction of what is popular and Danish (Lundgreen-Nielsen 1992, Sanders and Vind 2003, Østergård 1984, 1992b), Grundtvig embraced enlightened absolutism, which could provide all parts of the people equal access to civil rights (Kaae 1986). Discussions on parliamentarism or royal rule flared up in Denmark when a procession of peasants went to the Swedish king in Stockholm in 1914 to demonstrate their will to participate in the defense of the country. Their action provoked a political crisis and the Swedish government resigned as a protest against the king's involvement in politics (for Danish reactions, see the clippings in the Swedish National Archive, Stockholm: UD's tidningsklip, series 1, vol. 23). In Denmark, a similar recent example of the king actively manipulating the work of the parliament came with the so-called Easter-crisis in 1920.

26 Rudolf Mrázek has written interestingly on technology and the development of such a sphere of communication in Indonesia in the late era of Dutch colonial rule (2002).

Holstein as well as all of Schleswig, political bordering was employed in favor of a referendum and the division of Schleswig into one part to be included in Denmark and the other part to remain German. After the First World War, a referendum was in fact held in accordance with Wilsonian principles. The referendum in 1920 resulted in a division of Schleswig, the northern part to be integrated into Denmark. After 1920, the Danish-German border dispute was more or less settled although intermittently, it was discussed in public. In 1936, for instance, the Dane Claus Eskildsen published *Dansk Grænselære* (Danish Border Doctrines) which was reprinted several times. In the book, cultural arguments as names, footwear, and the number of churches were used to prove that the Danish "people's border," according to the "blood-principle," clearly followed the inlet Slien and the historic fortifications *Dannevirke*.[27] Thus, Eskildsen found that the proper border was further south than the one resulting from the referendum in 1920. This southern boundary, therefore, was naturally the southern demarcation of Nordic space. Even if Eskildsen expressed complete support for political bordering, or the "disposition-principle" as he called it, it was clear that his thoughts on cultural bordering legitimized a changing of the border; "we" only needed to convince the bewildered that political and cultural bordering processes should coincide geographically.

> *The blessing of harmony between nature and disposition* is felt strongly by the men and women who were taught a German disposition, but through the great revival in the time of the referendum [in 1920] or later found the path to the people which truthfully is theirs. It is a blessing to be Danish and to possess the right to express this in southern Jutland!
>
> Therefore it is both our *right* and our *obligation* to strive for that, as many as possible of our German-disposed companions north and south of the state's border are guided to recognition of where they belong and where they, therefore, find their blessing as part of the people! (Eskildsen 1945:153)

Through the dissolution of the Swedish-Norwegian personal union in 1905 and the changing of the border southwards in southern Jutland in 1920, a relatively accurate coincidence between the demarcations of the Scandinavian countries after blood- and disposition-principles, or according to both political and cultural bordering, was established. With comparatively homogeneous populations within the demarcations and relatively small minorities outside, the Scandinavian nations had no problems through the following era to establish themselves with good, political conceptions of themselves. The role of Scandinavia, and especially Sweden, as the Conscience of the World in the period after the Second World War was not only a result of the welfare state, the *ombudsmand*, and high codes of

---

27   Dannevirke is a line of fortifications across the Jutland peninsula, south of the present Danish-German border. Its earliest parts are prehistoric. The early system of walls was connected with Hedeby, which was a place for international trade in the Viking era. The walls were used most recently as military installations in the war 1863-1864.

ethics, but also a consequence of the absence of contradictory bordering processes, which in the same period triggered and/or legitimized conflicts around the globe.

Cracks in the image of this homogeneity appeared from the 1970s with public debates about the presence of immigrants and refugees. During that period, debates on immigrants' lack of will or ability to express themselves in Scandinavian languages can be perceived as a continuation of the discussions on the relation between language and state. Still, this cultural bordering discussion is not presented as an argument for the changing of outward borders, but for the establishing of ditches within. Not all who live in the countries perceive of all other inhabitants as culturally and objectively qualified participants in the nations (see for instance Hervik 1999, Månsson 2002, Pred 2000). The Scandinavian nations, thus, are no longer perceived of only as political communities that are the objects of a personal reflection. The hegemony in the public ideas and practices of bordering has shifted towards the cultural albeit, as will be discussed later, with differences between the countries. This transformation signifies that far more exclusive and threatening foundations for identity formation built on bordering practices are again in use from the 1980s (Hauge 1991). Political bordering continues to be of decreasing concern because it is not insistent on cultural and unchangeable ditches within and without.

Svend Poulsen and the Others – Novelized

According to Benedict Anderson, "print-capitalism" was a decisive part of the technological and economic preconditions for nations (1983). Newspapers and novels, which came from the printing houses, recounted events in contexts and with characters that created imaginations of community across time and within particular spaces or spheres of communication. The reader became an omniscient participant in the events, both knowledgeable and creator of the context (Varpio 1980).

The literary genre of historical novels is a magnificent tool for the dissemination of imagined communities, as novels are in general. The idea of community clenched additional dimensions as (hi)story forged both the precondition for and the explanation of the causes of community as it developed until the time of the author.[28] Through the texts, the reader is united with generations, which passed away long ago and who might be the couriers of a particular story as well as with generations to come who will be fellow readers. As a genre, historical novels are most often built on a foundation of knowledge about real historic persons and events. On top of this, the author freely composes the story. Through dramatic events that throw the figures of the text into fictive situations and complications, and through their involvement in romantic love-stories, the characters are tested for appropriate and unsuitable feelings for the nation, which are set sharply against each

---

28 Lukács finds interesting connections between the birth of the genre of historical novels in the time after Napoleon and the status of the nation in different countries (1969).

other. The genre's usual focus on single figures and their subjectivity has made it most effective as a vehicle for political bordering. As the texts focus the characters' conduct, there is no reason for filling them with notions of cultural difference and no reason to attempt to define the peculiarity of the nation that is taken for granted in a double sense – both as existing and as already known to the reader. Thus, "in the proper historical novel the past appears as a mirror for the present through 'the great dead stock, who will rise from the graves to teach the people of the present superb feelings and noble conducts'" as it has been formulated by Erland Munch-Petersen (from Agger 1992:11).

Confrontations between the Danish and the Swedish kings around Øresund create perfect backdrops for composing historical scenarios with fictitious occurrences. Novels with motifs from the seventeenth century's moving of the border and peoples' reactions have been utilized to incorporate generation after generation into political bordering processes situated between the two imagined communities. A part of these texts draw their motifs from the world of the snapcocks and most snapcock-novels are published in Sweden. Many of the authors display an interest in local history for audiences in the area. Some texts, though, rise beyond this local existence, or they contain such direct references to events already dealt with in the previous chapter that it is natural here to involve them and their images in the analysis. In Denmark, the first attempts at this genre is a story by Peer Ditlev Faber from 1820 (1965) and in Sweden, a novel by the pseudonym O.K. from 1831.[29]

The first to write a great historical and national novel about the snapcocks was the Danish author Johan Carl Christian Brosbøll. Under the pseudonym Carit Etlar, he published in 1853 *Gøngehøvdingen* (The Göinge chief), which together with a sequel from the following year has been one of the most popular Danish novels (Etlar 1991a, 1991b). The story of the novel is built around a number of figures of which Svend Poulsen is the most important. Apart from Svend, the characters and events have no or a very limited foundation in history. The story takes place mainly in the winter of 1657-58. With shrewdness and courage, Svend and his faithful companion Ib succeed in transporting a treasure through the occupied landscape, from Vordingborg in the south of Zealand to King Frederik III in the besieged Copenhagen.

Three themes are interwoven in this story, which are also represented in Etlar's other, numerous writings (Agger 1992) – Etlar was the Danish author who during the nineteenth century, had the most titles in print (nearly twice the amount of the internationally and much better known Hans Christian Andersen). The first theme, the transformation of a military defeat into a moral victory, embraces the whole story of the novel. The beating received caused the best, and in a few cases the worst, qualities to be offered by Danes. Even those who commit severe missteps can prove

---

29   See the reviews of such novels in Blom and Moen (1987:215-224), Busk-Jensen et al. (1985:150f), Edvardsson (1977:163-168), and Wiggers-Jeppesen and Boisen Schmidt (1981:145-152).

through later achievements their human qualities and deep political commitment to the nation. The figure Abel, who deserts the Göingers and Svend by leaving his post, illustrates this. He instantly regrets and later fully compensates for his error, as he sacrifices his life in the final encounter with Manheimer, the villain of the story (Etlar 1991a:80-92, 333-338). With Denmark's chronic lack of military progress since the publication of the book, this theme has made the novel comforting reading for generations of Danes.

A political unification of the people and the king against the nobility, who fail in the moment of defeat and think only of salvaging themselves and their wealth, is the second theme of the story (Etlar 1991a:25, 52f, 241f). Thus, national qualities exist among commoners, and not among the privileged. To stress this, Ib and the king characterize Svend time and again as "noble" and "chivalrous" (1991a:116, 240f, 260). The text has a social sting against aristocracy; this is also typical of Danish (scholarly) history-writing from the middle of the nineteenth century and into the 1970s (Kjærgaard 1979).

Situating Denmark between Sweden and Germany is Etlar's third theme. The text was written just after the Triennial War 1848-1850, which had confirmed that the Germans regardless of whether they came from the duchies or further south, were that time's enemies of the Danish – both the One-State and the imagined Danish community. Etlar himself had participated in the war and later wrote about his experiences. In the novel, the brutal and wicked German mercenary, the officer Manheimer, represents the real enemy while the Swedes are amicable, even as enemies. This is illustrated through the story of Ib's sister who is tortured to death by German mercenaries while the Swedish captain Kernbok tries to prevent this wicked deed (Etlar 1991a:94-97, 264-282). The Scandinavianism of the nineteenth century made Danish-Swedish contrasts light-hearted. Thus, the book, to an unusual extent, could inform the readers about their time's enemies as it clearly illustrated the necessity of political bordering.

The story Etlar composed is, in a few ways, modeled after the story Peder Ditlev Faber published in 1820.[30] Otherwise, the differences between the texts are as conspicuous as the similarities. In Faber's text, Svend Poulsen's most prominent enemy is also a German. "I serve my king. You hire your limbs to an alien," Svend says to him (1965:415). Besides that, the story is without any sting towards Germans. In Faber's text, Danish nobility is not depicted unfavorably. Only the acclamations of the noble characters join the two texts. Faber let Svend share with the poor and needy what he snatches from the enemy (1965:383, 394, 397f), and in the end, Svend promises to continue serving the king (1965:421). Faber's story, thus, foremost advances state patriotic fidelity to the king even if Danish-German political bordering is noticeable.

---

30   One example is the following: when Carl Gustav arrives in Zealand, Svend disappears, but his horse(s) return with a blow over the eye and a bullet graze on the neck (Faber 1965:385 and Etlar 1991a:53). As early as 1807, Faber had a few popular legends about the Göinge chief printed (Agger 1992:27). For writings on Svend, see also Wagner (2003).

Figure 2.1   Frederik III and the *Gøngehøvding* meet, by Poul Steffensen (1866-1923)

In Etlar's novel, there is also some state patriotic rhetoric. Loyalty to the king as well as to family and nation is decisive for characters to pass the moral tests inserted into the story (1991a:90, 259f, 283-307). The book conveys the same mixture of family, people, and monarch as a private in the Triennial War passes on in his records of the war and his representation of what the Danish soldier sang

Notes to Figure 2.1: Frederik III and the *Gøngehøvding* meet as equals. Both, consequently, can offer the other his service:

> "I knew His Majesty would recognize me, when first we got to talk with each other," Svend said with an undisturbed self-confidence, which for the third time made the king smile.
>
> "And which kind of favor do you request from us? Demand what you want with no fear or hesitation; let me hear your heart speak."
>
> "Well, I recognize now that His Majesty is mistaken, since my reason to appear here tonight was not to receive a favor, but to offer you my services." (Etlar 1991a:27)

Source: Illustration by Poul Steffensen (1866-1923) for Carit Etlar's historic novel; photo courtesy of Björn Andersson, Historiska Media.

(H. Rasmussen 1992). In Etlar's story, proclaimed loyalty to the king means something different from the same in Faber's writings. When the king and Svend meet each other as equals in the introduction to the novel (Etlar 1991a:26-30) the people is crystallized as an acting subject (though, of course, the deserting nobility and traitors are excluded). Together with the loyalty to the king, this creates the message that Svend cannot "desert people and fatherland" (1991a:102). Nevertheless, Etlar does not need to describe the cultural peculiarity of the nation in order to perform political bordering; he only needs to testify to the cruelty of German mercenaries and identify Swedish officers as gentlemen. When Etlar describes the people as a community and turns against the hereditary privileges of the nobility, he relocates his text far away from contexts of the seventeenth century. He regards the relation between a sovereign nation and the state in the same way as the first democratic Constitution of June 5, 1849 outlines it (Damsholt 1995).[31] A democratic sequence is even included in Etlar's text – in a critical situation, Svend makes his position as a leader available to his followers. However, he is of course immediately re-elected by his Göinger as their chief because he represents the right feelings towards the nation (1991a:90f). However, noble feelings for the community do not necessarily lead to personal success. Svend gains no earthly wealth from his endeavors and he loses three times in the game of love in the story and its sequel.

In Etlar's novel, the message is clear for both the author's time and posterity. The Danish political community has earlier been threatened, but the right feelings for the nation and the political unification of people and king have carried the country through severe crisis and ultimately toward progress.[32] After Etlar, the

---

31   June 5th was chosen as Danish Constitution Day because it was the day of the only victory worth mentioning in the first year of the Triennial-War. Ironically, the victory came after the German general Wrangel had attacked because he wanted to give the king of Hanover a victory on the king's birthday (H. Rasmussen 1992:51).

32   The three themes as a complex are also represented in the sequel to the novel. The unification of the people is mentioned prominently in the introduction to *Dronningens vagtmester* (the Queens Officer; Etlar 1991b:17).

Göinge chief, as a figure for a novel, seems to have been drained for possibilities in Denmark. The three themes in his text appropriate so crucial parts of Danish imaginations and their relation to the Germans that it would be very difficult to develop an arena for a new and different interpretation. Thus, the next novel about Svend Poulsen is from Sweden.

Carl August Cederborg wrote several historical novels with stories played out in snapcock milieus. Cederborg's authorship belongs, as did Etlar's, to "quantity-literature" (Agger 1992:8ff). Many of Cederborg's novels appeared first as serials in newspapers (Fernebring and Robertson 1993). Because of this, Cederborg is largely disregarded in Swedish literary history. The central elements of Cederborg's novels predictably are not the same as those found in Etlar's books. Not only does the vantage point from the other side of Øresund affect points of view but also the historical context is different. Still, there are evident similarities between the two authors – both writers use the historical novel to frame their stories with Cederborg recognizing the genre in ways that agree with what has been discussed above.

> The historical novel has a special assignment, namely to strengthen the youngsters' love for their native land, provide them with reverence for the efforts of their fathers in their times' culture, and teach them the often forgotten truth that the present has is root in the past. (from Fernebring and Robertson 1993:111)

Simultaneously with his debut as an author of novels in 1899, Cederborg revised his political affiliation. He defected from the liberals and went to the right, as he was opposed to what he saw as the liberals' cosmopolitanism. Moreover, he was also scared by the sprouting Social Democrats, "a rolling avalanche which in due time will destroy our ancient culture" (from Fernebring and Robertson 1993:90).[33] Cederborg's novels reflect his stance on the situation in Sweden during the years immediately before and after the turn of the century, which was a time when processes of industrialization and urbanization marked Swedish society. The dissolution of the Swedish-Norwegian union in 1905 and worrying societal developments were seen as signs of troubles within the Swedish imagined community, and were highlighted by a comprehensive investigation, starting in 1907, on causes for the sizeable emigration (*Emigrationsutredningen*). Additionally the forces with whom Cederborg joined hands reacted and made resolute attempts to revitalize Swedish nation building processes; the Exhibition in Stockholm 1897, the competition for the design of a Swedish national monument in 1907, the so-called Great Youth Gatherings, and authors like Verner von Heidenstam pursued the national cause.[34] The aim was to rectify aspects of society that did not

---

33   Niels Finn Christiansen has discussed relations between socialist thoughts and the national question in Denmark (1992) while Povl Bagge has offered a broader review of developments around 1900 with regards to different positions on nationhood (1992).

34   See von Heidenstam (1920), J.O. Nilsson (1991:64, 1994:172f), and Ehn, Frykman and Löfgren (1993:22ff, 47ff, 141ff).

function well so the Swedish nation could prosper. Cederborg employed historical novels for the promotion of this message and therefore, with varying intensity, he highlights his time's wickedness in Swedish society.

Cederborg wrote four novels between 1899 and 1921 that deal with the transfer of Skåne to the Swedish crown and the Skånian War. In his debut novel *Göingehövdingen*, 1899 (The Göinge chief, 1948a), Svend Poulsen is murdered. Two subsequent novels, "The Last Snapcock" from 1900 (1945) and "Snapcock-blood" from 1913 (1948b), expand on different aspects of the story found in the first book. The last snapcock novel, "The Priest of Önnestad" (1921), is not associated with the others (1948c). The books, maybe because they appeared first as serials, are built as sequences of depictions around a gallery of historic persons who are put through different trials in war and love that test their subjective commitment to the nation.

Despite declarations of loyalty to the crown, the element of state patriotic rhetoric that was included in Etlar's work is nearly absent in Cederborg's texts; where Etlar partly conceals his national message written into the composition of his text, Cederborg clearly accentuates his point of view, which reflects his stance in party politics. One of the reasons why Cederborg left the liberals was his support for a strong defense force (Fernebring and Robertson 1993:89). In his books, he glorifies those who are willing to sacrifice themselves. A young man begins his plea to his father to let him become a soldier, "The native land is in danger, we all know that. [...] I am young and strong. Why would I sit in the corner of the kitchen here in Ousby, while other young men are preparing themselves to fight for the native land? I certainly have a debt to my country and I will pay it" (Cederborg 1948c:267). He thus naturally becomes a soldier.

In this history, Denmark was the opponent. However, according to Cederborg, the real threat to the Swedish nation came from further afield. As is most clearly stated in "The Minister of Önnestad" (1948c), it originated from Russia. Etlar situated Denmark between Swedish chivalry and German brutality while Cederborg found Sweden in the path of history framed by a friendly Denmark and a hostile Russia, and given Cederborg's politics and the publication date of 1921, Russia's obviously more wicked successor, the Soviet Union.

The Swedish nation was also threatened from the inside; subjectively founded unity is essential for the survival of the nation, which Cederborg emphasizes in "The Göinge chief." Where unity is strength, the snapcocks lose because they are divided and immoral. The snapcocks are destroyed as a warning of what will happen if the people do not gather behind the political community. They do not represent the right feelings for the nation and they are not exponents of a "real popular movement" (1948a, I:212). Their own anti-social acts disassociate them from their popular anchorage. Svend Poulsen expresses it while talking about the evil witch who later arranges his murder, "Pus-Else is not a woman, not a human being. She is the impersonation of the evil spirit of the present snapcock movement, the spirit of strife and self-importance, the spirit of robbery and treacherousness, the spirit of degeneration" (1948a, II:97f). For both Cederborg and Etlar, the

Figure 2.2   *A Night of Misery and Terror* by Alexander Langlet (1870-1953)

Notes to Figure 2.2: In the novels by Cederborg, the snapcocks' moral deficits in Skåne are revealed as they gather the local population and plan to use it for human shields during an assault on the fortified manor, Wanås:

> It was for the locals from around Wanås truthfully a night of terror, this night July 1, 1677. For a long time, it was remembered by the people and even today it is, albeit only vaguely, kept alive in legends.
>
> Everywhere on the farms, people were pulled from their beds and in large groups on the roads they were herded towards Wanås. The snapcocks were especially interested in women and children, but also men were taken, and often the wild snapcocks did not provide their victims enough time to even get scantily dressed. Everywhere one looked, misery; cries were heard from the groups that were pushed and herded as cattle on its way to the slaughterhouse. And the cruel snapcocks found it especially joyful to scare the shaken victims by depicting how they would have to walk forward through the shower of bullets and build heaps of corpses, behind which the snapcocks themselves would be safe. The misery and fear of dying among the people was assumed to affect the defenders of the fortress and consequently, everything was painted in the darkest colors. (1948a, I:238f)

Drawing by Alexander Langlet (1870-1953) for the first edition of Carl August Cederborg's historic novel *Göingehövdingen*.

Source: Photo courtesy of Björn Andersson, Historiska Media. Credit: © 2009 Artists Rights Society (ARS), New York/BUS, Stockholm.

male characters of their stories are able to make good or bad decisions because they subjectively side with or oppose their imagined community. Thus, males are allowed to make bad decisions for the right reasons, subsequently correcting themselves and again becoming loyal to the correct nation. Both authors find that females are the real evil within imagined communities. In the few cases females are important enough to be included and are allowed to make bad decisions, they do so for the wrong reasons of self-importance and greed. This, of course, reflects the perceived property that the different genders were afforded at the time during nation building processes.

Like Etlar, Cederborg directed his guns against divisions of power favoring people who are "of noble descent and own a fortune" (from Fernebring and Robertson 1993:49). If the Swedish nation was to solve its major problem (of the time of the author), other and more popular forces were needed in power to meet the material needs of the people. In Cederborg's texts, the poverty of Sweden and its background appear and leaders of society comment upon it. At the beginning of the twentieth century, he puts the words in the mouth of a folksy Carl XI:

> The most dangerous enemy of Sweden is not the Danes, or the duke, or the Dutch: with the aid of God, France, and the Swedish military machine I will be able to handle them and win a tolerable peace. No, its most dangerous enemy is the big, oppressing poverty. That is what paralyzes every tendon in the body of

society, corrodes the marrow, and hinders us from using the power invested in us by God. From where then stems this huge poverty? Well, from the fact that nearly all the lands of Sweden, which are most desirable, are in the hands of a few rich families... (Cederborg 1945 I:65)

Two decades later, Cederborg continues embellishing the same theme but many things had changed. Instead of renewing his accusations against those who in reality had in the meantime lost most of their privileges, he let the governor general of Skåne Magnus Stenbock express his high hopes for the future:

It is the hard, impenetrable wall, as an unfavorable nature and old and new sins have erected around us...That wall is the Swedish poverty. However, one must not be blue. Perhaps it is exactly that wall and its perpetual resistance against any enterprise that has laid steel in the Swedish minds and hardened the Swedish swords... (1948c:40)

Cederborg, thus, successfully crafts his novels as vehicles for historical knowledge, letting his real and fictional characters bring the youth of his time that which he wanted the nation to remember. After Cederborg allows Svend Poulsen's assassination, the chief of the Göingers seems to have been depleted for possibilities in Sweden too. Thus, Cederborg found other figures from the world of the snapcocks through which he could cast his messages.

Similarly, later authors, without the help of Svend Poulsen, have been able to let the historical novel convey the burning political questions of their times. For now, it will be left open whether the Swedish nation really was effectively reaffirmed through the efforts of Cederborg and his contemporaries. At least the salvation of the nation was not the problem for subsequent Swedish authors writing about the snapcocks.

In 1968, Artur Lundkvist wrote a novel in which the activities of the snapcocks are depicted as a popular insurrection against a Swedish, imperialistic superpower. The people is fighting against German occupying forces, cavalrymen who plunder and rape, irrational authorities, and the draft of men to enter the war south of the Baltic on the sides of the imperialists (1968:25, 28, 77, 237). Lundkvist distinguishes between those who took to the woods to fight for themselves, as the repellent Hök-Elis who kept the spoils himself, and those who, under the leadership of Prästa-Nilsen, became irregular soldiers with communal ownership and even ran a sort of literacy campaign (1968:46, 88ff). To Lundkvist, the moving of the border to Øresund is the historical backdrop against which he paints an image, arguably a caricature, of the relationship between people and authorities. Of course, the reader is positioned on the side of the oppressed people. His text relates with his time's leftist rhetoric and engagement, siding with independence movements in Asia and Africa and rallying against the US involvement in the war in Vietnam. The book promotes anti-imperialism and international solidarity as ideals while hardly touching any national questions. The reader's position is made so much easier as the historical situation is painted in gaudy colors. The poor people of Skåne are so enervated by the oppressive authorities/imperialists that they take to the roads where they, in starving hoards, behave like wild animals. Soldiers slaughter infants and old people while noble irregular soldiers are unable to do anything to help their companions (1968:140f, 221).

From the time and world of the snapcocks, the armed activists have perpetually caught the interest of the authors of historical novels; when indivduals are fighting, it is easier to establish a person's feelings towards the nation and in turn, compose an intrigue that carries the depiction forwards, painted in black and white. In three books, Sven Edvin Salje has reversed this basic trend in the genre of historical novels. He mainly describes the problems of the times of the snapcocks as how they could be perceived by the settled, non-combatant population (1958, 1960, 1968). In his texts, he focuses on uncertainty, doubt, and the pressured situation between two fighting lords who both claim absolute loyalty. In some episodes, the difficult position between lords leads to double standards and the crumpling of humanity in the communities. The message of the texts is much wider than earlier attempts, but also less clear-cut as the figures are depicted much more in reflection, no longer only in black and white but now, all in shades of gray. The stories of Salje carry a general humanistic message, which might make them more credible in relation to the world of the seventeenth century than their predecessors. However, in this message the nation plays no role. The same humanistic message is also apparent in a novel by Birgitta Trotzig in which she discusses the vulnerability and solitude of marginalized groups. However in her book, the war and the nations are rather inconsequential backdrops to the story of poor, insane, women and children (1997). Knud H. Thomsen later tried to translate such a humanistic message into a Danish story from Zealand and at the same time he tried to reuse Etlar's Svend and Ib as minor figures (1989). The author's humoristic style though, makes his attempt far less sincere than the profound stories by Salje.

A Space for the (Hi)story

The historical novel's affiliation with nation building processes makes it natural that authors set their stories within the territory that in their own times, constitutes the space of their nation. The texts' anchoring in territory is such an important element that depictions of landscapes are often what readers remember the most from these stories (Fernebring and Robertson 1993:180-196). Therefore, Etlar can use the "real" story about Svend Poulsen from the Carl Gustav Wars in Zealand (though he moves the story from the second war to the first), while Cederborg follows Svend across the Øresund in the Skånian War. Interestingly, fiction writers follow in the national footstep of authors who are discussing the "real history" of the conflicts. It is possible to argue that the nationality of the author, to a large

Figure 2.3 **The *Gøngers* attacking the Swedish king's dragoons,**
by Poul Steffensen (1866-1923)

extent, decides whether they find snapcocks to be anti-social elements of the forests or maintain that they were a part of the settled population that fought to remain Danish. While Swedish writers seem to prefer the former, Danish authors tend to lean towards the latter.

Notes to Figure 2.3: Cederborg had to regard snapcocks as a threat to the national unity within the borders of Sweden. Etlar, on the other hand, promoted the snapcocks as the true defenders of the nation on Danish soil. It is the snapcocks and their leader, the *Gøngehøvding*, who fight for the unity and survival of the nation. Here, the brave freedom fighters are obstructing the Swedish king's soldiers as they penetrate the Danish beech forest:

> Immediately after firing their pistols, the dragoons made a turn towards the unfinished part of the obstruction in order to attack the defenders from behind. But the gøngers displayed great composure. They turned the long and sharp blades of their scythes against the horses' chests while marksmen hidden behind the trees fired volley after volley at the men. (Etlar 1991a:68)

Source: Illustration by Poul Steffensen (1866-1923) for Carit Etlar's historic novel; photo courtesy of Björn Andersson, Historiska Media.

What Benedict Anderson calls the "map-as-logo" (1991:175) is not only an iconographic representation of the nation as territory, but also a metaphorical space that limits the appropriate confines of a novel and national imagination in a very perceptible way. This is illustrated by the fact that no author has tried yet to write a historical novel about the conspirators in Malmö. Øresund has been a ditch, which seen from the Danish side has made the events appear as if they took place abroad. Moreover, that the coup depended on troops from Copenhagen made it difficult to handle for Skånian and Swedish authors. And likewise, that the coup is associated with Corfitz Ulfeldt, who is regarded as a traitor against both sides, has made it even less appealing. Thus, this (hi)story is far from qualified as a vehicle for the simple message of the genre of historical novels and this border-penetrating historical novel is consequently still unwritten. As an effect of this territorialization of history, Danish authors have been dispossessed as they cannot use the world of the snapcocks in Skåne as a basis for their work. Consequently, the snapcock novel remains mainly Swedish national property. Other domestic space motifs from the Danish-Swedish wars of the seventeenth century have been available, however, to Danish writers. The rebellion on Bornholm and the conspiracy to take Kronborg and turn it over to Fredrik III have tempted many.

Etlar very accurately formulated his messages about the unification of king and people into a Danish nation that is squeezed between friendly Swedes and hostile Germans and managing very well without its nobility. His depiction is so precise that every later Danish writer dealing with seventeenth century's Danish-Swedish relations has followed directly in his footsteps. The first successor was H.F. Ewald. For his motif, he took the failed conspiracy at Kronborg (1867). His text was inspired by the recent military defeat in 1864; its message did not focus on the loss but instead, concentrated on finding its reasons in the weaknesses of Danish society.

> The reasons that the kingdom came at the brink of the precipice were far deeper
> and must be sought in the actual condition of Danish society. The exorbitant
> egoism of the ruling caste had dried up the sources of prosperity, subjugated the
> people, and choked the genuine love for the fatherland. (1867, I:93)

Through the defeat (in 1758, as well as 1864 it is assumed), the people and the
king united together win a moral victory that must be converted into a new order
of things in order to propel the nation forward (1867, II:304). Although equally
skeptical towards poor peasants and privileged nobility, Ewald praises the
enlightened civil servant who without regard to his own situation, fights for the
wellbeing of the whole people and the fatherland. Master builder Steenwinckel,
who in the text is executed because of his faithfulness towards Frederik III and
Denmark, symbolizes this type of civil servant. Consequently, Ewald depicts state
patriotism as the framework for a progressive, political development for the entire
people. His book can be understood thus, as an attempt to disassociate himself
from the Danish Constitution of 1866, which allotted the large landed interests
increased political influence as compared to the Constitution of 1849. However,
Ewald also dissociates himself from democratization as such. Eighteen years after
the first liberal constitution, he is committing himself to an enlightened absolutism
but this, of course, was not to be re-constituted in History.

A later writer, Kelvin Lindemann, took up the case of Bornholm. His book
mixes the time of the novel with his own age so that Bornholm/Denmark is situated
between Sweden and Germany in a way that the bordering message of the text can
be hardly misunderstood, "that there was more of the right spirit of fighting and
love for the fatherland between those who were underground than between those
who were above" (1943:179). Lindemann cautiously identifies the underground as
nisser, or gnomes known from Danish popular folklore that are today associated with
Christmas, as they help, to some extent, the later import of Santa Claus. However,
the reader in 1943 would immediately understand that Lindemann was writing about
the Danish resistance-movement, which as elsewhere in Nazi-occupied Europe, was
known as "the underground." In the text, the Bornholmers felt that they must fight
for the right freedom, even if this means that they need to contradict what their
elected leaders believe in (1943:9ff, 105, 229).

> There was freedom in the old way which had to be paid in blood. That, they had
> now yielded as the price was too high.
> Then there was the other freedom, that which the authorities always mentioned,
> when it admonished the people to keep an orderly manner. This ought to be
> a fine freedom, nearly not distinguishable from the former, and then it was
> inexpensive, and one could pay it in small installments and gradually.
> But as time passed, they found, well, that added up they paid an exorbitant
> price – and that freedom of the old quality and for the old price after all was the
> cheapest. (1943:145)

A bordering political conception of the nation could not be more elucidated, at least not in a Denmark under German occupation where established politicians continued to work in parliament and cooperate with the occupiers. The book, with its concealed yet very clear commentary on the contemporary situation, was only available in bookstores for five days before it was prohibited. Ironically the author had to flee to Sweden, while his book continued to be distributed illegally (Agger et al. 1984:572, Lindeberg 1985:182).

In 1993, both the (hi)stories about Kronborg and Bornholm was interpreted anew (Amdrup 1993, Cordua 1993); these texts also follow entrenched paths even though they undertook an even more radical rupture with the historical novel as a genre than did Salje nearly a generation earlier. To Salje, doubts and uncertainty led to a broader, less precise but not less emotional, humanistic message disassociated from the nation-as-framework for comprehension and, therefore, there was little bordering potential (1958, 1960, 1968). Contrary to this, the two authors from the 1990s evidently celebrate the nation and its struggle for independence as a schemata for history, but at the same time they invalidate political and emotional aspects of this struggle. As they make nationality a natural and objective precondition, their characters are not subjected to tests that prove the qualities of their personality and leadership to the reader, and the texts do not depict superior and inferior feelings towards the nation. Consequences of thoughts and deeds are not drawn. The writers, thus, do not assign their contemporaries solutions to problems – the bad and good, the black and white found in the universe of Etlar, Ewald, and Lindemann; these characters are all turned more or less gray but we are not told what to think about that. To Jeanne Cordua, the Bornholmers do the right thing through their successful uprising; Bornholmers have always been and will always remain Danes. The novel communicates an objective naturalization of the national, letting it remain ambiguous whether the victory also implies the progress of the involved individuals' personalities.

Erik Amdrup is a far more complicated author who succeeds in keeping an unquestionably national, but otherwise ambivalent message drifting over 350 pages. When the people and the king are united (while, of course, the nobility is deserting), it is unclear whether this is a promising development. "Maybe king and people for a short moment believed that they had come closer to each other and that the voice of the people would be recognized in the future?" (1993:126). The lesson Amdrup distils from history is that the common people are nevertheless always let down by their leaders (1993:282, 342). By providing each chapter of his novel with quotes about German occupation and Danish resistance 1940-1945, Amdrup postulates similarity between two situations which are otherwise difficult to compare – the accomplishments of the nobility in the seventeenth century and the cooperation of Danish politicians with German occupiers in the years following the occupation of April 9, 1940. However, Amdrup is not extracting any political and moral guidelines for his characters from this comparison. Lindemann transformed the same calculation into the whole message of his novel, and consequently he had to flee to Sweden. But in the case of Amdrup's novel, it leads nowhere. Amdrup portrays the national as threatened from within by leaders who collaborate with foreign powers. And, from the outside, the hazard is more diffuse – partly represented by big powers in the proximity, which is a comment on his time's Danish referendum on an EU treaty (1993:162). The histories of the seventeenth century and of the 1940s confirm that this threat can be very tangible, but Amdrup is neither identifying it with neighbors in the south nor in the east. There are, though, other and more distant imagined enemies. However, the ambivalent message of the novel does not make it clear how the reader ought to relate with these; regardless it is evident that cultural bordering is seen as a necessity. The novel's only fairly sympathetically portrayed character, the non-historical mercenary Tor Mickelsen who knows that the leaders will desert the people, is both widely traveled and far-sighted when asked:

> – Do you know anything about the Swedes, Tor? How are they?
> – Absolutely as you and me. The Nordic peoples resemble each other like
> brothers and sisters, so they will never be able to reconcile.
> – Do you believe that? Could we reconcile better with those down in Italy or
> should we go as far as Turkey or Arabia?
> – That will not work either! In southern Europe they are so Catholic in their minds
> that they strongly desire to fry us in our own grease. The Mohammedans
> cannot tolerate us at all, and especially not a woman like you. They have
> their hands full with believing in and worshipping Allah, pass alms, fast,
> and undertake pilgrimages. In spite of everything, I believe that you will
> like the Swedes the better. (1993:288f)[35]

National Conjunctures

The single writers and their contributions to the historical novel as a genre have been considered here individually. The messages of the books reflect that at any given time, questions posed to the nations have informed the authors' texts. When the imagined community and its context changed, so did the actual message to the reader. In Denmark the three themes of Etlar's novel caught the national so exactly that they could be traced all through the genre. Subsequent generations have in his book been able to find guidelines for where political bordering should be directed in their own times. This must be the reason that the novel has preserved its topicality and popularity; it has been sold in incessant new impressions, and it is natural that the *Gøngehøvdingen* in addition became kind of a model for later writers.

---

35   As it will be discussed further in the postscript to this book, this passage mirrored a growing discomfort around the millennium with Others that increasingly shaped public debate as well as different kinds of executions of power in the Nordic countries. See for instance Hervik (1999) and Pred (2000).

These books can, though, also be approached differently. Instead of looking at what the single novel in the time it was written can tell about questions posed to the nation, all the novels and arrangements of them can be viewed as a discursive field. The intensity within this field reveals that discussions about the nation have had changing conjunctures. The appearance, reprinting, and transformation into other media of a novel, which in this context is so popular and exact as Etlar's *Gøngehøvdingen,* can be used as an entrance into considering the movements in the whole field. That Etlar's (hi)story has been transformed into and propagated through many media has not prevented the novel itself from being issued again and again, both in reprints and in new editions. It is not possible to follow the development year by year as numbers for all impressions do not exist.[36] However, it is obvious that the book is printed in huge numbers; presumably it is the most printed Danish novel. With the conjunctures for Etlar's text as an entrance, it can be shown how new editions of that text and the appearance of the other mentioned historical novels form into a picture of the movements in the field.

Etlar's novel became popular from the outset in connection with the conflict between Danish and German in the duchies into which it was directly written in 1853. Instantly after the following defeat in 1864, the text was, in 1865, for the first time, adapted for theater by Michael Wallem Brun who grew up in Norway. It was performed on a very popular scene, *Folketeatret*, in Copenhagen. The piece made the warrior of the seventeenth century look like the soldier of the wars in the mid-nineteenth century. Spectators flocked to enjoy "the tone of the fatherland which runs through the piece" (from Dyrbye 1990:93). The Danish nation could on the stage, border itself successfully against the southern neighbor. The staging by Brun contained only images of German enemies; any hint of a conflict with the Swedish was gone. In relation to Etlar's novel, this emphasized furthered good Scandinavian relations. Two years later came the novel by Ewald about the conspiracy at Kronborg, which approached the national differently. In agreement with his state patriotic foundation, loyalty to the king was essential while bordering against the Germans was fairly restrained. A Danish multi-cultural One-State still appeared to him as the solution so Ewald found it sufficient to dissociate himself from German "neatness and roguishness" (1867, I:81). On the same foundation he similarly removes himself from a Scandinavian association. Even if he lived in a time where "peace and reconciliation ruled between the Nordic countries" (1867, II:460), he did not trust Swedes, "The fox will kiss the goose before the Swede will be our good friend, and never will a wound be healed that the scar will not be recognizable" (1867, I:110). His anger might have been caused by the Swedish-Norwegian king's promise to help the Danes in the just concluded conflict; a promise with which the governments of Sweden and Norway did not agree to comply. The other Scandinavian countries did not send any troops to aid

---

36 Information on impressions and to some extent on the size of each impression can be found in the Danish list of books (*Dansk Bogfortegnelse*) and in the Royal Library's database in Copenhagen.

the Danish army in Jutland in 1863. Etlar did not like the staging by Brun even if it promoted the sale of his novel. In 1896, Etlar himself completed a dramatization of his book, and he even obtained a court ruling that Brun's staging was no longer to be played.[37] Despite its popularity, Etlar's novel was until 1896 not printed in more than 31,000 copies and until 1925 in 125,000 (Agger 1992:41, 44).

The period from around 1905 until 1930 seems to have been the glory days for *Gøngehøvdingen*, and they coincide with the period in which Etlar's staging was widely used; it was also the time in which the question of southern Jutland was again on the political agenda. As mentioned above, in 1920 the populace in the northern part of Schleswig, through a referendum, decided to be re-united with Denmark. From 1920 until 1931, nearly 100,000 copies of Etlar's novel were printed so that in 1931, 179,000 copies had been sold. This glory period also embrace the times during which Cederborg, east of Øresund, energetically worked on the recovery of the Swedish nation through his novel about the snapcocks. The 1940s experienced, as one would expect, several large impressions of Etlar's story. During the German occupation of Denmark also came Lindemann's book on the uprising on Bornholm (1943). A staging of *Gøngehøvdingen*, which was played from 1942, became very popular and toured the country into the 1950s.

The period after the Second World War, and especially from the mid-1960s until the mid-1980s, was by and large the lowest point for Svend Poulsen and the others; a reprint of *Gøngehøvdingen* from 1961 netted 249,000 to 256,000 copies. The decline corresponds with the absence of the national theme in the Swedish novels by Lundkvist, Trotzig, and Salje from the same period. This points to a time when many people understood politics as an international issue. A parody on *Gøngehøvdingen* was staged in 1971; later, another performance was on tour to Danish village halls (Dyrbye 1990:97).

From the late 1980s the old book again gained growing attention. Between 1989 and 1992, Etlar's novel was printed in at least eight new impressions and editions, two each year. And it was also transformed into other media. In 1988 and 1989 two different comic strip versions were commenced of which one since has appeared in a series of volumes. In 1990, the story was reworked for a ballad opera by very well known Danish artists. Thus, interest in *Gøngehøvdingen* and other stories that cross-examine the national fluctuates significantly. Each time Etlar's and other authors' historical novels have been issued frequently, it indicates times in which the national has been a topical issue that was debated in society.

---

37 The staging by Brun continued to be played, at least until 1903 (Dyrbye 1990:95ff, Agger 1992:25f). *Gøngehøvdingen* was staged not only in different versions for theater; it became a popular amateur-piece, a staging especially for kids was made, and the story was also transformed to puppet theatre. Different impressions of Etlar's staging, and the staging for kids by his wife, Augusta Carit Etlar, were printed at least between 1906 and 1918 and in a new version in 1989 (appear in the Danish list of books). The puppet theater-version is mentioned at least 1916 (Wiggers-Jeppesen and Boisen Schmidt 1981 #831). In 1981 it was played on a theater in Copenhagen and at the city's museum (Agger 1992:26).

A modern medium like film has naturally not been blind to such popular themes. Several times the novel by Etlar was transferred to the silver screen. The first attempt does not exist any longer. The silent movie *Stormen paa København den 11. Februar 1659 og Gøngehøvdingen* (The Assault on Copenhagen February 11, 1659 and the Göinge Chief) was made for the 250th anniversary of the event in 1909, coinciding with the novel's second glory days. On the first page of the program, the film, which was starring several of the most well known actors for the time, is presented as "by far the most costly living picture yet executed in Denmark." The film was very short and the plot had not much in common with anything from Etlar's text.[38]

In Sweden snapcock motifs were also filmed. The history about the *Uggleherrarna* was the basis for an otherwise freely composed story, *Snapphanar* (Snapcocks), filmed in 1941. The famous Skånian actor Edvard Persson played the leading role. The film's tribute to land and home makes it one among the many pictures depicting the common man, which Swedish film was bursting with at the time (Qvist 1986, 1995). The film was well received; it was remarked that it worked as a reminder of the contemporary Swedish dislike of taking sides in the war. And, it was additionally, nearly in the spirit of Cederborg, an injunction for unity within the Swedish political community. "I had nothing against Swedes when this war started. Swedes are by the way a people I like. They govern well and fight well. It is a people I like fabulously well," as the main character says. *Göingehövdingen*, the novel by Cederborg, was the source of a Swedish film in 1953 with the same title. As a fairytale film it was received quite well by the critics. A renewed attempt in 1957, partly filmed by amateurs, called *Krut och kärlek* (Powder and Love) is mostly known because it has been deemed the poorest Swedish film ever made.[39]

In 1961, Etlar's novel was filmed again for cinema. The film was fairly faithful to Etlar's text and in accordance with its message although the intrigue was somewhat simplified for the media. The film appeared, as Munch-Petersen expressed it, "strangely flaccid" (1990:321). Even so, a sequel on Etlar's second novel was filmed two years later. Most recently *Gøngehøvdingen* was filmed in 1992 for a series by Danish and Swedish national television in cooperation. It is said to have been the most expensive Danish production ever with many well-known artists participating. The series was much debated in the press, seen by many people, and nevertheless regarded as misbegotten. Even though a sequel built on Etlar's *Dronningens Vagtmester* was already contracted, the production was stopped. The TV-production relates freely to many points in Etlar's novel. The series seems to express a development that became increasingly clear from the

---

38 The program for the film, a later newspaper article about it, and a data sheet are in the collections of the Danish Film museum in Copenhagen.

39 In 1969, a TV-series in six parts, called *Snapphanepojken* (The Snapcock Boy) was created by Max Lundgren and Bo Sköld. It was later published as a book. On the films see Blom and Moen (1987:225f) and the encyclopedia *Svensk filmografi* (Swedish Filmography).

filming in 1961-1963. New editions of Etlar's work remain faithful to the central themes but they exclude the political bordering performed by Etlar's protagonists; the consequences of the tests of person's qualities and dispositions are forced into the background. The story, thus, appears more or less as a parody or a fairytale without a point. The producer of the television series went as far as to present it as an "action-series," and it was stressed that the best international instructor was contracted for the fighting scenes. The message of Etlar was partly pulled out of the story through an emphasis on the supernatural; the witch from Etlar's story was comparatively offered much more attention. From this development one can sense the same loss of obligatory agreement between genre and message as could be observed in Cordua's and Amdrup's adaptation of the historical novel. This development was not due to lack of talent among the artists. It is characteristic that the different appearances of Etlar's story constantly have involved some of the best Danish artists. The explanation must be instead that the admonitions of political bordering as the right feelings for the nation, which are present in Etlar's text, no longer appeal to the producers or perhaps, to the public?

Etlar's story contains principally a message for the political community. The characters who do not join the nation, do not do so for subjective reasons; they are even offered the possibility of rejoining the community if they change their minds and pass an enclosed test in connection with their decision. As it has been pointed out above, this political depiction of the imagined community seems to have been displaced by prominence for the cultural community. Attempts to employ the historical novel as a vehicle for this message do not work. The genre's focus on subjective choice and effort, represented as individual decisions about joining a political community, does not allow for an unquestionable understanding of belonging. When persons' feelings towards the nation are pulled out of the stories or are transformed into an objective circumstance, the historical novel is changed either to (rough) action or it becomes empty morals and appears anachronistic.

Of course, particular attempts can also be debated in different words. Both the script edition of the story and the performances by the actors in the TV-series *Gøngehøvdingen*, which was very much shaped by action and fighting scenes, were evaluated by reviewers as advanced dilettante.[40]

The discussed anachronism opens for unmotivated assertions of cultural community that are bordered from new, more distant but also present images of foes. The anachronisms confirm the existence of the nation but released from the political subjectivity and moral behavior of the characters.[41] In the first part of the TV-series, *Gøngehøvdingen*, Queen Sophie Amalie follows in the footsteps of Amdrup as she points to enemies further away than in Sweden or Germany. She says to Frederik III, "why hasn't someone, long ago, assassinated him [Corfitz

40 See Munch-Petersen's account of the different recent adaptations of the *Gøngehøvdingen* (1990).

41 Different times' images of foes are treated by the authors in Kristiansen and Rasmussen (1988).

Ulfeldt] and choked his loathing wife. – Sorry! Leonora Christina is your sister, but there has always been something Arabic to the way she dresses."

As it will be discussed below, the national has from the 1970s with increasing intensity been under pressure by alternative political levels – both by a larger EUrope and by lesser regions. At the same time the assertion of cultural communities has increased at the expense of political communities. As is obvious from the above, this understanding is associated with cultural bordering against Islam/the Middle East. This foe, however, in the texts remains stated as an objective fact with no representation and no discussion. It is tempting to see a connection between these tendencies and the growing anachronism in the close relation between the nation and the genre of historical novels. This picture, though, is simplified. The development is complex. At the same time as ambivalence is becoming more prominent in this relation, new impressions of Etlar's original novel are continuously printed and bought.

In the Modern Maze

From the mid-nineteenth century, the appearance of the historical novels signifies that nations are given tangible expressions as they illuminate important questions to and developments of political bordering processes. On the one hand, their stories draw important lines towards other political units whose existence is recognized as fact. Nobody who reads Etlar will doubt that Swedes and Germans are real, and that they ought to be allowed to remain what they already are – not Danish. The genre, on the other hand, does not illustrate very well how the national community from the end of the nineteenth century dug ditches around itself and filled them with imaginations of cultural difference.

From the end of eighteenth century, within peoples' ways of thinking and practicing the nation is established as *an idea of spatiality* that can explain and legitimize practices while at the same time it can also generate practices. The imagined community constitutes a self-referential system of bordering practices and ideas; practices and ideas that presuppose each other and possibly are changed together simultaneously. The reciprocity between explaining and generating is associated with a material, geographical demarcation on the map and on the ground – the country where the people, who constitute the nation, live.

Michel Foucault is recognized for his work on transformations of connections between ways of perceiving and performing cultural practices. In his studies, he discusses how power is generated within *discourses* (Foucault 1972, 1979, Krause-Jensen 1978).[42] Foucault studies how discourses associated with madness,

---

42   The works of Foucault have been much debated, and the understandings of the works are numerous. His thoughts are not unchanged through his authorship. There is no reason here to debate the method of Foucault, which has also been called an anti-method – that has been done within other fields (B. Svensson 1993, especially pp. 48-69). As other

punishment, freedom and individual rights, form cultural ideas and practices within modern society. Discourses constitute parts of a modern *episteme*, where naturalness, not as in an earlier pre-modern era divinity, constitutes frames of references.

> Where the concept of *discourse* designates a number of expressive fields (spheres for what, down through times have been expressed), the concept of *episteme* designates connections of links between sciences, formations of theories of knowledge, formations of informations, and discursive practices. (Krause-Jensen 1978:51)

Within the modern episteme, discourses are formatted in variable ways. The same naturalness, to which the nation refers, is expressed in, for instance, the nature of the mad.

I employ here *ideas of spatiality* instead of the concept of discourse because with so many perceptions of the latter, it has become more meaningless than meaningful.[43] An idea of spatiality refers to the mutual constitution of cultural thoughts and practices, of which the imagined community is just one example. Additionally as a conception, *an idea of spatiality* has the strength that it points out that transformations of thoughts-practices have spatial consequences. Bordering as cultural practice takes on and is afforded obvious significance; in the same way Foucault's works demonstrate how transformations of discourses appear as spatial reorganizations – madhouses, prisons, and clinics. Discourses within an episteme are not unambiguous. They do not develop in step; they are conceptualized in changing ways and therefore, they have shifting and contradictory consequences. For instance, in modern society it is difficult to determine how to categorize a person who has both committed a crime and is insane (Foucault 1979:19ff). In a similar way the national is conceptualized through cultural processes as both a political and a cultural community, and these comprehensions are joined with and appeal to other modern spatialities such as language, race, ethnicity, even if they do not necessarily coincide with or logically depend on any of them. In continuously changing circumstances, the nation is joined together with and separated from other ideas, which do not immediately appear to be in a similar way geographically demarcated; included in this are gender, age, and class. Often it is assumed that the spatiality of nations is appealing to the spatiality of ethnicity; for example, this assumption is the basis of logic within the previously discussed work of Anthony D. Smith. While in a lot of analyses it appears as if it is possible

---

scholars have argued, I doubt that Foucault thought of his work as an outline for others to follow. Rather, through using his studies and concepts as inspirations much more can probably be gained (Österberg 1995:39).

43   Foucault himself is not unambiguous in the way he is using the concept. He uses it partly as debates and the mutual elaboration of ideas, partly as talks or discussions – possibly scientific discussions (Foucault 1972).

to clearly define language or gender, the relation between nation and ethnicity has continued to be very difficult to disentangle. It seems impossible to define nation and ethnicity so that they appear as both reasonably tangible and analytical sophisticated concepts (Calhoun 1993, Hylland Eriksen 1993). That language, gender, and other ideas within the modern episteme are not as self-evident as they might appear and thus, have at least that in common with nations and ethnicities, is another matter; ethnicity is still often seen as a principal matter of concern in relation to nations.

The modern episteme, with its references to naturalness, constitutes in this way a maze in which each idea of spatiality is associated with certain others ideas through direct or meandering paths.

> The discourses must be treated as discontinuous forms of practices, their different manifestations sometimes coming together, but just as often unaware of, or excluding each other. (Foucault 1972:229)

Through each manifestation of a particular cultural practice, new possible conceptions and coherences open in the modern maze, while others close behind the practicing group or individual. What is suitable in one time-space appears inappropriate in another. Moreover, at least in theory and with references to history, it is possible to turn back, to try to open previously passed links, or to repeat the whole journey. As Benedict Anderson stresses (1983), it is in this context important that each subsequent wave of nationalism learn from all its predecessors.

It is not until the imagined community is opened as ideas and practices that people gradually can begin to consider which bordering practices ought to fill its spatiality and where they ought to be situated so that they can be contemplated and filled by the nation's naturalness. To think and practice bordering offers new possibilities that require time to assign trustworthy and recognizable forms. Uncertainty in the selection as well as different interpretations and positions constantly create new compositions and possibilities. Thus, there is not one development. Times of anxiety, in which ideas and practices are tested and subsequently accepted or rejected, are long and complicated (Hobsbawm and Ranger 1992). There is no logical necessary relation between the idea of the imagined community and any other particular space in the modern maze. Some national projects have been fixed in times when certain relations between spaces in the maze were hegemonic, while other national projects have been fixed at other times when other ideas and practices were privileged. This explains why some nations need one, and only one, language, religion, and ethnicity while other nations are able to be as strong with more of these different ingredients.

The opening of the nation as an idea of spatiality is a process though which the continuation and termination of earlier cultural practices are woven together. In this process, an earlier, pre-modern comprehension of the demarcation of state and society as a system of mutual obligations and privileges between lord and subjects was eventually and slowly liquidated. State patriotism could create and explain constantly fewer practices and thoughts. In its place, the in-itself-sovereign-*imagined community* was expanding its explanatory space. As argued above, first this transformation focused the object of political loyalty, which was then replaced. An elevated individual, favored by divinity, was being substituted with a horizontal, secular collective. This transformation necessitated reconceptualizations of the state, its role, and the privileges and obligations of the citizens. But at first, a political structure continued to be crucial for the state's substance and demarcation. Gradually the nation-state was transformed into an idea of spatiality in which the citizens could participate. Within this idea, the nineteenth century saw the state creating many new relations with and within its population. Different practices were tested, some were proven to be relevant in relation to the nation while other practices turned out to be inappropriate and were again terminated. Conscription was introduced, newspapers spread throughout the country, and later mail, trains, political parties, cinemas, telephones, radio, television, tax returns, sewage pipes, and the daily weather forecast knit the nation into one. In completely new ways, the spheres of the private and the nation-state were integrated (Hobsbawm 1990:141ff, Löfgren 1990). Simultaneously, connections to other similar nations were established and elaborated. In 1838, regular steamship traffic across Øresund, between Copenhagen and Malmö was started.

Building on this transformation, an invented cultural community joined the political community during the latter part of the nineteenth century, and gradually achieved hegemony. Both political and cultural conceptions of the nation were continuously present, though. Even if one of them had hegemony, developments are expressions of how they play together with and against other ideas of spatiality in the modern maze. Developments are therefore always expressions of complex coherences and processes (Hannerz 1992:113). By being associated with the national, many practices are given substance and legitimacy. But these are not unambiguous developments. Gellner illustrates how the nation has had an astounding ability to gather and maintain many different, and not infrequently opposing, cultural, political, and economic processes (1983:124f).

Because of the way the idea of spatiality worked in the modern maze, all contemporaries did not get the same opportunities in relation to the nation. For instance, it was unclear for a long time which part the female could play. To Malling, it appeared that in state patriotic society, men performed virtues. Later the cultural formation of genders was transformed and naturalized under the impression of the practices and ideas of bourgeois groups, which during the nineteenth century were also leading the development of the nation. The male became associated with energy and the public sphere. As a conscript he would place himself at the disposal of the nation and as a breadwinner, he would do the same for the family. The female, though, became associated with fervor and morality as evident in the wife and mother becoming associated with the emotional side of the nation. The female would socialize the next generation to have the right understanding of the nation, which was an enlarged version of the family. However, the female obligation to reproduce the nation, which was to be executed within the four walls of the home, tied simultaneously the domesticated, feminized individual to the public sphere and society. This also made females dangerous. As it was the case in Etlar's and Cederborg's novels, the real danger to imagined communities stemmed from particular females who neglected their obligations and betrayed their men by being immoral (Linde-Laursen 1999a).

In the early 1840s, when a radical political movement such as the Scandinavianists sought to demonstrate their power-potential, they made a huge effort at the 1845 students' meeting in Copenhagen by including female spectators at the arranged participants' reception in the harbor of Copenhagen. To obtain a permission to use the biggest meeting hall in the town – the Riding house at the royal palace – they were forced however, to promise that they would only allow a male-only audience. Consequently, no females were present when Orla Lehmann (who later became a minister in the Danish government), as one of the more than one hundred speakers at the occasion, made the participants take an oath, with direct reference to the oath of the Ballhouse at Versailles in connection with the French Revolution, on the unity of the Nordic countries (F. Nilsson 1996:43ff).[44]

Still, the confusion of homely fervor and responsibility towards the national public opened up the possibility for a women's prolonged advance into the national public, and political, arena. From the very start they manifested themselves through embroidering colors and knitting stockings for soldiers to bear (and die for); subsequently they demanded and got civil and later, perhaps, equal rights to the imagined community.[45]

A range of representations that every nation required developed through such an intertwining and separation of thinking and practicing; if this had not happened, each nation would not be recognizable to others. The development started with the early American nationalisms that brought songs and flags (B. Anderson 1983:78). Later a still growing list of representations, which every nation must have or invent, emerged: its own language, landscape, money, stamps, national museums, and much more (Löfgren 1989:7ff).[46] For instance, after the Napoleonic Wars

---

44   See Stråth (2005) for a short review of different Scandinavianist ideas from the 1800s into the 1970s.

45   This development is discussed by Berggreen (1991), Colley (1992), and Mosse (1985), see also Alonso (1994:384ff). On women who bought prizes and donated colors to rifle clubs in the 1860s, see Kayser Nielsen (1993:102f). The state patriotic virtues could figuratively be manufactured as allegorical, antique female characters (Kryger 1986), while the female embodiment of the nation (in this case Denmark) from the mid-nineteenth century became "*Mor Danmark*" (Mother Denmark), who often appeared ready to defend the feelings and colors of the nation (Adriansen 1987, 1990).

46   How well established the list is can be illustrated through (more or less humoristic) attempts to create miniature nations who employ the symbols (Strauss 1984). It can also be illustrated by the fact that the first things the Baltic states got themselves when breaking away from the Soviet Union were their own flags, stamps, and money. The Baltic states had been deprived of symbols internally and externally for half a century, but with their re(new)ed, legitimate sovereignty they regained access to such symbolism.

in 1818, Danish officers who were among the occupation troops in France got a contest arranged at home; they sorely needed a "national song" to employ when the British sang *God Save the King* and the French sang *Vive Henry Quatre*. The contributions to this popular contest for an official national anthem illustrate great uncertainty regarding how an anthem like that ought to appear. Many contributors fell back on the secure program of state patriotism drawing on king and fatherland. The winning text is no longer generally known in Denmark (Conrad 1991). Later this genre became so stereotypical that everybody easily can recognize it – the different nation-states are represented by their particular, but also so similar, melody at international sports competitions and the like.

As Hobsbawm notes, the loyalty of the citizens in the nineteenth century was not given to the state but to a particular version of the country, which was invented and selected though processes like this (1990:93). Or, as Ernest Renan states, "It is good for everyone to know how to forget" (1990:16). As inventors, scientists who were interested in the people and therefore the nation, played a large role. This has been described as a rediscovery of the people's culture (Burke 1994) or as a creation thereof (Eriksen 1993). Regardless which perspective is employed on the efforts of ethnologists and folklorists from the mid-nineteenth century, an ideological construction of the people's culture, with which the nation could identify, was created from criteria as age, source, distinctive character, exoticism, and beauty as well as economic and political applicability.[47]

Before 1800, it was the political borderlines of the states that were threatened when for instance, Stockholm and Gottorp repeatedly joined against the Danish king. Later, political bordering as evidenced and carried by historical novels, pointed to the importance of individuality and subjectivity within the imagined community. Cultural bordering, subsequently, identified new understandings where often the ditches that demarcated national cultures seemed to be at stake. Therefore, ever since the nation as an idea of spatiality was opened, the nation has been perceived as being threatened by ideas, commodities, capital, and people that are penetrating its borders (Löfgren 1989:17ff, Tingsten 1979). It is in this particular context one must understand the dread, which was felt by Cederborg and like-minded, of social democracy and the cosmopolitan inclinations of the liberals. However, this is also the context for understanding the same persons' and groups' inability to see that similar efforts to protect and develop the national particularity

---

47 Most often the selected elements were not typical of the whole. For instance, the most Swedish landscape was found in the relatively socially, homogeneous Dalarna, which on the other hand was an ideal illustration of the idea of the undivided, imagined community (Frykman and Löfgren 1987:59ff, Rosander 1986). The selection process took many paths. The big, international exhibitions became an arena on which elements were tested (Ekström 1994, Stoklund 1994). Some things were moved directly from the exhibitions into museums to be preserved for the nation and the future (H. Rasmussen 1979:89ff). On the contributions by folklorists and ethnologists in the Nordic countries, see Alver (1989) and Honko (1980).

were taken in countries all around them, in their own time and through cultural bordering. Bordering processes consequently most be understood as balancing acts between efforts to perform both border penetrations and border maintenance; and this should be understood that often, controlling border penetrations through selecting certain practices is an imperative for border maintenance (Andreas 2000). Within these frames each generation establishes its comprehension of the imagined community and its expressions and representations. Some elements are taken from the parental generation, others are prepared by themselves, or most often borrowed from others. Yet the older generation regards these "loans," which do not differ from what their own generation did earlier, as a sign of the decay of the cultural community. New practices on public arenas therefore often lead to discussions and conflicts, placing groups against each other over what are possible and suitable national and bordering practices. In these instances youth is often ascribed the role as "the good enemy," which ill-timed practices can elucidate the foundations and values of society.[48] Anxiety for the survival and power of the nation is, thus, often directed against youth, who are at the time trying to establish their own conception of the imagined community. The school system and other didactic institutions and measures are consequently regarded as important instruments for the preservation, often conservation, of bordering thoughts and practices associated with the nation (one illustration is Hirsch 1987). Thus as the nation became available to more and more people throughout the nineteenth century, new schoolbooks were issued that introduced children to the imagined community. In schools, children were taught their own language, were informed about their own history, and were trained in other abilities that were requisites for their participation in modern society. Through the history books they learned about the deeds of their forefathers. In geography classes they gained knowledge about the blueprint of countries through lists of rivers, towns, and other important parts – their own and those of others (Feldbæk 1991b). And in this way thus, borders around the nation were drawn and legitimized.

---

48   See Stanley Cohen's innovative work with the term "moral panic" to describe this (1980), and the development of this theme by Jonas Frykman from a Swedish material (1988a).

Chapter 3
Bordering Narratives

Statement by the Swedish national school council
May 1922
Through the efforts, which have been carried out by the Nordic Associations on different areas to increase and deepen the feeling of solidarity among the Nordic peoples, it has naturally appeared as an important aim, that the instruction, which in the schools is conveyed on the circumstances in the neighboring countries, is of such a character that it corresponds to the closer relations between the peoples, which especially during the latest years has developed, and the demands for increasing mutual knowledge and understanding, which have been the result.

From this, it is natural, that the associations' attention has been directed among other things to the teaching of history and geography.

(*Nordens läroböcker i historia* 1937:15)

Who are the Others?

The Eider is as natural a border as Øresund. (N.F.S. Grundtvig, May 26, 1849)[1]

As was the case a century earlier, when the commoners were instructed in the transferred provinces at Øresund and elsewhere, attention around 1800 still mostly focused on religion and reading. Religion can be understood as that time's political science. Knowledge ought to strengthen the state patriotic relations between lord and subjects, as Malling imagined and through his reading book, suggested to practice.

From the early nineteenth century instruction in history seems to have expanded in schools for commoners (Feldbæk 1991b). Written history had been employed from the early Middle Ages to prove the antiquity and dignity of dynasties with no concern over the "verity" of the information in a scientific sense.[2] Knowledge about the realm of the lords and events related to the development of the states/royal houses could better than anything else legitimize the demarcations of kingdoms. History explained how it came to be that those living in a certain territory happen to share a state; consequently, this knowledge also could be employed in political bordering. The propensity of the Enlightenment to demythologize and explain coherences was continued in the field of history in the latter part of the nineteenth century as a professionalization and scientification of the subject (Floto 1985). With critical scrutiny of the historical causes as its means and objectivity as its

---

1 Quoted here from Lundgreen-Nielsen (1992:108).
2 See, for instance, the *Chronicle of Denmark* by Saxo or the Swedish *Chronicle of Karl*.

aim, history ought to be written to illuminate *"wie es eigentlich gewesen,"* ("how it really happened") as German Leopold von Ranke, one of the creators of history as a modern science, phrased it. This development predictably meant that myths of kings and royal space were eliminated meticulously from history. However, earlier times were simultaneously subjected to the projection of the idea of spatiality of the imagined community, which was hegemonic in the time of these historians (Østergård 1991b). This nationalization of the past established a close association between the subject of history and the nation, which was materialized as the "historicity of a territory and territorialization of a history" (Poulantzas 1980:114). Consequently, history is often mentioned as the principal vehicle among the subjects taught in schools for the invention and maintenance of nations. Furthermore, history could also be employed to legitimize the dominating political system of the time (Tingsten 1969) as it had earlier supported the royal houses.

Contemporary political cross-border relations and potential conflicts over the drawing of borders on maps were decisive for both the politics of the nation-states and for which demarcations on the ground were claimed as important to history. As a result, studies of the relation between Danish and German within the One-State and in a broader perspective became the main approach for Danish historians to understand questions of nationality.[3] In a similar way, the establishment of Swedishness from the late eighteenth century must be understood in light of its relationship with Russia and the 1809 lost Finnish part of the Swedish realm as well as its relations with Norway, which in 1814, it entered into a dynastic union with Sweden. In these processes, relations between Denmark and Sweden played a comparatively inferior role. The political and military developments around the Baltic had erased the geographical-antagonistic relation between Øresund and the Mälaren region; it had been cooled down by international powers, a process on the Danish side financed by the Øresund toll, and thus, had become of relative lesser importance. At the middle of the nineteenth century, the border in Øresund had become natural, as Grundtvig explicitly articulated it when he argued that the river Eider (the southern geographical demarcation of Schleswig) was the ideal national border between Danish and German.

This relaxation of Danish-Swedish political relations was also obvious when the Nordic Associations, from the 1930s, initiated attempts to create consensus regarding how events of the intertwined history of the different Nordic countries were to be presented in schools. However, despite such efforts, a joint Nordic and non-national approach to history was not employed. Rather, attempts to sanitize conflicting understandings of the past out of schoolbooks further emphasized the imagined communities as the natural basis for the state since antiquity; what was

---

3   This interest marks, as it has been pointed out above, the articles of *Dansk identitetshistorie* ("History of Danish Identity," Feldbæk 1991-1992), but also numerous other writings. See for instance Adriansen (1987, 1990), Pontoppidan Thyssen (1980), Rerup (1980), and Yahil (1991). As demonstrated above this perspective also dominated the approach to the questions developed in the historical novels.

asked for was that the depiction of earlier conflicts between the Nordic states be fair and objective.[4] This objectivity had to support the existing distribution of territory, and studies in and teaching about history had to legitimize the demarcation of the imagined community. The course of history, therefore, had to confirm the naturalness of this idea of spatiality as well as the actual distribution of territory. Thus, as Danes and Swedes had always been exactly that, history could explain their fights and conflicts but it could not, at the same time, question the objective reality of their natural differences.

As illuminated below, this identity void propelled a new understanding of and interest in the border in Øresund. As the cultural community during the latter part of the nineteenth century gained hegemony as what explained and sustained harmony between nation, state, language, ethnicity, and other cultural categories, cultural bordering was launched. In such processes, Sweden and Denmark found that they, in crucial ways, shared fate and perspective on identity formation. The two Scandinavian nations could not profile themselves as different from Germany and Russia respectively, as both the latter were too big, too culturally complex, and too acutely present in their political life as potential adversaries. Furthermore, it would be difficult to deal with the many Danes and Swedes who through historic conflicts had ended up within the borders of those big neighbors (in northern German and in the Russian grand-duchy Finland). For obvious reasons, it would be improper and difficult for Sweden and Denmark to practice bordering against nations with which they actually shared history – Finland, Iceland, and Norway. Nevertheless, they could border themselves from each other. Nearly equally small, nearly equally many, and nearly equally distant from the centers of Europe, what was Danish could be reflected in what was Swedish and vice versa. Moreover, both nations could look back on "great days of yore, When worldwide renown was valour's guerdon" as Richard Dyback in 1844 wrote in "Thou ancient, thou Freeborn" (later, the Swedish national anthem). The old multi-cultural kingdoms of Sweden and Denmark became reduced through time (around 1905 and 1920 respectively) to approximately their natural dimensions as cultural communities. Consequently, they appear as nation-states with an extraordinary singularity between *nation, state, land, area in which the languages are spoken, and people*.[5] This singularity between

---

4   See: *Omstridda spörsmål i Nordens historia I*, 1940; *Omstridda spörsmål i Nordens historia II*, 1950; *Omstridde spørsmål i Nordens historie III*, 1965; *Omstridte spørsmål i Nordens historie IV*, 1973, and *Nordens samhällen i Nordens skolor*, 1983. On the background for these efforts in the times following the First World War and on similar international efforts, see *Nordens läroböcker i historia* (1937:3-14).

5   In imagined communities, internal similarity and external difference are accentuated in attempts to measure up to this singularity. Nevertheless, it only exists in very few cases. Walker Connor mentions fourteen states that do not contain considerable national minorities. In half of these cases, considerable national minorities exist outside the state's borders (1973:1). He does not discuss his criteria, but he counts Denmark among these ideal nation-states while Sweden is not on the list.

political and cultural communities is reflected in the languages spoken. Most often, it is possible to use nation, land, people, and state as synonyms. The singularity also explains the positive and untranslatable content of the Danish word *folkeligt*, which propagate Danes as "a people – small, friendly, and with respect to democracy all other peoples indefinitely superior" (Østergård 1984:89, see also Østergård 1992b). "Folkeligt" often can be translated as "popular" but this translation does not convey very well the Danish word's strong positive connotations. In some cases, "folkeligt" also can be translated as "petit bourgeois culture," which in Danish is not necessarily understood as something negative. It is this singularity between the political and cultural community that made it possible to frame statements in terms of inclusive political bordering while at the same time maintaining a concealed and exclusive cultural bordering message. This singularity, as pointed to above, became important especially after the Second World War when cultural communities were in disrepute and the Scandinavian nations found an easy path towards becoming the conscience of the world. It is important to maintain that this singularity is only extraordinary in comparison with other states (Tägil 1995). In both Sweden and Denmark, the political demarcations embrace cultural minorities with a long history in the territories of the imagined communities (Samis, Finns, Faeroes, Greenlanders, and Germans). Likewise, both states today embrace comparatively new minorities of immigrants and refugees. In addition, both Swedish and Danish minorities are found outside the borders of their states (Swedes in Finland, Danes in Germany/ South Schleswig, and emigrants on more distant shores).[6]

The natural border in Øresund was an ideal expression of this singularity. The changing of the border occurred during a historic window of opportunity when borders could be shifted without creating later disagreements between political and cultural conceptions of the nations. On the one hand, it occurred after both Lutheran centers of power had commenced transforming towards the European central state. As discussed above, the Swedish state was strong enough that it could implement political and administrative uniformity within the acquired territories even if the measures were extraordinary at the time. Consequently, there was no space like the one for which Ulfeldt fought – a republic ruled by nobility in Skåne and associated

---

6 This singularity also actively obscures the Danish abandonment of its colonies, as this is most often omitted from understandings of Danish History. It would be easy to establish an analysis of Denmark's (or Denmark-Norway's) colonial politics from the 1600s through the 1900s. However, as there is no representational space in the national narratives for Denmark as a colonial power, it is difficult to analyze Danish narratives as post-colonial. Instead of active denial, there is nearly complete ignorance. Denmark sold its possessions in India, Tranquebar (south of Madras) and Serampore (at Calcutta) to the British East India Company in 1845. Its forts and possessions in West-Africa, in today's Ghana, were sold to Great Britain in 1850. And the Danish West-Indies (the Virgin Islands of St. Thomas, St. John and St. Croix) were sold to the United States in 1917. Greenland and the Faeroe Islands are still part of the Danish Commonwealth. However, they are not part of the European Union and are gaining more and more independence even if in the foreseeable future they seem likely to stay within a Danish Commonwealth with at least a joined monarch and defense.

with Sweden in a personal union (Heiberg 1993:167ff). Furthermore, in Skåne, no historic legitimization or other basis for political provincialism existed, as was the case with for instance, Schleswig and Holstein. On the other hand, the moving of the border to Øresund occurred before the imagined communities had been opened as an idea of spatiality; no national ideas and practices and therefore, no political consequences of culture, existed in the 1600s. Consequently, the moving did not leave any Danish national minority east of the new border. Similar windows of opportunity, which allow for a border later uncontested in national terms, must exist at all other borders. As discussed above, Sahlins depicts how the Spanish-French border became naturalized as locals learned to employ national authorities as allies in their mutual conflicts (1989, 1998). That border moving in the valley of Cerdanya and at Øresund happened at the same time. However, this does not indicate that such windows of opportunity existed in the mid-1600s and not earlier or later. Local contexts and historical processes differ; processes such as colonization and de-colonization have immense impact, affording earlier and later waves of nationalization different potentials and problems. Yet everywhere, the same singularity between processes of political and cultural bordering processes would constitute such a window of opportunity.

As cultural bordering processes gradually achieved hegemony in the late nineteenth century, nations increasingly needed to consider how they were understood as natural units by their surroundings and how they themselves conceived of other nations as natural units. Consequently, they no longer demarcated themselves only from their immediate neighbors with whom they shared political lines. In theory, they had to perform cultural bordering and dig ditches towards all other existing nation-states. As mentioned above, Michael Harbsmeier has discussed such relations between cultural communities (1986) and has developed a perspective on the schismogenetic character of modern, national identities in which the mutual recognitions and acknowledgments of cultural differences between nations is essential:

> Nations...seduce each other to recognize themselves in the images held by others. Not from that to become another than one is, but to become the one which one in reality always has been. (1986:53)

However, a nation's concept of itself is of course dependent on in which other nation's image of itself it chooses to recognize this self. Sweden appears differently if compared with Denmark or Finland:

> Seen from the inside, there are always relatively many possibilities; an arbitrary choice between foreign countries' images. Seen from the outside the number of alternatives is reduced. Nevertheless, in all circumstances national identity is a question about being able to recognize oneself in the images of one's national peculiarity, which other nations already hold, or which it is realistic to believe one can manipulate them to accept. (Harbsmeier 1986:50)

Thus, a schismogenetic identity does not place an imagined community in opposition to all others; identity appears as a relation between two elements – one nation and a significant Other.[7] This explains that general attempts to determine a nation's form and content to itself or through implicit or explicit comparisons with several or many others get stuck in variety:

> And each time it is as if everyone knows it in advance. Nationality works both perfectly and precisely in jokes, in conversation, and in entertainment. But each time someone attempts to realize it, each time someone deliberately attempts to write about it under the auspices of sciences: anthropology, folklore, ethnology, sociology, psychology, then the otherwise so living, sparkling sensation for the differences between nations and each nation's peculiarity vanish like dew before the sun. The history of nations can easily be written, but their peculiarities and differences are difficult to pin down. (Harbsmeier 1986:54)[8]

Thus, it is for instance difficult to determine what is Swedishness. Is it what differentiates Swedes from all others, but not necessarily is very essential? Is it an essence, even if it happens to be shared with others? Is it what has been Swedish for a long time? Or, is it Swedes' conception of themselves? (Arnstberg 1989:15). Jonas Frykman has described the national as a fairly determined narrative. Yet, he notices that the content of this narrative is ambiguous. He finds that this variety is caused by elements adopted and integrated into the narrative at different times (Ehn, Frykman and Löfgren 1993:120-160). One consequence of the argument above is that the great variety in the narrative about for instance, Sweden, is as likely to originate from elements of recognition from many other nations' different images. Any significant coherence is therefore never to be expected within bordering narratives. Furthermore, it should be expected that bordering narratives should be ambiguous otherwise they would not be able to include everyone within the imagined community who will continue, of course, to belong as well to many other ideas of spatialities such as gender, generation, age, class, sexuality, race, religious affiliation, and so forth (Linde-Laursen 1999b).

A viable path to avoid ambiguity and confusion in the study of national identity is to approach it with the perspective that it always has a schismogenetic character. By focusing on the relation between one nation and one Other nation, much variation is eliminated while the narrated differences that should be expressed through cultural bordering processes are emphasized. The historic changes and

---

7   As B. Anderson expected, some other cultural collective identities are theoretically constructed in the same way. See the analysis of the development of European identity (in Boll-Johansen and Harbsmeier 1988) and the investigation of the Orient by Edward Said (1978).

8   For a contemporary attempt to single out what constitutes Danish in a scientific way but without maintaining a comparative perspective with another imagined community, see Gundelach (2002).

contradictory developments within the nations at both ends of the studied relation should also be illuminated It is furthermore important to study not only the relation but also how the relation is part of societal developments at both its terminal ends. If such an approach is not implemented, the Other remains an exotic and basically unknown contrast. With this perspective, the following is an analysis of Swedish and Danish cultural bordering processes, analyzed through the relation between the two nations. This perspective highlights the first characteristic mentioned by Wilson and Donnan; a border "simultaneously separates and joins states" (1998a:9). Naturally, this will also lead to discussions of which impact this relation has had on the creation of the two nations at different times. What is possibly specifically Nordic about the welfare (national) state is also included – the linking of class interests, the Social Democracies, and the well-developed bureaucracies. The Danish in this study does not only represent something different, an Other, in relation to the Swedish, and vice versa. Differences and similarities between the two cultural communities must also be explained through analyses of developments within both Denmark and Sweden. In this sense, this perspective represents a *genuine comparative and historical cultural analysis* (Østergård Andersen 1992).[9] This argument does not imply that there exists one Swedish way or one Danish development on all arenas. However, there are *differences* in ways of regarding and solving problems of all kinds but not necessarily different solutions to all problems within all arenas. Because the analysis is specific in this way, it only illuminates the relations between the parties who recognize and acknowledge the others' cultural, objective differences. Thus, when something is described below as Danish, it is only Danish in relation to something correspondingly Swedish. And thus with this, it is not claimed that the particular element and development are uniquely Danish or Swedish. As James Clifford emphasizes, the ethnographic truth is inherently partial, both committed and incomplete (1986:7); the crucial aspect of this is then to make an explicit decision to how this ethnography will be committed and incomplete. Something Danish can be identical with something German without therefore becoming less different from the corresponding but dissimilar Swedish.[10] Studying bordering then, is trying to understand the intersections between similarities and differences as "a dialectical or two-directional journey examining the realities of both sides of cultural differences so that they may mutually question each other, and thereby generate a realistic image of human possibilities and a self-confidence for the explorer grounded in comparative understanding rather than ethnocentrism" (Fischer 1986: 217).

Cultural bordering over time is filled with contents through narratives that explain a specific relation with anOther. Thus, nations take positions to secure themselves representations through which they can both contemplate all other communities and be inspected and imagined by them. In this process, relations

9 For similar views, see Fischer (1986), Marcus (1998:68ff), and Ortner (1999a:8).

10 The attempt to understand bordering in a more specific sense is similar to what some students of the US-Mexican border have attempted (Vila 2003).

with some significant others come to play prominent roles. Common history and geographical proximity are elements in this selection process, but it is not decisive for the importance of any specific case of cultural bordering. Translating geographical distance into cultural distance ignores that certain geographical arenas attract more relational attention than others. Most other nations, for instance, have narratives about America (rightly, the US) but Austria means little or nothing to Danes and Swedes even if it is geographically and politically closer (than the US) through EU membership. Besides, as will be discussed below, cultural identities are often built from ignorance of society and developments within the other part of the particular relation (Østergård Andersen 1992:89).

As historical novels could shed some light on political bordering and its development, narratives about differences between Danish and Swedish can be employed to illuminate cultural bordering. The focus must be on narratives about Danish and Swedish that fill the ditch between them with differences. Since the national idea of spatiality is in constant flux as a consequence of how it associates with and disassociates from other spaces in the modern maze, thoughts and practices that express difference and peculiarity must be sought on many varied arenas. Thus, an effort in multi-sited ethnography is needed (Marcus 1998). Furthermore, studying developments of the relation between Denmark and Sweden is, at the same time, an archaeological reconstruction of a part of the modern maze. It is this complex of conceptions of, developments in, and employments of the Danish-Swedish relation that are of interest. Below, the discussion follows the approximately one hundred year period when cultural bordering obtained and held hegemony.

Since the Enlightenment, people's conduct has been explained as a consequence of regional differences in their natural environments. For example, travelers in Sweden such as Jacob Bircherod in 1720 and Carl von Linné in 1749 described differences in houses and clothing they observed (see G. Christensen 1924, von Linné 1959). Generally, plains-dwellers were considered sluggish while people living in the mountains were characterized as loving their freedom and energetically defending it. Within the modern maze, such differences gradually were edited to accommodate imagined communities. Nineteenth century travelers' accounts and debate literature were filled with interpretations, which suggested the national form as political borders as well as provided conceptions of how the other community objectively differed as a culture (G. Christensen 1923). Eventually, a new kind of knowledge became necessary for people's ability to understand and navigate a world filled with cultural differences. While still important, instruction in history that could legitimize political communities became insufficient. Ideological imagineering of cultural communities was promoted through journals and exhibitions and from the latter part of the nineteenth century in Danish and Swedish schools, a new topic – geography – was added to the curriculum to introduce knowledge about the borders of the state. The subject clearly had a national aim from the very beginning, as it was important to provide the students a general view of the nation and its territory as well as the nation's surroundings (Olsson 1986:59ff). In 1866, authorities advised on this topic in Sweden:

In this way, the teaching of geography must both fulfill its national aim, to provide an accurate knowledge of the present state of the fatherland as a condition for any national education. Moreover, it must correspond with its general human significance, which is to open the mind and understanding of the apprentice for the great whole, of which the fatherland is an organic part, for our part of the world and the great European society, for mankind and our whole world. (Olsson 1986:64)

From around 1900, geography was assigned more class time and new educational materials were introduced. Schoolbooks that were published between the end of the 1880s and the First World War were often edited, reprinted, and sold well into the 1960s, and used in schools even longer (Linde-Laursen 1988, Olsson 1986:75ff). The same authors wrote books for both primary and high schools, and apparently they also readily borrowed materials from each other (Tingsten 1969:256). The compositions of the books, including the short stories that animate most cases in the texts, are hardly distinguishable from one print or author to the next; even if the texts were later revised, the structure of both books and education were mapped out prior to 1914. Cultural bordering was outlined thusly at that time, for generations of teachers and pupils.[11] Through geography lessons, the cultural community was adduced and explanations to contemporary differences between nations were disseminated. Within this perspective, not only were Denmark and Sweden located facing each other across the political border in Øresund on the big map in the classrooms, but they were also countries with different cultures. Authors of educational materials and teachers in classes induced imaginings of how people lived and how society was organized on the other side of the Øresund ditch. From around 1900, narratives about these different lives and this dissimilar society were employed in debates about how things were or ought to be on "our" side of the border. Through comparisons, the desirable and the unwanted were provided tangible forms and the advantages and insufficiencies of one's own nation were illuminated and, therefore, possible to debate.

---

11   Even if authors probably employed foreign models, it illuminates the connection between geography and the cultural community that each nation arranged its teaching materials to be written by domestic writers. Translations seem not to have been used to any extent. It is peculiar that geographers seem, until the present, not to have involved themselves in the study of how the topic has been involved in nation-making, but rather geographers have entrusted such investigations to students of other sciences (J. Anderson 1988, Johnston, Knight and Kofman 1988). Lena Olsson discusses how new educational materials from the 1960s partly introduced new approaches in teaching geography that on the surface appear less nationalizing. Instead of descriptions of other nations' populations and cultures, today the focus concentrates on, for instance, GNPs or the number of doctors or telephones. Thereby, the depiction of the northwest European nations as the technologically, politically, and economically most developed was reproduced, albeit from a different perspective, on the grounds of statistics and science and thus, objectively and without obvious prejudice (1986).

Geography Lesson

> In primary school, when I was obligated to comprehend that I had a fatherland,
> my geography-book informed me in the lesson about Iceland, which was in a
> separate volume, another was on the Scandinavian countries. The rest of the
> world was in a third book. From the beginning of the loss of innocence and in
> learning, Scandinavia was a remote world. And the lesson on Iceland started
> like this:
>   "Iceland is a small country, far away from other nations."
>   And we, inevitably, had to learn this glittering passage by heart. And at every
> test we solemnly answered to the question "What is Iceland?" that "Iceland is a
> small country, far away from other nations." And there was no alternative if one
> desired a good grade in geography. (Bergsson 1988:63f)

When Icelander Gudbergur Bergsson later traveled the world it appeared that also
other people understood Iceland as something far away. Sweden and Denmark were
not as distant as Iceland from other nations. In Swedish and Danish geography-
books from the late 1880s, the other country's peculiarity and development was
accentuated as something about which it is important to be informed. This was
in accordance with the division of the world, which also was used in Iceland's
geography books – the fatherland, Scandinavia/the Nordic countries, Europe/the
West, the Rest.[12]

In *Skolgeografi* (School-Geography) by Ernst Carlsson, which for a long time
completely dominated Swedish schools (first impression 1887, last 1948, see
Olsson 1986:86ff), "The Kingdom Denmark" was among other things described
in this way:

> Denmark is one of the smallest European states, barely bigger than [the Swedish
> landscape] Jämtland but quite densely populated. [...]
>   With their well-tilled fields, plentiful meadows, closely situated farms, and
> lovely beech-groves, these parts of Denmark [eastern Jutland] very much
> resemble Skåne. [...]
>   *The Danes* are, as the Swedes and Norwegians, Scandinavians. They acclaim
> the Evangelic-Lutheran teachings and talk a language, which much resembles
> ours.
>   *Education* is well developed. There is a university in Copenhagen (since 1479)
> and many other schools in the towns. Noticeable is also the numerous folk high
> schools in the countryside for both common young men and women. [...]

---

12   This section is built upon an examination of the collections of educational
materials in the Danish Pedagogical University's Library in Copenhagen (Linde-Laursen
1988). A similar study from material in The National Psychological-Pedagogical Library in
Stockholm was made in connection with a BA thesis written under my supervision (Rysén
1993). The material is presented in more detail in these two references.

The constitution is *limited monarchy*. The king has at his side a Parliament, divided into *two chambers*. [...]

Agriculture and *cattle raising* are the principal businesses in accordance with the nature of the country. Of Denmark's area more than three fourths are fields and meadows. And the number of cattle in relation to the population is bigger than in any other European country. Especially dairy is highly developed so that much butter and livestock is exported, mainly to England.

Trade is favored by the location at Öresund, which is a path for traffic between the North Sea and the Baltic. However, *industry* is, while Denmark from nature lacks coal and iron, less important. (eighth impression, 1904:69ff)

This description is nearly word-by-word similar to simultaneous impressions and other books' accounts. Although different authors emphasized somewhat different elements, or the same author did so in books for different grades, this is very close to an ideal presentation of the little, flat, and fertile country with waving fields of grain, cows in the meadows, and sows in the pigsties; industrious, prosperous and (through the folk high schools) educated farmers exported products through their own cooperatives mainly to England for which reason trade was comprehensive. The population was competent in business although not always perfectly trustworthy. Industry, which was gathered in the biggest city in the Nordic countries, Copenhagen, was less important due to the lack of minerals and ore. The people shared a petit bourgeois culture that attached importance to helpfulness and evenness in a true democratic spirit.

This presentation was supported by Swedish writers of the time in its entirety or in parts. Thus, similar imaginings are repeated, regardless if the particular writer tries to depict their tiny neighbor's tourist sites (Velander 1906), characterize its people (Nyblom 1900), or offer a more comprehensive image of its whole society (Sprengel 1904).

Through cultural bordering, what was Danish complemented what was Swedish so that what Sweden had plenty of, Denmark lacked, and vice-versa. Danish geography-books depicted Sweden as an immense country. In the north, there were huge forests often identified as the "gold mines" of Sweden. Together with raw materials from its underground, timber from the north formed the major part of Sweden's exports. In the south, there was agriculture but only a lesser part of the country was cultivated due to boulders and a sparse layer of tillable soil. Even if Sweden could export dairy products, its less developed agriculture had problems with keeping the country self-supplied with cereals, which was compensated through the import of grain and the "export" of people, foremost to North America. Skåne, however, which resembled Denmark, was fertile – and by the way, was former Danish soil. It was also mentioned frequently that Swedes had learned a lot of modern agriculture by studying Danish farmers' folk high schools, organization of their work, and their cooperatives. In the much-used *Geografi for Folkeskolen* (Geography for Primary School) by C.C. Christensen and A.M. Krogsgaard, the Swedes in 1909 were described as:

> The Swedes are industrious, and the common Swedes are thrifty and far more frugal that the Danes, but they are not as educated and in fights they far too often reach for a knife. Many emigrate to America or Denmark.
>
> Sweden is as Norway, a sparsely populated country. In addition to *agriculture* and *forestry,* Sweden also has *crafts* and *industry.* At the mountain *Gellivara* far away in Lapland are the most plentiful deposits of iron…Also in central Sweden there is much iron and at the town *Falun* there are copper mines. (Third impression, 1909:38)

Often Danish texts noted that even if there were more Swedes than Danes, their foreign trade was lesser than Denmark's. In conjunction with this, it also warned that ruthless exploitation of raw materials from forests and mines would soon deplete Sweden's resources. However, it also mentioned that industrialization, which began around 1880 and was rapidly progressing, could eventually improve the utilization of natural resources. "*Norrland*, rich in lakes and forests, might soon become Sweden's land of the future, where big towns grow from forestry and mining along the roaring rivers" (C.C. Andersen 1909:33).

When Danes described Sweden, it was a land of opportunities, but also a land of contrasts and injustice – where a minority lived in lavishness and splendor while the majority lived in poverty. This economic inequality was mirrored in the political structures of the state, where only the king, the nobility, and the rich had influence while the right to vote was strictly curtailed:

> The king shares the legislative power with the Parliament, but Sweden's constitution is not nearly as free as Norway's. Only plutocrats can be elected to the Upper House (first chamber). In addition, the right to vote for the Lower House (second chamber) is so limited that only one out of four of the men of age within the state hold the right to vote. While Norway is mostly a state of peasants and tradesmen it is contrary the nobility and wealthy classes that hold the predominant influence in Sweden. (C.C. Christensen 1899:88)

The portrayals of the Other in the geography books are of course quite undifferentiated. And on some points, the result was not very flattering for Sweden when measured by the Danish standard. Around 1900 however, little interest seems to exist among Danish writers to write more in-depth descriptions of Sweden. Swedish writers directly notice and wonder about this apparent lack of interest (Nyblom 1900:9, Sundbärg 1911:95). This, though, did not indicate that the two partners in this relation lacked agreement with regards to how people at that time could know the other part and how they could narrate their impressions.

Knowledge about the other was, in many cases, established through the comprehensive migration of Swedes who, for a season or longer, went to Denmark to find work because until around 1900, wages were substantially higher on the western side of Øresund. Migration in the other direction was far less although later the situation gradually reversed. Most of the emigrants from Sweden to Denmark came from the southernmost regions, especially from Skåne. Both male and female migrants were generally unmarried and came from poverty. The migration began in the mid-nineteenth century, culminated in the 1880s, and slowly vanished after 1900. Both the seasonal workers and those who moved across Øresund more permanently mostly found employment among the lowest paid and most taxing jobs. As among the poorest, persons born in Sweden were very well represented in Danish statistics on minor crimes and prostitution (*Emigrationsutredningen* 1911, Willerslev 1983, Bloch et al. 2000). Feelings between Swedes and Danes were usually amicable and many Swedes married Danes. Nevertheless, problems materialized now and then in regions and trades in which many Swedes were employed; this was not least due to the use by employers of Swedes as blacklegs during strikes. Swedish authorities were not, at least towards the end of the phenomenon, very pleased with this migration. Swedes unable to provide for themselves and girls who became pregnant were returned by the Danish authorities and thus, became a burden for the Swedish social institutions.

When the Folklife Archive in Lund (Sweden) in 1952 issued a questionnaire on "Danish and Swedish" (LUF 84) many of the 66 respondents could recall this migration.[13] They themselves or others from their home communities had gone to Denmark because there, they found employment, shorter work hours, better pay and board – and, aquavit (*snaps*) was inexpensive. Many of the responses contain elements that remarkably agree with the narratives from the geography-books. Relations between Swedes and Danes were recalled as good.

> The general understanding of Denmark has always been, although more so earlier than now, that Denmark is a country that flows with milk and honey. It has a developed culture, and the Danes are more humorous and shrewder in business than we Skånians. (LUF M:12016)

Moreover, some would tell that Swedes were much more industrious and serious than the lazy and comfortable Danes (LUF M:12145, 12148, 14343). The narratives of the relations found in the geography books illuminated a few central themes, which with variations, were repeated in the responses to the questionnaire – the character of the country and the people, the state of politics, and economic life.

Some elements were judged differently in the many descriptions and experiences conveyed in the handwritten answers to the survey. Was it really true that a Dane was lazy or diligent? If the depictions became too detailed, their attempts to provide cultural bordering could easily be disproved. But while keeping to the broader brush, the narratives generally acknowledged the other part's cultural dissimilarity and provided answers to who and what one was. The relation between Denmark and Sweden was thus, established and narrated so that the depiction of the other part was in agreement with that which the other part

---

13   For a discussion of the general use of the questionnaires from the Folklife Archive as ethnographic sources, see Sjöholm (2003:18ff) and Frykman (1988b).

itself could identify. Danes and Swedes mutually recognized and acknowledged their differences, and the relation therefore could be employed to emphasize what one found important about oneself.

The light in which Danes saw themselves was refracted in a Swedish prism, which underscored essential Danish peculiarities – *peasantry, family-enterprise, and democracy*.[14] These Danish qualities became associated with the name of N.F.S. Grundtvig. As a man who managed to extract crucial tendencies in his own time, he later obtained the status of an assembling symbol for a broad cultural and popular movement that had a tremendous impact (Sanders and Vind 2003). In 1865, Enrico Dalgas introduced the reconstruction of the Danish imagined community after the defeat in 1864 and the dissolution of the One-State, under the vigorous motto, "What is outside lost, shall be inside won!"[15] It became a starting signal for a successful campaign to reclaim extensive bogs, especially in Jutland. This campaign (and the image resulting from it) became an expression of the Danish cultural community especially concerning how strongly Danes identified themselves as peasants. With the surrender of Schleswig, Holstein, and Lauenburg in 1864, the most industrialized parts of the One-State were gone. The loss of this industrial potential was not compensated through an aggressive policy to push the development of Danish industry. Rather, a legion of peasants-in-the-making redeemed the loss as they threw themselves onto the bogs, reclaiming lands that are today, largely deemed marginal; with monetary compensation from the EU, it is expected that they be pulled out of normal agricultural production. Simultaneous with this campaign, Danish exports were reoriented. Earlier, agricultural products from Denmark were mostly sold on the markets in northern Germany. After 1864, the goods were instead sent to England, which during the following century became the principal market for Danish butter, pork, and other agricultural outputs. To support this reorientation, a new and rapidly growing export-harbor was established in southwestern Jutland, in Esbjerg. The bog land campaign and the successful reorganization of Danish agriculture, in conjunction with the establishment of a network of cooperatives for processing and export, confirmed that the peasant was the principal element in the Danish nation. This could successfully be mirrored by the poor results of Swedish agriculture and that country's wasted uncultivated areas. Thus, the Danish cultural community was reflected by and described through its peasants' achievements. Family owned and

---

14   Instead of peasantry and family-enterprise, other writers employ concepts such as "petit bourgeois culture" – a traditionally Marxist concept (usually with negative connotations but as noted above, in Danish, not necessarily so as it can in some instances be understood as "*folkelig*"), "the life-mode of the self-employed" (T. Højrup 1983a, 1983b), or "agricultural-capitalism" (Østergård 1984, 1992b).

15   The words of Dalgas were not without foreign inspiration. The Swede Esaias Tegnér had after 1809 talked about, "Within the borders of Sweden to win Finland back." The "inside/outside" metaphor in itself is telling about how such ideas contribute to bordering.

Figure 3.1   Danish 500-crown note, *plovmand*, in circulation 1910-1945, by Gerhardt Heilmann

Notes: The peasant, as a manifestation of agriculture and family business, was a central feature in narratives about the Danish from the end of the 1800s. He (and it was ordinarily a he) was also depicted on the most valuable note in circulation. The Danish 500-crowns note, colloquially known as the *plovmand* (the man plowing) was crafted by Gerhardt Heilmann. The *plovmand* was introduced in 1910 in a blue version. In 1945, an exchange of all Danish notes was implemented to detect illegal earnings during the years of German occupation in the Second World War. At that exchange, an identical version in red was introduced and was in circulation until 1964, when a *plovmand* with an entirely different image was introduced. From 1974, the peasant disappeared from the notes of the nation.

run farms became the model for business in every branch of Danish economic life. Narrated as the nation's basic element and favorably considered in legislation, the organization of the farm spread to fishing, crafts, services, and industries. Thus, the economic, ideological, and linguistic forms, which are associated with the family enterprise can be employed throughout most of the 1900s to analyze large parts of the Danish population's ways of organizing their lives (T. Højrup 1983a, 1983b).

The peasant not only was the model for the hegemonic way of thinking about economics, but also for thinking about politics. From the late 1800s, the self-reliant peasant who owned his own farm became important, recognized as the opposite of the feudally oppressed and deprived peasant of eastern Denmark in the 1700s. This focus existed both within the political debate as well as in historical studies where the historically freer peasants from other parts of the country did not get much attention. Peasants (from eastern Denmark) thus became the principal objects for Danish history from the 1840s and far into the twentieth century. As a result, peasants became identified simultaneously with the Danish political community (Kjærgaard 1979); the *free* peasant became the country's principal political agent.

This interpretation was confirmed in 1901 with the so-called Change of System (*Systemskiftet*) and when parliamentarism was adopted. Representative democracy based on a dynamic perception of freedom, which accepts conflicts and extensively takes the political (and economic) rights of the minority into consideration, became a prominent part of Danish self-perception and was accepted within a broad spectrum of political tendencies (Østergaard 1984, 1992b, Stenius 1993). This peaceful, political evolution of the people's united interests with the state was perceived as something particularly Danish. And it held true regardless of whether this perspective was applied to the study of the introduction of absolutism in 1660-1661, the phasing out of absolute monarchy in 1848-1849, or the contemporary political debate; it of course, also materialized in Etlar's historical novels. This Danish perception of democracy could strengthen itself by looking at its reflection in a Swedish mirror that profiled a country where the right to vote was limited and where the rich held supremacy. In this mirror, Denmark became, as expressed by Grundtvig, the country where "few have plenty and fewer are lacking." Although these words most often are understood as a depiction of economic conditions, they can also be adopted as a depiction of the political situation after the Change of System. Consequently, through cultural bordering, economic, political, and national ideas of spatiality were united within the modern maze; this intertwining afforded the free peasant the principal role in Danish society.

From the other end of the relation, Swedes could similarly strengthen and understand their cultural community. Their country was poor, somewhat backward, and lacked properly developed democratic structures; it also was depicted this way in Cederborg's historical novels. Embedded in this narrative though, were possibilities for doing so much better. Sweden increasingly imagineered itself as a project for the future.

Exactly because they touched central modern circumstances of political and economic life, the bordering narratives could be employed in debates about the conditions and developments of society. The relation between the Danish and the Swedish no longer involved the division of power in the Baltic realm or approached questions of moving borderlines. Through cultural bordering the relation dealt with establishing and developing ideals and oppositions between the two nations' internal political discussions.

In 1904, the young Swedish critic David Sprengel (1880-1941) traveled to Denmark and found the elements that his contemporaries discovered during their geography lessons. Not least through his admiration for the Danish literary critic Georg Brandes, he interpreted these elements as so attractive that he accentuated them for his fellow Swedes as an example. To Sprengel, the Danish was a modern development of an ancient culture that was shared equally by all. It was in opposition to Sweden where contradictions were embedded historically; in earlier periods, only the aristocracy was cultured and therefore, in his own time, the people was cut off from any kind of traditions. Sprengel depicted the modernization of dwellings, the free schools, the growing Social Democracy, and other matters as positive, Danish elements. The idea of cooperatives had in the late nineteenth century saved Danish agriculture from a market crisis and at the same time, had transformed the occupation. "It is truly a factory. All what was known as rural idyll and rural loafing, rural slowness and rural sleepwalking, has with the transformation of the system been tossed away; the work has become considerably harder" (1904:170). This change had been beneficial for the intensively working smallholders and that, Sprengel found, was Denmark's fortune. The smallholders would affiliate with the idea of socialism: "The smallholder will master the Danish countryside, as the industrial workers the Danish towns..." (1904:178). And, the tradesmen would take care of selling. "For the businessmen of Copenhagen, his work at the office is nearly the same as for the American: life and poem, love and poetry. To the Swedish businessmen and in particular to the Stockholmian, his occupation during office hours is much more often a tedious story" (1904:127ff). All groups in Denmark partook in the development of this modern society. "All these writers are associated with their time, stand with both their legs down in the whirlpool, no-one appear as ignorant of, excluded from, what is going on" (1904:376f). Sprengel was not oblivious to problems in Denmark. He depicted among other things prostitution in Copenhagen albeit without mentioning the participation of poor, Swedish women. Sprengel's advice – to find solutions to Swedish problems in an imitation of modern Danish democracy, agriculture, rationalization, and trade – was not met with undivided enthusiasm in Sweden where his book was considered controversial (Sprengel 1904:470ff).

Turning Bordering Upside Down

Through cultural bordering, a narrative of differences was created and later transformed or recreated. This narrative pointed to a series of elements and was usually value-laden, as it contained relatively evaluating assertions and symbolic inversions (Ehn and Löfgren 1982:38). The other part's peculiarity was most often perceived as either superior or inferior to one's own. Usually though, it was possible to turn interpretations of such symbolic inversions upside down so that what was regarded as superior could be depicted as inferior from another polemic or political point of view.

A few years after David Sprengel presented his thoughts to the public, Gustav Sundbärg, the Swedish statistician and leader of the 1907-1913 investigations into the reasons behind the huge Swedish emigration (*Emigrationsutredningen*), presented in an addendum to his investigations his understanding of Danish-Swedish relations to the public. The addendum was subsequently published as a book that became immensely popular as evident by the no less than a dozen prints in its first year (1911). Sundbärg found that the lack of economic development of Swedish society was the basic reason for the emigration. But he also pointed out that this was accentuated because the character the Swedish people had some serious deficiencies in "the lack of human understanding, of national instinct, and of economic calculation" (1911:100). What Sundbärg found as the error of his contemporary compatriots also appeared to him as their grandeur – they were far nobler and less manipulative than other peoples.[16] Sundbärg's text is thus, one of many similar (contemporary and later) descriptions of national characters in European countries. Herbert Tingsten has pointed out that these descriptions are amazingly analogous; the author's people possessed great virtues which due to the lack of national pride, individualism, and jealousy as well as the admiration of the foreign, were never developed in the best interests of their own nation (1979).

At crucial points in his analysis, Sundbärg employed bordering against the Danish. In comparison with Sprengel, Sundbärg came to a completely opposite interpretation of the existing narrative of difference. The economic life, which Sprengel had celebrated as nearly American, Sundbärg found significant in terms of the Danes' constant assaults on Swedish interests. According to Sundbärg, from the Kalmar Union over Scandinavianism and to his own time, Danes, with their developed sense for psychology and desire to promote the interests of their own nation, had enriched themselves at the expense of Swedes:

> The Dane's ability to persuade and convince is immense and the Swede's resistance is exceptional small; mostly as he lacks the support, which *ought* to come from the national instinct. (1911:89)

If the Swedish was thus weak, it was magnificent in its weakness. Denmark's grandeur on the other hand was founded on a fragile and egoistic basis:

> Strictly speaking, the Swede is very often tremendously talented as a thinker, and nearly always, his perspectives are *greater* than the Dane's. Just look at these in Sweden, so common huge, high brows – real thinker-foreheads. They represent (the possibility for) thinking in great, clean lines, not in the Dane's odd tortuous loops. Contemplate how this latter way of thinking is drawn in the Dane's always – one could nearly say painfully furrowed and twisted features! (1911:13)

Such negative interpretations of the narrative about Danes invited counterattacks. And they came.[17] While all writers agreed on how cultural bordering between the Danish and the Swedish could be narrated, they differed significantly in their evaluation of each of the elements of the narrative. Predictably, the responses emphasized democracy, petit bourgeois, and the cultivated (understood as both

---

16   Sundbärg's book is still well known, but of course he was not the first who depicted the problems of Sweden in this way. See for instance the short reflective address by Andersson (1881).

17   Levin (1911), Nielsen (1912), and Olrik (1912). The contributions by Nielsen and Olrik came first in Danish in journals (respectively *Ugens Tilskuer* and *Nordisk Tidskrift*). The text by Nielsen then came simultaneously in Swedish and Danish as a book while Olrik's only came as a book in Swedish. The book by Sundbärg was not published in Danish but both the central Danish replies were available in both languages.

the arable and what is regulated by the authorities) as core characteristics of the Danish. At the same time, they stressed Denmark's leading role in the economic life of the North. While Sundbärg had seen this as a consequence of Danish betrayal, the replies stressed the impact of history and development on the Danish:

> Thus, there is really a contrast, the whole contrast, which is between a people [the Danish] which through centuries has been used to obey laws, and a people [the Swedish] which to a great extent until recently has lived as settlers, usually in their solitude tempted to and forced to take the law into their own hands, whereas a Dane had a representative for authorities nearby. [...]
>
> We [Danes] were not only the geographical small and strategically open country. We were also the fertile, easily developed, and civilized country... Besides the problems of war in its most difficult form we had also earlier than Sweden the burden of peace on our shoulders; the considerations of trade and business, the influences of satisfaction and culture; and gradually as the events disowned our political-military instincts the civilian views and considerations seeped more and more into our resolutions. [...]
>
> The example [Danes as employer, points] also to something particular Danish, namely our propensity, for better and for worse, to be humane, fair, social, liberal [...]
>
> We knew [in 1905, at the dissolution of the Swedish-Norwegian royal union] by and large nothing about Sweden, but that it politically was "backward" compared to us, i.e. fewer had the right to vote... (Nielsen 1912:28f, 35f, 40, 45f)

The contributions from the Swedish side praised Sundbärg's character sketch of the Swedes but tried, at the same time, to mitigate his severe sentencing of Danes (Nielsen 1912:5f, 25). More recently, Jonas Frykman argued that the aphorism of Sundbärg was in agreement with other depictions of the time, and that they have had a great impact on how Swedes have been perceived since then (Ehn, Frykman and Löfgren 1993:143). Thus, it was the judgment of different particularities that Sundbärg and his contemporaries influenced, rather than the narrative about the Swedish as such. By pointing to deficiencies within the Swedish nation, it was emphasized as a project for the future – something that could be improved through an effort. Sweden was, according to Carl Laurin, the country with "the greatest past, maybe the greatest present, and certainly the greatest future" (1911:365). Sundbärg's depiction of the Swedes was, as Frykman has called attention to in another context, in agreement with perceptions within a large group of people who were socially mobile in his time (Frykman 1989:34f). Sundbärg himself called his times, "the era of social motion" (1911:142). Personal progress, achieved by individuals who worked their way into the growing middle class, was legitimized as efforts made on behalf of the Swedish project. The lack of a past, about which Sundbärg repeatedly complained, was not a national defect. On the contrary, it could be regarded as an essential element in a national project that fought to become reality – in the future.

Unlike the Swedish project for a greater future, the Danish cultural project was narrated as already realized – as it was steadily anchored in a historical dimension. "Is there a more splendid sign of culture than the combination of a lively feeling for and keen work in the life of the present with reverence for and a continuous connection with the past? Tradition, in the most beautiful understanding of that word, exists in Denmark," Laurin appraised (1911:367). Thus, east of Øresund, a Swedish middle class was about to establish practices and ways of thinking which secured them control over the development of their cultural community (Frykman and Löfgren 1987). This process of legitimization developed in ways similar to how peasants and others had been able to legitimize their comprehension of Denmark. The Danes referred to the success of agrarian reforms, their cooperatives, and to Grundtvig as a unifying symbol for the peaceful evolution of the whole nation and the state (Østergård 1984, 1992b). Cultural bordering reflected the different historical circumstances of the two countries. Denmark and Sweden walked hand in hand each their own way towards modern society; the Danes almost walked backwards. Scrutinizing the development of farming, they were able to explain how they had approached a glorious present. The Swedes apprehended their everyday life with their gaze toward the future and the dream of everything better to come.

Figure 3.2   Danish 5-crown note, in circulation 1912-1945, by Gerhardt Heilmann

Notes: In the 1800s, prehistory, like all other history, was nationalized. Consequently, Danish peasants every day worked in and managed the national heritage. In the country, where the terms stone, bronze and iron-age were first coined, prehistoric remnants in the landscape (as well as these relics' association with an imagined or real society of free peasants in the past) had a significant influence on national identity processes. Archaeology became something of a national science and pastime for high and low (see M.L.S. Sørensen 1986). Prehistory thus appeared on items in everyday use, such as on this Danish 5-crown note, which was in circulation from 1912 until 1945, crafted by Gerhardt Heilmann.

In his contribution to the debate following Sundbärg's aphorisms, Laurin wrote as introduction to his considerations:

> The psychology of people often affects those who practice it as hashish affects the Arabs. It is said about this somewhat hazardous stimulant: "One have the feeling, that the earth disappears under one's feet, and one have a not disagreeable sensation of hanging in the air." (1911:361)

Cultural bordering as a practice demanded that the contemplator towered above the earth so that only the general images appeared. The not disagreeable sensation probably was caused by the certainty one could achieve from that distance with regards to how to paint the differences between nations. Still the contemplators disagreed on which conclusions and consequences should be drawn from these distinctly marked bordering narratives. While the narrative was firm, the understanding of it could be turned upside down, and was done so often repeatedly!

Culture and Identity

Sundbärg and his contemporaries did not agree on how to evaluate the relation between the Danish and Swedish. Nevertheless, they agreed that there, between the two nations, was a ditch of differences (Febvre 1973). They also agreed on the content of this ditch and thus, on how to narrate cultural bordering. As the participants in the debate evaluated their nation by comparing it with their neighbor and by employing a certain edition of differences, they emphasized their understanding of their own nation. Thus, their disagreement was limited to how specific elements of difference ought to be considered from their own vantage point.

To describe their cultural communities, Denmark and Sweden, through the relation between them, they accentuated difference at the expense of similarity. To understand the content of this relation could be a satisfying result in itself. A proper comparative analysis however, as mentioned above, must be carried further. Through analysis east and west of Øresund, differences, which have been pointed out through the relation, can be viewed in their proper context. Through this, it is possible to appreciate why these differences are of importance, and it becomes possible to analyze how they have developed. It appears that difference can result from similar processes and that similarity can result from different paths.

Thinking in difference and similarity is embedded in today's debates both in the public sphere and in cultural sciences.

> No discipline manages well without its members' efforts to define its identity. This can be carried out through comparisons with and demarcations from other disciplines. It can also be carried out through central concepts, which drag a complete cluster of associations which together depicts the peculiarity of the discipline. Within ethnology, "culture" is such a concept. (Gerholm and Gerholm 1989:8)

Culture from the 1970s became a concept that was employed in every possible and impossible context – also outside the cultural sciences. In many situations it explained and often subsequently legitimized differences and misunderstandings between individuals and groups. Thus, the concept was employed to establish differences between groups of people (Ehn 1992). During the same period, a kind of resentment towards the concept spread among cultural scientists. Fewer and fewer used the concept of culture, most often arguing that the problem was that culture described homogeneity – similarity within a demarcated field. Culture came to depict something unambiguous that could be demarcated and which constituted a social entity. This thinking in similarity was at least from the late 1970s under pressure in discussions within the cultural sciences. Analysis of world-systems and global patterns made clear that demarcated social entities did not exist. Furthermore, discussions on the role of the scientist gradually clarified that the similarity, the culture, was a construction developed by the student and transfused to the studied (see for instance Clifford and Marcus 1986, Hastrup 1988, Stocking 1983).

Scientists it seems abandoned culture (for a reaction, see Ortner 1999b). In its place the concept of identity proliferated in scientific discussions. This change of concept was associated with a programmatic shift from interests in similarities to a focus on differences.[18] "Every thinking in identity is thinking in difference, and every thinking in difference is thinking in identity," as Hans Fink emphasized (1991:218). This shift, of course, had consequences. In modern societies many agents fear a loss of identity because of social dissolution, cultural change, and other cultural processes. Therefore identity is, as Hans Hauge called attention to, also a threat (1991). Who is not encompassed by the demarcations? Who is different, has another culture?[19]

Ole Høiris has summarized the development in the field of social-cultural sciences as "a change of the formulation of problems from a focus on what is similar despite differences (the tendency of social sciences) to a focus on what is different despite similarities (the tendency of cultural sciences)" (1988:95). Both ways of analyzing imply in similar ways that the student establishes an understanding of the studied. As writing about culture brings about wholes, writing about identity brings about difference – and consequently the possible threat to exclude (Damsholt 1993:82). It is through the analysis and the writing of ethnographic objects that similarities and differences are acknowledged. This, of

---

18   It should be noted that there is a simultaneous shift in the scope of the anthropological object. Where usually a topographically quite limited community used to be in focus, as in community studies (Bell and Newby 1971, Redfield 1960), (much) larger objects, such as nations or the city, gender, class, race, diaspora, and so forth are now being studied as comprehensive systems. This in itself has propelled a new conceptualization of the anthropological object; see for instance the discussions in Marcus (1998).

19   My object here is not the many different understandings of the concept of identity. The concept is, as is the concept of culture, ultra complex (Brück 1984, Fink 1991).

course, does not necessarily imply that those who perform as objects and provide the materials are aware of differences and similarities in their everyday life. "We [the cultural scientists] write cultures, as they are not texts; and our authorship implies a composition developed from their reality. The lives are theirs, but the words ours," as Kirsten Hastrup has emphasized (1988:133).

Thus, studying national imagined communities does not imply that culture can or ought to be considered as definitely different from country to country. The differences, which can be demonstrated, between Denmark and Sweden are not more important (if that could be measured, which is doubtful) than many other differences that relate with for instance gender, age, generation, religious beliefs, sexuality, race, ethnicity, class, or for that matter, any other difference anchored in other ideas of spatiality in the modern maze (Hobsbawm 1990:123f). Differences founded on these circumstances can be crucial and play a more vital role for persons in more situations. Other geographical matters, as for instance affiliations with region or locality, similarly play their role for the formation of identities. An individual's identity is a complex process within which each person, in various but always tangible situations (what anthropologists talk of as everyday life), establishes and often also represents and narrates conceptions of him or herself and of his or hers surroundings from one of or a composition of these different ideas of spatiality. Therefore, identity must always be considered an open process in which the formation of identity takes place along many borders, on many levels, and through continuous transformations, or as what Marcus calls "multiple overlapping fragments of identity" (1998:63). To break out one piece of this process and assert that this particular bit is more important or deeper, more fundamental or crucial than one or any of the others is unproductive (Hannerz 1992:68-99).[20] Bordering between imagined communities is not neutral in relation to other differences but compositions of different groundings for identity must be investigated as tangible formations. To try to catch the whole of a person's or group's identity will either lead to the study of individuals and psychology (from where the identity concept originates), or lead to an endless argument in which each foundation for identity continuously can be claimed to be outdone by another; the very reason that nationality, as Harbsmeier found (1986:54), works well in jokes, but dissipates under the scrutiny of science. In both cases the idea, that some people under specific circumstances really can share insights and experiences, is liquidated; and consequently cultural analysis is severely limited.

Its anchorage in conceptualizations of differences suggests that the concept of identity has ontology separate from the concept of culture, which usually conveys coherence, harmony, and entity. Fusing the two concepts as "cultural

---

20   This is not the place to discuss the impact of metaphors (Lakoff and Johnson 1980). Nevertheless, it is obvious that adjectives used in conjunction with identities – deep, shallow, superficial, invented – have a tremendous impact on the study of identity as such. Considering adjectives and metaphors, thus, is an important element in the study of identity.

identity," common in socio-cultural sciences, thus, seems paradoxical. However, the argument here is, to the contrary, that it is exactly by combining the two concepts and considering their combined perspectives and potentials that some of the problems with each can be solved. Culture is first of interest when it can be explained as a real or imagineered similarity, which is founded on specific political, administrative, ideological, economic circumstances; as mentioned above the different frameworks that Ulf Hannerz describes, everyday life, the market, the state, and social movements are of interest here (1992:40ff). It is from this point of view that culture captures social relations between people. Culture becomes similarity of ideas and practices resulting from historical developments that are continuously being recreated or transformed. Identities help to locate such similarities through historical sequences, across territories, and between social groups.[21] Or as Donnan and Wilson writes: "much can be learnt about the centres of power by focusing on their peripheries" (1999:xiii). It is by letting differences locate when and where similarity expires or starts that the study of culture becomes credible. That is why the identity analysis must be contrastive, and why in its written form it will focus, exaggerate, and cement differences (Hastrup 1988:125-128). As discussed above, the recognition that no particular difference is of ultimate importance is what provides for a plausible examination of an identity. Similarities and differences exist in many overlapping contexts and situations, each relating to sets of ideas of spatialities and executions of power; without asserting any theoretical or hierarchical coherence, people's everyday life is what unites this multitude of perceptions and experiences. The analysis of culture and of identity, therefore, constitutes two readings of the same object; implicitly or explicitly the one is always represented in the other. The analysis of culture focuses the core to expose what is distinctive, what characterize the entity – similarity. On the contrary, the analysis of identity focuses difference(s). It is where these two readings intersect, exposing how related and yet different similarities meet, that cultural identity can be recognized and finds its expressions. To renounce the concept of culture conceals rather than exposes this.

When two imagined communities border themselves and mutually recognize and acknowledge their cultural differences, many different agents participate in creating their ditch. The study of the national, consequently, must be initiated as an analysis of identities. However, to be able to understand nations as entities it must be continued also as an analysis of cultures. The ambition must remain through cultural history and analysis to be able to create sufficient context to be able to explain both differences and similarities.[22]

---

21   To study cultural variations associated with time, space, and social milieu is, in this connection, still the object of the science of ethnology; ways, methods, and theories have changed considerably during the latest decades (Bringéus 1976).

22   This understanding of the study of culture is not different from that of Hugo Matthiessen and other cultural historians from the latter part of the nineteenth and the beginning of the twentieth century (Linde-Laursen 1989a: especially 73-112). It is also an

The play between similarity and difference, and the latter's coincidence with the political division in Øresund, constitutes the basis for the study of cultural bordering ideas and practices. Hovering above the water, Sundbärg and others exactly recorded the coincidence between the political border and their imagination of the cultural communities. This clarity provided them with the certainty they sought in their efforts to establish identity where they thought they identified a lacked of such. However, just because such processes have been brought to fruition and therefore exist as narrations of differences does not indicate that they will remained unchanged.

A Converted Reality

> That, which was a dream in 1905, was in 1940 reality bright as day. (Kjær Hansen 1941:76)

The narrative about Sweden contained from around 1900 a project for the future, something for every(Swedish)body to pursue. Industrialization was obviously imagined to be the path forward. Especially the natural resources in Norrland, the ore, the forests, and the "white coal" in the rivers, should contribute to success. A project so essential to the nation caught Sundbärg's interest. He found that Swedes were predisposed for engineering and natural science; fortunately, Swedish engineers and technicians could meet the huge challenges. Sundbärg, though, complained in general terms, that the development of industry was obstructed (1911:82f). These obstructions evidently were overcome. The potential, which was demonstrated from the late 1800s, was brought out after the turn of the century and eventually changed important elements in Danish cultural bordering against the Swedish. In the course of time, such conversions were registered in geography books. Nevertheless, as they came in continuously new, only slightly changed editions and as old editions were in use for long periods, they presented (after the first editions) an often antiquated imaginary of nations. The transformations, though, were highlighted in other media. From around 1930, a series of Danish accounts of every aspect of Swedish society signifies increasing interest in the eastern neighbor:

> In Sweden, we find pleasure and satisfaction in the increasing interest in our country, which not least during the last year has taken on so many expressions in Denmark. (Sweden's ambassador in Copenhagen, Baron Hamilton, in his foreword to Stangerup 1941)

---

approach, presented by, for instance, the Swedish source of inspiration for many Nordic historical anthropologists, Börje Hanssen, who wrote of supplementary processes and the significance of tangible examples (1973a).

In some Danish books, Swedish transformations are presented as the reason for writing, in others they appear through observations that depict the general background. Irrespectively if the works are travel books, contributions to debates or something else, the changed Danish recognition of Sweden is noticeable.[23] Danish observers noticed progress in all fields. Agriculture had made a great leap forward. Not only was it now able to provide for the domestic market, it also produced a surplus for export. With inconceivable haste, an extensive and expanding export industry had been established which used and processed domestic raw materials. The formation of a series of Sweden's later world-famous groups, as for instance Boliden, LKAB, Saab, Skanska, SKF, and Volvo, started in this period:

> Modern Sweden, as it has formed itself in the period from 1905 up till now, has, from being an indifferent fringe in Europe, become a powerhouse with crucial impact on the economic world-view of our time. [...] We know that Sweden today is also the economic leading state in the North. [...] The starting point for understanding modern Sweden is...the industrialization of Swedish economic life. (Kjær Hansen 1941:76f)

The basis for this rapid development of Swedish economy was explained as a certain type of "controlled capitalism," where employers' associations and labor unions acted together in the interest of the whole society (Kristensen 1938:221). Sweden entered "a new period as a great power," in which it "around the world... has become a myth, a brilliant example, and through that anew a factor in the history of the world" (Kristensen 1938:20).[24] Included in the transformed Danish narrative were no real envy of the development and the economic results of Sweden, which were built on very different natural resources. On the contrary, one finds a quite extensive admiration for the achievements. But still, the different structure of Swedish economic life, with big and growing corporations, which processed what was dug up of the underground or torn out of the forests, stressed family enterprises and the work of the peasants through hundreds of years with cultivating the soil as particular Danish features. In relation to the densely built and through history developed and cultivated Danish soil the intensity in the development of the new, expanding Swedish economic life, which lacked a retrospective dimension, was regarded as exotic: "Sweden is however an odd alien country; we are brought to think of America" (Nyrop 1929:13). Thus, it was

---

23   See books as: Nyrop (1929), Spager (1935, 1937), Kristensen (1938), Mielche (1941), Stangerup (1941), Vogel-Jörgensen (1943), and Bergsøe (1946).

24   "Around the world" ought in this context not least to be read as a responds to the interest which followed Marquis Child's book: *Sweden, the Middle Way* (1936) (on the debate see: UD's tidningsklipp, series 3, vol. 344 B, the Swedish National Archive, Stockholm). Child's stressed the importance of the cooperatives in the development. His book contains a whole chapter on Danish agriculture and cooperatives; although this is rarely recognized and not what made the text famous.

Figure 3.3   *When Gustaf Carlsson was on his way to the station in Örsjö, he stopped at Grythult and thought: this place is Heaven on Earth, but I probably have to sell the cow*, **diorama by Ola Terje**

Notes: Swedish nature with "disorderly" pine forests and open meadows is an established component in representations of the Swedish. Ola Terje, photographer and diorama builder, catches this Swedish imagination in his art. In this diorama he additionally includes aspects of how poor and backward Sweden was before it was transformed into spearheading Modernity.

Source: Photo by Ola Terje; reproduced with permission from Ola Terje.

no longer Danish economic life, which, within cultural bordering between Danish and Swedish, distinguishes itself by being compared to the achievements of the great country of emigrants in the west.

Sundbärg mentioned, how Swedes' ability to organize partly compensated their other deficiencies (1911:84f). Within the Danish narrative, the thorough organization of state and everyday life from the 1930s became an explanation for the force and speed of the reconfiguration of Sweden:

> The Swedes are, regarded as a people, overall more inclined to order, discipline, and solidarity than their western neighbors, that is what, in the perspective of history, has supplied it with the power to achieve, what it decided to undertake. (Nyrop 1929:19)

Sundbärg's emphasize on the particular ease with which Swedes could be organized was vehemently rejected in 1911: "for our people [the Danes] can commend itself for the same qualities, which presumably are jointly Scandinavian" (Nielsen 1912:12). Twenty years later, Danes readily acknowledged that this trait was something different and particular Swedish. Around 1900, Swedes were depicted as rough, knife fighting, and less cultured because their state was not sufficiently developed. The opposite was true after 1930; their state was then over-developed. Commentators explained that while democracy in Denmark was also founded on a humanistic principle, in Sweden it was limited to the political system, without any relation with everyday life.

> But it is again this odd thing with the Swedes; even if they in many instances are more progressive and democratic than we [the Danes] are (they obtain for example the right to vote at age 23 and have the right to bring prams and dogs into streetcars) the population is peculiarly tyrannized by authorities and public offices; they all also agree to be tread on by ordinances and conventions. (Bergsøe 1946:106)

Sweden was around 1900 described as undemocratic due to the limited access to its political institutions. About one generation later it was the subjugation of the people by the modern state which was regarded with skepticism. The development had not even achieved the erasure of social barriers: "The drawback here is the undeniably larger social difference between the lower-paid and the top, compared to our conditions" (Holst 1941:42). Hansgaard summed up this realization:

> The frugal, hardworking Swede saw in the state something elevated, which well-being and right was far more important and had priority to any other right. The Danes, as always, not very providently, but casually, in general jovially, found that the state ought to provide a good example, and that human rights must be promoted and protected by the state. (1956:95, without italics)

Even if Hansgaard interestingly hereby intended to characterize the situation around the Skånian War, his conception of differences between Dane and Swede was truly embedded in his own era.

From the 1930s, bordering illuminated what was particular Swedish and that merged with the understanding of Sweden as a fundamentally Social Democratic project. The party and its leaders, first Hjalmar Branting and later Per Albin Hansson, were looked upon as the driving forces in a process which aim was to establish *Folkhemmet* (the people's home). It was observed that both the aims and means of the party were developed not only in collaboration with the federation of labor unions but also in close cooperation with a line of prominent intellectuals. From the late 1910s, the Social Democratic party was the biggest in Sweden, and it occupied the post as prime minister from 1932 until 1991, only interrupted by right-wing governments in 1936 and from 1976 until 1982.[25] Through cooperation in parliament, in the 1930s, 1940s, and 1950s between the Social Democrats and the right-wing peasants' party (*Bondeförbundet*), later with *Vänsterpartiet* (the Communists, literally "the left party"), a tradition was institutionalized in Swedish politics for governments with actual functioning majorities in the parliament, as the government was promised by their partners not to have a majority against them. For many years, many Social Democrats therefore voted in elections for *Vänsterpartiet*, as they would ensure this party, which had promised never to toppled a Social Democratic government, representation in parliament; with a doubly interesting metaphor this was called: comrade voting. The result was that formally minority governments most often were able to realize their political aims and means.

The Swedish project for society and economic life was not only founded on political institutions and the exploitation of raw materials. The development was also founded on the promise that continuously larger parts of the population got a stake in and enjoyed the results of these processes in their everyday life; as the Social Democratic father-figure of the country Per Albin Hansson swaggered: "What sustains the faith in us [the Social Democratic party] and consolidates our progress is…that we present practical suggestions which interests the people in its everyday life" (from Kristensen 1938:290).[26] As Frykman suggested, the reserve and lack of psychological insight, which Sundbärg found among Swedes, were in accordance with patterns among socially mobile persons. And while continuously larger portions of the Swedish population gradually got the possibility of participating as culture builders in the national project, Sundbärg's characteristic of his compatriots continued to be topical also after 1930.

Cultural bordering between Danish and Swedish had changed. Nevertheless, Danes could still in crucial aspects receive recognition for their own peculiarity when they contemplated their reflection in the cultural community on the other

---

25 From an American perspective, one might use "non-socialist" instead of "right-wing." However, in Scandinavia in general Social Democrats are regarded not as socialists but as pretty much at the center of the political specter. Non-socialist, thus, is an incorrect description. The term "right-wing," on the other hand, is not totally correct either as that is usually in Scandinavia used about the far right. The literal translation of the Scandinavia term: "*borgerlig*" is bourgeois, but that too gives some very incorrect connotations. Thus, "right-wing," as the least problematic term, has been preferred. The whole problem of the impossibility of translating exactly the terms of Scandinavian politics into English (an especially into American English) points to differences in the conceptualization of politics which ought to be investigated but that is beyond the scope of this book.

26 See also Frykman's discussion on the relation between everyday life and the national narrative (Ehn, Frykman and Löfgren 1993:161-201).

side of the Øresund. The controlled capitalism and the ability to cooperate and pull together within the Social Democratic people's home was in the narratives from the 1930s emphasized as a particular Swedish version of solidarity; "in the character of the Swedish people is an ability, proven through its development, to rise above the fear of today and the uncertainty of tomorrow to realize the requirement of the whole and to act fortunately in the spirit of this realization" (Holst 1941:51). In the hegemonic Danish narrative this assessment, though, was turned upside down. The image of Sweden was most often not depicted as an admirable ideal of solidarity; rather, it was regarded as a deterring caricature of the execution of power by the state onto its citizens. The Swedish project was from the 1930s cemented as modern; and the narrative, therefore, could be employed in Denmark as a depiction of all the evils that would befall this nation, if it adopted modernity as an ideal.[27] The (hegemonic) narrative about Sweden was in the Danish debate not employed to reject democratization, urbanization, industrialization, or other processes, as far as the results were improvements of infrastructure, dwellings, means of production, and many other things. However, it was employed to demonstrate that the results of modern society's influence on everyday life would be the dissolution of traditional social networks. The state would assume and control tasks which were earlier carried out within families or local communities and, consequently, tradition would be liquidated as an important foundation for identity formation.

Narratives about the consumption of alcohol became the arena which to the greatest extent symbolized the different societies' control of their citizens' everyday life. In the earlier Danish narrative, Swedes were rough and less cultivated, they drank too much, and then they became violent. The Swedish state's attempt through dry areas and the alcohol trade monopoly, *Systembolaget*, to regulate the citizens' drinking was never totally successful. Swedes, who drank, were by Danes recognized as both less formal and less cultivated. In the narrative, this soon became symbolic – Swedes by drinking also emancipated themselves from the modern project and their state's undemocratic interventions into their everyday life albeit temporarily. Thus, by drinking Swedes became more human. The Swede Neander-Nilsson offered a psychoanalytical explanation for his compatriots' transformation under the influence of alcohol. He found that the Swede basically was a child of nature who only when drunk was able to recognize himself or herself as an *I* (1946:15-22).[28]

---

27   The close connection between the modern and the Swedish national project from around 1930 became a pivot for much historically oriented cultural analysis in Sweden from around 1980. See for instance Frykman and Löfgren (1985, 1987), J.O. Nilsson (1991, 1994), and Ehn, Frykman and Löfgren (1993).

28   In this part of the Danish narrative it is completely ignored that temperance movements also in Denmark celebrated triumphs until around the First World War; see Porskær Poulsen (1985, 1986) and Bundsgaard and Eriksen (1986). Many Danish municipalities were dried up after local referendums. The best known depiction of the battle of the booze in Denmark from this period is the novel *The Fishermen* by Hans Kirk from 1928 (1999).

Around 1900, Swedes found in their narrative about the Danish fuel for discussing what their country lacked and how the ongoing development ought to be managed. In the period until 1930, when the public sphere in Denmark did not demonstrate any significant interest in the Swedish, Danish solutions in some instances were turned into models for elements in the construction of the Swedish people's home. Per Nyström, who was a central figure in the Ministry for Social Affairs in the 1940s, is of the opinion that basic features in Swedish social politics (and in the English social reform, the Beveridge plan from 1942) were modeled from thoughts developed by the Danish Social Democrats C.V. Bramsnæs and K.K. Steincke in the 1920s (Nyström 1983:221-230). Denmark, thus, became on some points a model for Sweden in "the latest generation's continuous development towards greater political and economic justice" (Wendt 1941:73). However, similarity between the nations on single points does not overshadow that there were and are differences between the two states' version of modern society; for instance in areas of education and the fight against unemployment they diverged (Knudsen and Rothstein 1994).

While Danes' interest in the Swedish people's home from the 1930s was increasing, and while the narrative of the Swedish as an abominable, modern contrast to the Danish was developed, Denmark gradually disappeared from public debate in Sweden. The number of books in Swedish on the Danish became smaller than the reverse.[29] The Swedish narrative on Denmark confirmed the importance of history and agriculture: "every part of the country carries the marks of being tilled by humans and of the historical development" (Numelin 1935:68). In Swedish geography books it was noted that most Danes by now were employed in industry but it was in enterprises producing from agricultural raw materials. Continuously it was true that "Agriculture is unquestionably the life nerve of Denmark" (Swedberg 1948:21). To Swedes it was explained that industry, which processed and exported foremost products from animals or which manufactured machinery for agriculture, developed from other conditions than Swedish industry based on raw materials. Thus, they understood that Danish industry had a different structure: "It...is not assembled in big corporations to the same extent as the Swedish, and this is in accordance with that the Danish industry, with the exception of the provisions industry, is not an exporting industry to the same extent as Sweden's" (Swedberg 1948:59, without italics).

In travel books, and to some extend in geography books, Denmark became to Swedes in the period after 1930 foremost a holiday country with beaches and an exciting city with public entertainments (Copenhagen). The friendliness, openness, and talkativeness of the population made the country appear nearly in a southern wrapping: "In Danish mentality one finds something southern, something continental, which does not belong in the rest of the North but maybe exactly is what has contributed to making the whole culture here so graceful" (Numelin 1935:108). The more southern and unrestrained frame of mind also left its mark

---

29   Single examples are Åkesson (1934), Numelin (1935), and Krantz (1951).

on the relation between state and citizen, as the rights of the individual were highly respected. In Denmark the individual, not least, took responsibility for his or hers drinking and smoking. Swedes were aware that authorities in Sweden on the contrary meddled in everyone's affairs:

> For them [Swedes] it is unthinkable that the citizen could act in accordance with his own judgment and responsibility. He must have rules and regulations as guides while walking through this vale of tears. (Krantz 1951:45)

Did the narrative on the Danish gradually vanish from public discussions in Sweden, this was not because the Swedish thereafter turned inward and rested in itself. Relations to anOther stepped forward. A Denmark not resolutely seeking modernity was no longer of current interest to cultural bordering and as a comparison for the Swedish future oriented community. Instead Sweden was correlated to other conscious modern projects. Until the Second World War, Germany played a role. After that, as discussed by Löfgren, the relation to the huge settler society in the west gradually obtained an immense influence on the conception of the Swedish (Ehn, Frykman and Löfgren 1993:65ff; see also O'Dell 1997, J.O. Nilsson 1994, 2000). This orientation towards the US was obvious to observers:

> I wonder if any European population or capital today resembles so much the US as the Swedes and their Stockholm? The similarity is probably to a certain extent due to a forceful admiration for the US, but a not insignificant reason is likely to be the similarity of the two countries nature: the colossal distances, the abundance of nature, and the abnormal growth of the cities. (Bergsøe 1946:49)

Metamorphosis

The Danish narrative on the Swedish people's home was from around 1930 quite fixated for half a century. The narrative accentuated, that industrialization, urbanization, and capitalization from that time quickly transformed Swedish society.[30] As the earlier narrative, the converted Danish-Swedish bordering was employed to confirm that petit bourgeois and liberal democracy were essential to the Danes. Whereas Danish peculiarities earlier were stressed through the comfort in a nation of peasants, which evolution and results were more advantageous than its neighbor's, the relation after 1930 was contrasted against Swedish large-scale industry. Before 1930, Swedes were reserved and less cultured. Later they became cultivated, but simultaneously worked in so much state and so much Social Democracy, that it happened to ru(i)n their everyday life. While society before had been undemocratic, because the citizens were excluded from participating in the

---

30   How this was described by a Swede, see for instance Ludvig Nordström's book about Swedish national unity (1934) or other of this author's writings.

management of the state, it later became undemocratic because the state outgrew the people. The particular Danish, which was emphasized through this cultural bordering against Sweden, remained through this transformation of the narrative interestingly unaltered. The Swedish narrative about the Danish remained largely unchanged from the end of the nineteenth century and into the 1980s. However, the unchanged narrative was employable to profile a completely transformed Sweden. Nevertheless, Denmark remained countryside, small-scale business, and liberal democracy.[31]

Through cultural bordering, the Danish was marked as stability, while the Swedish was discontinuous and recreated. Perpetually, however, they mutually recognized and acknowledged the other part's cultural difference, and the narratives continued to be in harmony with the two parts self-images. Swedes accentuated with pride the industrial and social development of their country. Nevertheless, in agreement with the Danish narrative, Swedish voices eventually emerged that pointed to flaws in their society. From 1939, a group of prominent active creators of the modern project started to gather at intervals. They discussed how the schools, dwellings, families and labor market of the new industrial society could be organized in ways that raised individuals in the people's home to partake in the democratic system: "It is crucial to bring up, to educate people to be active, self-going, and responsible citizens" (Segerstedt 1944:35). The solution this group drew up (Segerstedt et al. 1944), which was founded on the participation of individuals and families in the building of the institutions, could however be turned upside down. Other debaters pointed to the same institutions as the very cause for passivity and all evils in modern society. In Swedish society, the individual now had problems handling "centrally directed development of opinions and a growing guardian society" (Nothin 1956:93). The Social Democratic project, thus, despite its initial merits, could easily end in a bad way:

> A majority-parliamentarism can easily slide into something that is dangerously near to a totalitarian society. A transition to a real dictatorship is, then, awfully near. (Nothin 1956:157)

Despite some disagreements about the consequences and the course ahead, the alliance between the Social Democrats and the academians prevailed. Though not agreeing on the evaluation, all concurred on which elements were part of the cultural narrative and agreed on the remarkable change, Swedish society had achieved during the time of the previous generation. When asking himself, if the

---

31   Elements of the relation between Danish and Swedish after 1930 are expressed in memories written by people who after the Second World War moved across the Øresund (Lund University Folklife archive A:4817 and Osbak 1993). The narratives also gradually appear in geography books for the schools, in the same way as they surface in mass media (see for instance a series of articles in the Danish daily *B.T.* in December 1961, UD's tidningsklipp, series 5, vol. 188, the National Archive, Stockholm).

Swede was still human after these societal changes, Neander-Nielsen answered that they as children of nature had trouble adapting to the new surroundings. The many divorces, thusly, were due to "the country-Swede's inability to settle with the new urbanized, rationalized, and industrialized existence, his deep dissatisfaction with the uprooted condition created by modern life" (1946:53f). This was exactly the conception, Danes recognized and could use as a mirror. Interestingly, these differences with regards to how modern society should be evaluated are precisely reflected in the two languages. While the word "modern" in Danish has been associated with something dubious, it is in Swedish related with something unquestionable positive.[32]

During this period, if evaluated with a different scope, Denmark and Sweden came out quite identical. Regardless of the fifteen years between two Indians' stay in the North, there are remarkable similarities in their accounts and attempts to characterize the two nations. Both Danes and Swedes kept to themselves in their nuclear families and did not socialize with their surroundings; parents and their children did not talk to each other, the private and the public were completely separated, and people in general were conflict-shy and had problems communicating feelings (Dhillon 1976, Reddy 1993). The two writers' presentation of the two countries is so general, that their accounts reasonably could be taken to depict any modern, western society. Seen from afar, similarities between Denmark and Sweden dominated the picture. Nevertheless, in their mutual cultural bordering processes it was differences that were evoked, narrated, as well as used in political discussions, in jokes, and in other contexts.

Through the intertwining of similarities and differences, the Danish and the Swedish emerge through their metamorphoses as two versions of the Scandinavian Welfare State. Both managed to synchronize political and cultural communities and both could maintain that their society embraced unusual homogeneous populations, socially as well as culturally. Regardless of similarities in the historical processes, the countries' different paths towards this homogeneity were clearly profiled in the relation between them.

The narrative about the Danish emphasized continuity from a pre-modern peasant society. The historical process took a central stage and these peasants' individuality, which was not to be mistaken for egoism, was the background for understanding how interests were joined in modern Danish society. The

---

32   This is at least the case after 1930. Realistically, there is no alternative to translate Danish "modern" to Swedish "modern"; this difference therefore makes communication across the border difficult under specific circumstances as listeners will receive semantically quite different messages, without being able to detect that. Furthermore, how the semantics of "modern" is evaluated in English of course complicates this point profoundly. Realistically, English "modern" will be understood remarkably different by groups within societies, not to speak of by groups from different societies that fully or partially employ English (for instance England, Wales, the US, New Zealand, Kenya, Belize, Nepal, Hong Kong, United Arab Emirates...).

strong Social Democratic party bended its understanding of progress within this hegemonic, liberal understanding of individuals' equal standing in peasant society and in history (Østergård 1984, 1992b). This was obvious also from the other side of Øresund:

> Few people are as the Danes willing to accept universal human points of view. The political contradictions are therefore nearly all of a social kind and not as bitter as elsewhere. (Sörlin 1951:158)

In Sweden, on the other hand, the narrative focused modern industrial society as a complete mutation from an earlier, exhausted peasant society. History was, thusly, not recognized as a continuous process but as a rupture. What happened prior to this discontinuity was regarded as an unhappy, poor, and dark chapter, in which the people had been divided over access to political influence and economic possibilities. Thus, it was the now and especially the future that represented the positive. The basis for this positive narrative was the unification of the interests of the blue and white color workers and the academians. In harmony these groups strived to realize this inclusive, new, light chapter, during which internal differences would be equalized. While progress was expressed through a broad Social Democratic movement, the national project became associated with this party.[33] Other groups in Swedish society had to relate to this hegemonic narrative, even if they, as illustrated above, were free to turn the evaluation of elements upside down.

The relation between Swedish solidarity, collectivity, and responsibility on the one hand and Danish independence, individuality, and freedom on the other became a point from where the relation between the dominating cultural bordering narratives crystallized. And this point was easily available and could be invoked in debates about what were reasonable and unreasonable developments in modern society on both sides of Øresund.

After 1930, the Danish-Swedish relation was additionally inserted into a more general pattern that sorted nations in Europe according to a northeast-southwest scale. The northeastern nations were more reserved, had more state, and corresponding less humanity. The southwestern, contrarily, were more individualistic, people were open and happy, and they had less state and governing. Narratives about nations towards the terminals of this continuum tend to be more coherent than narratives about cultural communities in the middle. The images of for instance Finland and (Southern) Italy therefore are more likely to be the same among different other nations, while narratives about Switzerland are much more ambiguous and difficult to agree on. From this perspective, Denmark has two

---

33 The Social Democratic movement was not only a political party. It reached far into and embraced other popular movements in society; these included the labor movement, sports organizations, a remarkable portion of the temperance movement, some revivalist Lutheran fractions...

northeastern nations as neighbors: Germans and Swedes, while its southwestern relations stretched to southern European and Mediterranean countries and possibly to England.[34] Sweden had northeastern relations to Russia and in some ways to Finland, was quite equal to Germany, while they had southwestern relations just across Øresund. Copenhagen became in this perspective the start of a southwestern "continent"; as it was in fact also often depicted in travel accounts and tourist advertising. While Sweden in such relations largely remained a northeastern country, Denmark was both the northeast towards most of the European continent, and the southwest towards the other Nordic countries. As argued elsewhere, cultural bordering, thusly, produced diverse narratives that situated Denmark perfectly in the middle of the world! (Linde-Laursen 1999b).

Cultural bordering processes between the Danish and the Swedish imagined communities thus from the end of the 1800s displayed both continuity and discontinuity. The form and comprehension of the Danish cultural community seemed stable. The Swedish, on the contrary, was fundamentally changed, and as a sharp divide was drawn between a before and an after 1930, both eras were afforded clear cultural identities. Every national project includes narratives about events that symbolize such transformations. Already happened or expected events in the history of nations are accredited extraordinary emotional and symbolic significance; they are turned into tradition (Hobsbawm and Ranger 1992). Among monuments, those which can be called "national" are often erected to commemorate faithfully such events or persons associated with them. National monuments and events must be ambiguous to allow multiple interpretations in order to similarly materialize the state, the nation, and a whole people, regardless of what competing ideas of spatiality might be available for interpretation (like age, gender, and so forth).[35] Such a grand, inclusive event was indeed planned to signify the transformation of modern Sweden. The Stockholm Exhibition (*Stockholmsutställningen*) from May 16 through October 9, 1930, was both in its time and later recognized as a symbolic representation for the birth of a new star among nations (see Pred 1992:3ff, 1995:97ff).

---

34   It is possible that the UK and Ireland, with their dualistic political systems and other peculiarities from this perspective are not at all European, but island nations. A realization of this might explain some continuous problems within the EU, where especially Britain obviously does not always share aspirations with other member states. While such differences can be and are dressed up as differences from national interests, it might be illuminating to approach them within a wider framework and analyze them as differences of cultural identities. Henrik Nissen has discussed the role of perceptions of Germany in Denmark around 1930 (1992).

35   I have elsewhere discussed *The Liberty Memorial* in Copenhagen as a national monument (Linde-Laursen 1994a).

De Nova Stella[36]

> Then, with support from this people, educated, cultivated, well-nourished, well-dressed, organized, orderly, free, fresh, and determined, *Svea Rike* ["the Swedish realm"] walks out in the world to peacefully conquer a spot in the sun, and this is done by manufacturing from Sweden's mountains, forests, soil, and waterfalls a series of the best quality products on the world market, the result of brilliant inventions, craftsmanship, and professional pride. (Nordström 1930:8, text for the part of the Stockholm Exhibition called "Svea Rike")

The exhibition unofficially was called "*Funkis*" (from functionalism, the Swedish term for the style that in the US often is called "international"). The aim of the exhibition was to show dwellings and furniture that most people could afford, of good quality, and with a pleasant appearance. It also presented suggestions for half or wholly public arenas as gardens, burial grounds, streets, and means of transportation. From the start, Funkis was associated with "the God of the new ear, functionalism." At the presentation of the plans for the exhibition, functionalism was presented as "an intellectual and morally clean conception of the artistic problems."[37]

A year after the exhibition some of the persons involved in the project published a manifesto that clearly expressed their aspirations: functional differentiation, the industrial production of dwellings, democracy (social equality), and development. They wanted to show that humans could master space in accordance with the rationality of the time:

> Creating from the era, the will to be inspired by its light sides, its outstanding technical inventions, its freer humans, is the only reasonable [thing to do]. This interest, this composing with the visage of the era as theme by itself establishes the demands that we try to fulfill. We cannot be inspired by the era, without feeling solidarity with it. We have to enroll as it servants, we have to help to solve its problems. [...] *accept*[38] the existing reality; which is the only means by

---

36 *De Nova Stella* ("the new star") is the title of a book from 1572 by the famous astrologist Tycho Brahe (1546-1601). The book describes the development of a super nova in the Cassiopeia formation. As an early example of empiric, modern, science in Europe, I find this a suitable title for this part on the emerging stardom of the Swedish Welfare State.

37 Quote from Gregor Paulsson, the executive director of the Swedish Handicraft Association, which backed the exhibition (*Stockholms Dagblad* October 26, 1928). A lot of material on Funkis from both Swedish and international press is gathered in the collections of the Swedish Ministry of Foreign Affairs (UD's tidningsklipp, series 2, vol. 835-837, the Swedish National Archives, Stockholm). Besides many small announcements about the exhibition, it seems as if most of the debate about the exhibition is written by a small number of very active writers.

38 *Accept* (Swedish: *acceptera*) was the title of the manifesto, it is in Swedish not spelled with a capital a.

which we might control it, to be able to change it, and create culture as an agile tool for life. (Asplund et al. 1980:14, 198)

To the organizers functionalism was not a style. It was a principle: "that, which is well suited to its purpose, is what is beautiful."[39] The beauty as decoration was substituted for its value as a thing: "the simple, the functionally natural and beautiful demands as much fantasy, as much artistic ability as the traditionally decorative" (Asplund et al. 1980:169). From everywhere, also from abroad, it was recognized that the exhibition represented something remarkable and new in society. Furthermore, it was not only the backers of Funkis who valued its exposition of the transformation of Sweden. Sigurd Schultz walked around and "felt that for once, he was equal to his era and himself" (1930:133). Finnish architect Alva Alto, who had followed the planning of the exhibition and was invited to the opening, enthusiastically depicted it as "the intellectual Sweden's hold on Mr. Average-Swede" (*Åbo Underrättelser* May 22, 1930).

However, there were also many, who forcefully disassociated themselves from the forms they met in buildings and displays. As border penetrations, transnational cultural expressions always have threatened ditches between nations; even if, paradoxically, the exact same forms that were regarded as threatening often subsequently were nationalized and were turned into parts of the always changing, established set of representations for the nation-state.[40] Thusly, functionalism was already prior to the opening of the exhibition being debated as a threat to the traditional and the national. The Finnish architect Eliel Saarinen regarded functionalism as something universal, which in the US was already going out of fashion. Opposite, the Swedish architect Carl Malmsten, who redrew from the exhibition, saw it as a style and as something new and threatening coming from Germany. One of the principal promoters of the exhibition, Uno Åhrén, however, ensured that "Functionalism...of course can...become national. In every country it takes on its slightly different forms, gets a national impression" (*Stockholms-Tidningen* August 25, 1929). Some people, though, were difficult to

---

39   The organizers tried as far as possible to avoid to mention the concept "style" all together, to stress that what they pursued was not a "style" but an expression of a more profound purpose (see Asplund et al. 1980:142ff, *Stockholms-Tidningen* August 25, 1929). The exhibition motto here translated as "that, which is well suited to its purpose, is what is beautiful" ("det ändamålsenliga är det sköna"), was created by Gregor Paulsson; an alternative English translation: "form follows function" does not provide the same intensive quality as the original Swedish. In general, a lot of the words/texts associated directly with the exhibition reads like slogans and are difficult to translate while maintaining both the rhythm of the language as well as the meaning. Realizing that something very important to the authors and the time is lost, my translations focus on the meaning. For a discussion of the Stockholm Exhibition and the relation of the presented ideals to Social Democrats, see J.O. Nilsson (1991:67-90, 1994:177-204).

40   See for instance Tom O'Dell's study of the Swedification of the American car (1992 or 1997) or Gestur Gudmundsson's study of rock music in Iceland (1991).

reassure. They for instance saw the exhibition as "an artistic bolshevism...The most unpleasant and at the same time correct that can be said about the 1930 Stockholm Exhibition is, that it has not been Swedish" (*Sydsvenska Dagbladet* October 2, 1930).

In general, the exhibition was reviewed positively in Danish newspapers. Even if they questioned functionalism as a style, they reviewed the social ambitions and the festive displays at the exhibition positively: "They mess about with steel and concrete, with glass and wood, paper and rubber, and the result is a saxophone-hymn to technique, which is the strongest impression of the Stockholm Exhibition" (*Nationaltidende* May 16, 1930). The organizers had realized that Funkis also had to be "an ordinary festival for the public. Here one has to employ the principal of Roman politics: 'bread and circus'" (Paulsson 1930:18).

Anker Kirkeby, reviewer for the Danish daily *Politiken,* was recognized as one of the only negative, European critics at the premiere (*Stockholms-Tidningen* May 18, 1930). It was not because he turned against the developments of his time:

> I think that we can expect a great revolution of the world...in front of our eyes, new humans are appearing. It is within ourselves that the revolution is in full swing. We eat differently, drink differently, dress differently; we think differently, make love differently, breath in a new way; we will be living a life, completely different from now. A new race is created. Yes, we are downright transformed into completely different humans with new suites of sensations and feelings. (*Politiken* June 22, 1930)

Kirkeby was familiar with the idea of a modern life, but was strongly critical of the architecture of Funkis. He found that earlier exhibitions of dwellings in Paris 1925, Stuttgart 1927, London 1928, and Copenhagen 1929 had presented "le Corbusier...the master-architect of the era" more pleasantly and harmonic (June 22, 1930). "The architects sin here [in Stockholm] against the very fundamental principle of the new functionalism: that just as well as a steamer must be a steamer, an exhibition must also be an exhibition" (May 16, 1930). To him, Funkis showed the democratic potential in industrial production. This potential provided a means to approach nature instead of withdrawing from it, as had happened through the history of architecture. However, in Stockholm it had become fare too much technology and correspondingly too little form and spirit.[41] Kirkeby propagated the nation as the proper space for all kinds of cultural processes. Consequently, he paid tribute to Svea Rike, the part of the exhibition that told the history of the

---

41   Leif Nylén has told that le Corbusier was invited to the premiere in Stockholm. However, by mistake he was given a second class ticket for the sleeping car; sulky he turned around on Gare du Nord and never saw Funkis (Pred 1992:49, 1995:160). The democratic ideals of the organizers could of course be taken too literally! Earlier exhibitions, the impossibility of regarding the modernist, functionalist style as one form of expression, and the development of a moderated Danish style is discussed by Carsten Thau (1992).

Swedish nation and its transformation in his time into "the leading country in the North." In Svea Rike, Kirkeby saw "a kind of giant advertisement for the firm, the modern private capitalist Sweden" (May 16, 1930), which Alan Pred much later has characterized as:

> An exercise in collective memory (re)construction in which the State as well as corporate capital is situated at the vanguard of technological innovation, is portrayed as a pathbreaker in the application of scientific rationality to the benefit and liberation of ALL. (1995:137)

It was in this part, all spectators should face that they had their past behind them and that each of them was the standard-bearer of modern society. Realizing that, they would much better in all the other sections of the exhibition understand the recommendations on how this effort ought to be practiced. The last part of the text from Svea Rike turned this intent into a slogan:

> if Svea Rike is to reach this goal, which is the whole intentionality of its history: to be regarded higher and higher by the world; that is dependent on each man and woman in this country.
>   Compatriot! Brother! Sister!
>   It is up to you to create the future of Svea Rike! (Nordström 1930:18)

However, some observers found that the image of progress and prosperity depicted in Svea Rike was not true to the everyday life of Swedes. Kirkeby found, that the exhibition did not mention "the gravity of the everyday, of the profound social differences, of the unemployment and housing shortage" (*Politiken* September 25, 1930).

Another Dane, who went to Stockholm, was Poul Henningsen; or as he is usually referred to by Danes: PH. He was man of many talents that always managed to be highly provocative for the establishment as an architect, inventor, and writer. What Anker Kirkeby found good, PH as his antagonist found bad, and vice versa: "This is the first time in the history of exhibits that it has been approach exactly right" (*Politiken* August 6, 1930).[42] When Kirkeby praised the planetarium and the aquarium, PH found them indifferent, banal parts: bread and circus. Correspondingly, PH did not like Svea Rike: "awash with visual instruction and tourist advertisement" (1930:81). However, PH was absorbed by the didactic aim of the exhibition. Funkis was not only entertainment supported by the state. The organizers wanted to educate the common people:

---

42   That Anker Kirkeby and PH did not appreciate each other is apparent from the introduction to one of PH's reviews. He constantly refers to their dispute: "He [Kirkeby] will consequently be the donkey which we will ride through the exhibition" (1930:81).

It is your taste, the exhibition will change, and you cannot resist. It will come sooner or later. [The exhibition] is like a swell across Scandinavia, an attempt to free itself from the pressure of the past. [...] Beauty is harmony. Harmony with itself, harmony with its surroundings, with its age. [...] It is for the *masses* we work...harmony will be created in the everyday life of the citizen, agreement between the general sense of beauty and the general political conviction. [...] I congratulate Sweden, that it has escaped the claws of the national art. [...He who] creates the ordinary bulk article, which makes life more comfortable or cheaper, actively participates in working for political progress. (*Politiken* August 6 and July 3, 1930)

Before the opening, an advertisement campaign in Sweden and abroad had marketed the exhibition. Advertisements, all approved by the chief architect Gunnar Asplund, were also a significant part of the impression the visitor had of

Figure 3.4   *The Corso* **at the Stockholm Exhibition in 1930, by Max Söderholm**

Notes: Chief architect E.G. Asplund, "The Exhibition thus tries, through its propaganda to support manufacturing that strives towards the mutual goal of through existing and innovative shapes to the products to provide the environment in which we live a striking character and quality as well as a new form in agreement with the foremost qualities of the time. [...] On the *gradiner* stands the 80 meter tall advertising mast from which lit ads and floodlights illuminate the whole exhibition area and are visible from most of the city" from *Almanack för alla* 1930. Stockholm: P.A. Norstedt and Söners förlag.

Source: Illustration from Paulsson 1930; photo courtesy of Björn Andersson, Historiska Media.

the exhibition (Pred 1992:22ff, 1995:124). Kirkeby, and many with him, found the ads appalling; they were annoying and confusing. Especially the lean, tall central advertisement mast was detested. PH, on the contrary, pointed to the advertisements as one of the exhibition's most important artistic contributions (1930:81f).

Even if they both had positive expectations to the modern project, PH and Anker Kirkeby as antagonists experienced Funkis in oppositional ways. Kirkeby saw the exhibition as a positive attempt to try to reorganize the national project, while PH with sympathy reviewed its message for a more modern, more beautiful everyday life for the masses in his own time. To PH, the exhibition wanted to

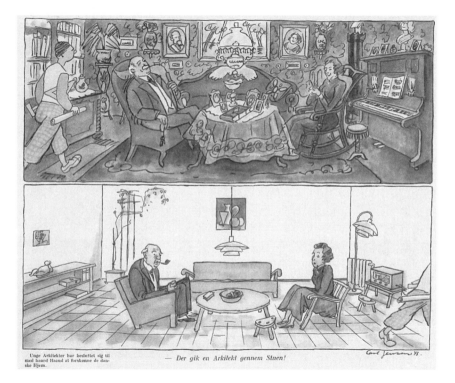

Figure 3.5   **"Young architects have decided with a heavy hand to beautify the Danish homes:** *An architect walked through the livingroom,***"** **by Carl Jensen, 1943**

Notes: As Lone Rahbek Christensen has demonstrated (1986), *funkis* (international style) was not necessarily comforting to those who from the beginning were the targets of the social programs of the architects. Drawing by Carl Jensen, 1943, for *Blæksprutten* 1943, a humoristic annual publication with comments on politics and everyday life.

Source: Reproduced with permission from Gyldendal, Copenhagen; photo courtesy of Björn Andersson, Historiska Media.

accelerate a development that could not be stopped: "It comes from Russia, it comes from being used, from everyday life, from economic conditions. It grows from the people" (Henningsen 1930:89). The two opposite opinions illustrate that in the modern maze there is no such thing as a one-way path, which all have to walk. Ideas of spatiality open and close to each other in complex ways. Within the cultural bordering narrative between the Danish and Swedish, however, Kirkeby and PH did agree. Also to PH, Sweden had become the leading country in the North: "We have constantly been after Europe, though ahead of Norway and Sweden…today Denmark is a backwater that receives its impulses from the north" (1930:92).

The Stockholm Exhibition was in its time and has later been regarded as an outstanding symbol for the brighter future of Sweden, and for substantial changes to everyday life. Kirkeby was right; a (r)evolution occurred; however, it did not happen instantly. In fact, the people, for whom Funkis was made, were not as interested in the message from the exhibition, as the organizers had imagined. The exhibition was visited by 4.1 million. However, only 63,881 guides and a humble 5,767 catalogs were sold (Pred 1992:42ff, 1995:151). The vast majority considered the Stockholm Exhibition more as an amusement park than a place, where the possibilities and developments for the nation and modern society were contemplated. It also lasted quite a while, before the masses experienced the ideas for dwellings and their organization, which were presented in the exhibition. Its time saw Funkis as a symbol; for entertainment and/or modern possibilities. It was only in its posterity that some people believed that the didactic attempts reflected everyday life in Sweden, 1930. In that same year, 48.2 percent of all apartments in central Stockholm consisted of one room with or without a separate kitchen.

As more new dwellings were built, the Stockholm Exhibition became a symbol, not for something that could happen, but for something that had happened; even if the tangible manifestations of modern advances often took quite different shapes than suggested in 1930. Funkis became a shining star, a *nova stella*, which clearly marked the transition from a pre-modern to a modern everyday life and society. In practice, more and more people became cultivated; not least through internalizing new ways of living and consuming. However, it was not furniture of bend steel tubes that came to characterize Swedish living rooms. Gradually, a Swedish functional style very much made out of wood developed. This style implied also a re-evaluation of the past peasant society from stinking poverty to a fragrance of cleanliness and naturalness. Peasant society remained a definite past. However, the distance to it became so large that in the narrative, this past was allowed to become exotically familiar. Reproductions from Carl Larsson's famous images in "A Farm" and "A Home" appeared in many homes, and such imaginings next materialized as IKEA warehouses (Löfgren 1993).

In Sweden, academians and Social Democrats joined in their efforts to created and disseminate new thoughts and practices for the developing ideas of spatiality. Simultaneously, PH and other intellectuals found themselves in opposition to a Danish Social Democratic movement that acted within the confining narrative

Figure 3.6   The contemporary Swedish family lived a modern life, from "The Citizen Book," 1950

Notes: "Thoughtfulness and personal courage to break with customary, boring and unimaginative habitual conceptions of what is beautiful and what is ugly – that is the essence of the notion of coziness" (from "The Citizen Book", *Medborgarboken* 1950:117).

Source: It is unknown who photographed the illustration for the book; photo courtesy of Björn Andersson, Historiska Media.

about a democratic, liberal peasant society. The Danish Social Democrats, consequently, could not or would not engage the cultural-radical tendencies in the period between the wars. Instead of being integrated with and leaders of the state, as in Sweden, Danish intellectuals were send on a weary hike in the dessert, which for many led to political radicalization. Even if cultural-radicalism and communism was far from identical, they often were mixed together in the public's understanding of the position taken by many public intellectuals.[43]

---

43   In Nolin (1993) different positions of the cultural radicals in Norway, Sweden, and Denmark are discussed. The complexity of the phenomenon is stressed by the fact that there, of course, also was radicalism without any party-affiliations in Sweden (as for instance the so-called Fogelstad-group of women gathering around the journal *Tidevarvet (The Epoch)*; see Eskilsson 1993). Morten Thing in detail has discussed the Danish intellectuals' relation to the

Funkis could become such a significant symbol for modernity and freshness because it originated from this interface between Social Democrats and academians in Sweden. In Denmark, the situation was very different. PH had already in 1923 in a presentation formulated, what later became the motto of the Stockholm Exhibition: "the most concise, the economic, the most practical, the most true, is also the most beautiful." The reporter added, that it was "a sentence that struck the listeners" in The New [communist] Student Association (after Thing 1993b:224). PH later became involved in a project, which immediately could have obtained the same shining star powers as Funkis.

Denmark's Metamorphosis

The Danish Ministry of Foreign Affairs early became aware of film as a medium that could be used as a vehicle for disseminating the narrative of the cultural community among foreigners and Danes abroad.[44] The initiative to the first official contribution to this genre was taken in 1923, and the result was presented in 1925: the silent pictures *The Denmark Film (There is a Lovely Country)*. By 1930, this film was already considered old-fashioned. A committee was formed and it decided to produce a picture with sound. It also managed to raise the necessary funds for such a project from various private foundations. It is unknown who among the committee members had the somewhat precarious idea to give the task of scriptwriting to always provocative PH. However, his name began to appear in the minutes of the committee's meetings in February 1932. In July the same year, he presented his first draft to the committee. PH envisioned "a complete modern geography," as he imagined the material itself would inspire him to find "the right modern form." Furthermore, he stated, it would "entail great expense to depict any history, especially the most cinematic kind, with Vikings and things like that" (from J. Sørensen 1980:30, 41).

It took two years to make the film, which was premiered simultaneously in Copenhagen and at the World's Fair in Brussels on April 29, 1935. The next day a nearly unison Copenhagen press grounded the film. The strongest critic was possibly – Anker Kirkeby:

Communist Party (1993a). What might be important to note for a non-continental European reader is that these intellectuals very much propelled public debate; they were listened to and discussed, even if people did not agree. They in many ways came to act as the generators of what the public could/should be interested in; and thus they paved the ways between the different ideas of spatiality in the modern maze. In the US, in can be argued, such a role for intellectuals and artists was actively denied, for instance through the McCarthy processes.

44   I have in two articles (1995b, 1999) more in detail discussed PH's film *Denmark* and another contemporary film *The King Commanded (Kongen bød)*. PH's film is also discussed in: Agersnap (1973), Hammerich (1986), Schepelern (n.d.), J. Sørensen (1980), and Thing (1993a:255ff). The history of the film is also illuminated here from material in the archive of the Danish Film Museum.

who would not in advance have pointed to him [PH] as the obvious man to make a modern film that could reflect Denmark's working everyday...And then the whole thing fell to the ground with a bang. [...] all the pictures appear shot in boring, overcast weather with dank winds. [...] Painstakingly, through most of the film, all beauty was avoided. Throughout, the Danish types were depicted as paunchy men and broad-bottomed matrons. And the shady sides of life were emphasized too strongly through slum and trivialities. It is very well to attach importance to the working Denmark. However, work also has a smile; we mostly were presented with its tiresome sweat. All these boring cranes, monotonous telephone lines, unlovely pylons, horrible wind wheels and throbbing locomotives are not Denmark, at least not essentially Danish, but just as much German, English, Italian, Palestinian, Japanese...It was Denmark seen from the rear. [... The accompanying jazz was] far too often just Negro noise...[The accompanying voice, which was PH's own, was] irritating dull and grinding jargon. [...The film accentuated] the dreary and unsavory sides to the fabrication of commodities. [...] A far too gray and depressing Denmark. (*Politiken* April 30, 1935)[45]

What the reviewers objected to most in the film was its lack of the proverbial Danish smile, the joy and enthusiasm – about being Danish. A confessed internationalist and modernist like PH so obviously lacked the right emotional qualifications to be able to portray Denmark: "Herein lies the fatal fault of the film. They chose a man without a fatherland" (*Dagens Nyheder* April 30, 1935). The conclusion was spontaneous, tangible, and certain: "it must be withdrawn or destroyed immediately. Under no circumstances should it be allowed, through the agency of official Danish authorities, with state approval, to bring our country into disrepute for its poor technique, tediousness, and gross lack of tact" (*Dagens Nyheder*, May 1, 1935).

The committee had closely followed PH's efforts and had expressed its satisfaction with the film. Representatives of government departments and businesses who had attended the premiere in Copenhagen had been favorable in their comments, although with reservations. The Secretary of Social Affairs, Mr. Steincke, found that: "The design of the film is a little too modern for my brain;" and the National Director of Tourism seconded, that "as a whole it will scarcely influence foreigners so that they will conceive of Denmark as a modern society with a large pleasant and modern city and with modern hotels at the seaside resorts." Some among these premiere-goers regretted a lack of coherence, of sunshine, and of pretty girls in the film. However, they pointed out that the film was not made for tourists, and endorsed it as successful "propaganda for Denmark as an enterprise" (*Politiken*, April 30, 1935). Only *Arbejderbladet* (the communist newspaper) supported the film. PH never was a member of the party, despite many points of contact between the intellectual cultural radicals and the communist party during

---

45 When a shortened version of PH's film was premiered in 1950, Anker Kirkeby again was very negative in his review (*Politiken* March 23, 1950).

the 1930s (Thing 1993a:242-265). Surely, least of all PH needed support from the cultural radicals, who in the public were lumped with the communists. He had already felt compelled to assure the committee that he was not a communist, was not a member of the party, and had never voted communist.

Faced with such a massive criticism, the committee was forced to act. All public showing of the film immediately was halted and further showings prevented. The committee asked PH to edit a new version of the film. Some scenes and songs were removed, others were enlarged or inserted. Some songs, commonly regarded as national, were added. At the repeat premiere in November 1935 in Odense (the major provincial town on the Danish island of Funen), the reactions of the press were considerably more muted. PH himself expressed that the film had become better.

Some Danish cultural radicals soon acknowledged the film as groundbreaking. In a review after the premiere, it was stated that: "Time will show that 1935 was an important year for film history." This understanding was echoed in 1948, when *Denmark* in an article was proclaimed the "total national film," unlike those which dealt with limited issues such as social security, the laws on paid holidays, or the care of the elderly.[46] Abroad the film in many places was received with enthusiasm. In 1938, PH himself showed the film in Stockholm, and it was "met with a long applauds" (from Hammerich 1986:277). Within the Danish cultural community, the film did not gain popularity. The company responsible for renting it to cinemas in Denmark imposed the condition that the customer should take *Denmark* when renting other major films. When the film was shown to Danish-Americans in San Francisco in 1937, some of the audiences were outraged. A local Danish-American newspaper editor sent an open letter to the Ministry of Foreign Affairs, which among other things stated: "There was no excuse for the fiasco. We were as ashamed as wet dogs and felt that our pride in our native country had received a blow" (*Politiken*, November 2, 1937).

What kind of gross tactlessness was it PH had committed, since he deserved such massive domestic criticism? PH himself found that he had been loyal to the task, he was given:

> – What tendency did you want to give the film?
> – Only to serve as a record! To paint an accurate picture of the geography and everyday life of present-day Denmark. However, I did think it was my duty to paint the most favorable picture of Denmark that I could, because it is and is supposed to be propaganda. One could easily have made a caricature of Denmark on film, but that would have been going beyond my contract. Therefore, I kept strictly to the idea that it should be a favorable, democratic, etcetera film. I believe that was my duty. (*Arbejderbladet*, May 2, 1935)

---

46   Review by Karl Roos in the magazine *Vi Gymnasiaster*, 1935 (a radical socialist periodical, quoted in Hammerich 1986:277) and Theodor Christensen: Problemet Danmarksfilm, 1948 (reprinted in J. Sørensen 1980:11-16). On the cultural radicals, Karl Roos and Theodor Christensen, see Thing (1993a).

Figure 3.7   *It will likely not be a huge success, as fiasco is too easily spelled*
with Ph, **1935**

Notes: The intense public interest in PH's film was expressed in many media. *It will likely*
not be a huge success, as fiasco is too easily spelled with Ph (which rhymes very well in
Danish!). From Danish daily *Dagens Nyheder*, May 5, 1935.

Source: It has been impossible to learn who was the artist, GF. Photo courtesy of Björn
Andersson, Historiska Media.

To be able to understand the reactions it is essential to regard the half-official film's relation to the narrative about the nation. The film as a whole is a kaleidoscopic picture of everyday life in Denmark. There are more than a hundred separate scenes in the 55 minutes of running time; however, no scene in itself is directly provocative. The interior and the private are not presented, and the social problems of the times, for instance rows of unemployed, were not included (Ernst 1973:8ff). What provoked, then, can be summarized as the film's depiction of modern society and the Danish woman.

PH's film, the only film he ever directed, was filled with movement. Many brief fragments and the absence of geographical coherence, the film jumps from north to south, from east to west, underline the mobility of the medium, and this affects the overall impression of the pictures. Regardless of where we are, everything is part of the same societal situation; place does not matter. It is the modern rhythms of the age, and of the city, which set the pace for the activity of the entire nation, whether at work or at play. There is little rest to be found in the scenes in the film. Rather, the action is moving, the camera is moving, or moving objects are filmed with a moving camera. This mobility assures that the lasting impression of the film is of panoramas rather than details. There are few portraits, and those are brief. There is nothing in the film that relieves the anonymity of humans in modern times. The intense rhythm, reinforced by the constant focus on means of transport, stresses change and shapes the image of the film. This rhythm is underpinned by the accompanying music, which is played by jazz musicians.

Some students of Danish films have considered the style of PH's film to be something significantly Danish. However, even if this enthusiasm for PH's contribution may be adequate in an isolated Danish context, the film is seemingly closely related with the theories and films of Russian and German directors from the 1920s. Dziga Vertov's *Man with a Movie Camera* (1929) and Walther Ruttmann's *Berlin: die Sinfonie der Großstadt* (1927) have been mentioned as films, which may have provided PH inspiration or models (Strunk 1978). According to Børge Høst, PH has explained that he did not know of these films while working with *Denmark*. Nevertheless, both PH's selection of motives and his composition have obviously an affinity with films by Ruttmann and other contemporaries and pupils. Thus, PH's film ought most reasonably to be regarded as a Danish rendition of an international film genre from the 1920s and 1930s, which subsequently exerted great influence on the documentary film as a means of expression (Barnouw 1983:51-81).

PH presented the potential of modernity *also* as liberation. Others in the Danish cultural community saw it *only* as the threat of lost context, individuality, biography, and history. For PH and the cultural radicals, modernity was a vision: a utopic promise of a new everyday life for the masses. Modernity was not a development of artistic expression, but had political democratic dimensions found in the social development of everyday life, in leisure, sexuality, dress, language, and forms of behavior; exactly as modernity was invoked by the organizers of Funkis. In this spirit, PH clearly regarded film as *the* modern medium. He found himself capable of forming film into a modern expression and through this exciting new medium, he could reach the anonymous masses with a political message. Consequently, the aesthetic and the political were integrated aspects of PH's and the cultural radicals' vision for modern society (Henningsen 1994). In relation to the hegemonic cultural narrative on Denmark as a liberal, democratic, historically founded country of peasants, PH's moving pictures coincided with the opposite of a modern utopia. To most, it depicted a dystopia that was only celebrated by a few intellectuals and communists. Democracy in 1935 was not only a concept signifying an appreciation of Danish peacefulness and societal evolution. To others, PH's understanding was an exaggeration directed against particular contemporary political tendencies. Accordingly, one of the financiers thought that the film was a "little too democratic" for the rest of the world. The film, then, must be viewed as a manifest continuation of PH's commitment to democracy and his stance against Nazism and fascism; and against expressions of these movements in Denmark and in everyday life.[47] What the reviewers and the public desired was a retrospect on the process and a tangible historical foundation for their present. From this point of view, it did not help when the Director of Tourism ensured that foreigners would not "conceive of Denmark as a modern society." PH's way of depicting the people as a modern and anonymous mass with no history was a direct attack on some generally shared understandings of how the Danish cultural community ought to be represented. Danes in general did not want to be portrayed as nameless parts of the modern crowd.

When nations become conceptualized as imagined communities, the respectability of individuals is tied to that of the whole. Consequently, ideas about the different proprieties of gender mean that their roles in the creation of the nation's respectability are different, and of course far from unambiguous. Gender is not a trans-historical concept, so male and female take on different shapes in different times and different nations (for instance Berggreen 1991, Colley 1992, Mosse 1985, Stokes 1998). Although there were many women performing in harmony with the national narrative in PH's film, his pictures were viewed as an attack on the morality of the national family. At the premiere of the film, the senior civil servant, who had taken the initiative to have the film made, explained that "the film was supposed to convey the smile of a mother to her child" – probably trying to invoke an image of Denmark sending a message to Danish emigrants abroad. However, one reviewer found the film rather to be "the smile of the modern mother, coming home from the jazz ball, meeting her children on the steps on their way to school" (*B.T.*, popular conservative, April 30, 1935). In the hegemonic narrative in Denmark in the 1930s, as in other contemporary northwestern European nations, women's contribution to the respectability of the nation concerned the group oriented, private, emotional, family side of the community and its internal morals. Men had to perform individually, outwards, functionally, and manifest – including acting as breadwinners.

---

47 See for instance the publication: *Hvad med Kulturen?* (What about Culture?, Henningsen 1933), or the article: "Du er selv Nazist" (You yourself is a Nazi, 1933, reprinted in J. Sørensen 1980:19-23).

Judged from this association between two ideas of spatiality in the modern maze, nationhood and gender, far from all women in PH's pictures performed respectably. The presentation of women cycling through the city and exposing themselves lightly clad on the beach had parallels in other films and that was at least welcomed as worth watching. These women acted a passive role, directed by a man, and as indicated above many reviewers wanted to see more of that; females were expected to smile nicely and stay in the background. In *Denmark*, however, one could see countless anonymous women working actively in public space, outside the domestic landscape where they should be found. They worked at cleaning herring, milking, or shearing sheep, and these tasks, which made them sweaty, smelly, and visible, made them unacceptable as representations for a female side to the nation. These women let down their nation by doing something wrong, at the wrong time, in the wrong place. These anonymous working women in *Denmark* were naturally not PH's free inventions. They illustrated what was an everyday reality for most women outside parts of the middle and working classes. It was just that these women were in places appropriated for masculinity in the national narrative. These suntanned, but definitely not sunbathing (Kayser Nielsen 1995), women in the pictures were an illustration of a residue of an old society which, according to the national narrative, had been done away with by the democratic evolution of society. They were peasant women who had not yet become housewives. Thus, they presented a threat to the perception of femininity and were a reminder of a part of Denmark's modern geography that still existed, even if it was important to remember to forget as a part of the present. PH's images of the Danish women, thus, were highly provocative. The respectability of the national family, its honor and dignity, amounted to nothing in the public's comprehension of the film. Furthermore, in the cinematic gaze, the misplacement of these women in the public landscape was underlined by pictorial components in PH's film. There are many phallic symbols, such as chimneys and telegraph poles, and many powerful objects swelling with masculinity, such as trains, tractors, and cranes. The choice of these pictures further reveals PH's perception of the modern world as a promise of power and speed.

Like a national monument, a national film must be able to relate to the hegemonic narrative, be able to symbolize the people, the nation, and the state alike, and be able to connect the individual with the collective. PH did not achieve that with his pictures. The comment by Secretary Steincke hit the nail on the head. It was not only in his brain, PH's exposition was "a little too modern." The national narrative was incompatible with a presentation of Denmark as a modern society, with no traditions and no foundation in the history of its peasant community.

In 1962, the film director Børge Høst discovered that the original version of PH's film had disappeared. He found that it was destroyed as a result of being re-cut by PH for the second version in 1935 and again by being used for the production of a shortened version of the film in 1949. Børge Høst, thus, undertook to reconstruct the film for the National Museum. After the premiere of the reconstruction in 1964, most people expressed their surprise that the film had been so controversial 30 years earlier. "Poul Henningsen's picture of Denmark is the most beautiful and most poetic I can remember ever seeing on film, a warm, atmospheric, unceremonious poem to the fatherland, put into pictures by a born film poet" (*Berlingske Tidende*, conservative, May 9,1964). The reconstructed version of *Denmark* soon became very popular.

This revolution of the evaluation of the film fits well into the generally accepted narrative of Danishness sketched here. Thirty years, one generation, had passed from 1935 to 1964, and those who saw PH's film in the mid-1960s or later had to see it in the light of their own biography. What people in the 1930s had seen as a modern dystopia for an uncertain future by the 1960s had become a utopia from a popular, democratic, familiar past. In just 30 years, a dystopic urbanized landscape pulsating to modern rhythms had not only been realized in many respects. It had been overtaken by development and had been transformed into a bygone rural utopia that explained the present; and that was precisely what Danes wanted to see. This modification granted the film value. Students of Danish film now found that the film mostly had "nostalgic and historical interest" (Strunk 1978:18). "With consistency, charm, and poetry and without any attempt to stodgy symbolism, the style of *Denmark* has the character of a suggestive choreographic pattern of a small world in constant and tranquil movement, in harmony with itself" (Ernst 1973:10). As a developing process, the modern had taken on new forms. Consequently, PH's earlier avant-gardist pictures had been loaded with nostalgia (Berman 1982:74). PH's film was transformed to what it had never been before: a national monument, which both materialized the people, the state, and the nation, and illustrated the individuals' biographies as parts of the history of the Danish cultural community.

The presentation of females in PH's film, similarly, was no longer a problem after one generation, because femininity and its association with Danishness had been reinterpreted. Women and the family were from the 1920s at the center of the whole debate about and later implementation of the (modern) welfare state. The portrayal of femininity in PH's film, which in the 1930s was seen as a threat to the morality of the nation, was in the 1960s fully compatible; not just with the real functions of society but with the whole idea of the liberation and equality of women, which was gaining ground. This new interpretation of femininity, which was accentuated in Scandinavian national narratives, made it possible, or even desirable, in the 1960s for women to establish themselves as visible in public space and to take part in work. Their visibility on the public stage was no longer threatening and immoral. At least it was not a problem as long as they kept out of what was now regarded to be male territories and chose to perform duties, which the welfare state had taken over from the family.

Denmark and its history illuminate how national narratives are produced, contradicted, and accepted through processes over time. When PH created his filmic geography, he composed a picture of a modern nation living to the beat of modern rhythms. The film's mobile style, presentation of women and men, and lack of ambiguity made it a controversial image of contemporary Denmark. The mobility made the past inaccessible to the viewer and made the future a threat. One generation later, PH's picture of the present had become a moving picture of the past. Within the reorganized historical context, the movie did not conflict with the generally accepted narrative about how the cultural community had become Danish. Had PH in his film emphasized history and established an alternative version of it, this reinterpretation of his film would have been impossible. The lack of history, which in the 1930s so obviously damaged the film's appeal to a broader audience, later insured that it could be incorporated into the narrative of the very path leading to the typically Danish. PH's "modern" geographical pictures were overtaken by the transience of the present and became part of the Danish nation's historical traditions, loaded with emotions for a past that had once been understood as a horrifying future. It became a newborn film star; a *nova stella* in the past.

The Swedish Developer and the Danish Lover

PH's moving pictures eventually became identified with the Danish past. Parallel to this process, Sweden as described above developed as a dystopic picture to Danes and as such in its own way substituted for the film. The Swedish Model often was characterized as a third way, in between capitalism and socialism. In the late 1960s and early 1970s, when the word socialism was used more often and with very different connotations than it was later the case (not least after 1989), people were inclined to identify Sweden mostly with the socialist way. This, of course, made the Swedish Model a both appreciated and despised object to some observers: "To an observer, who dislikes socialism in all its forms, the politics of equality appear as something very suspect and the description will show that. The same is true for what is written about politics on education, the labor market, the right of co-determination, and so forth" (*Sverige i utländsk press* 1972:4). However, it was a time, when many foreigners indeed took Sweden to be an ideal, even if a critical evaluation of the possibility of transplanting the model to other states gradually became increasingly common. This development is evident from the yearly reports from the Press Bureau of the Swedish Ministry of Foreign Affairs (*Sverige-bilden i utlandet* 1958, *Sverige i utländsk press* 1968-). Sweden felt the gaze of the outside world:

> The French journal *Nouvel Observateur* publicized in September an interesting investigation among French youth between 15 and 19 years old. To the question: "Do you want to transform the French social order?" 79 percent answered yes and 15 percent no. On the follow-up question, which social order they would prefer, around one out of three declared preference for "une socialisme à la suèdoise." No other system got more than seven percent. (*Sverige i utländsk press* 1972:8)

Most often, however, it was as a vehicle for discussions of the modern and not of socialism, that Sweden was introduced into debates abroad. "In many reports it is outright said that a study of today's Sweden is like gazing into the future" (*Sverige i utländsk press* 1970:1). The country represented something that would become reality also elsewhere (Ruth 1984).

The hegemonic narrative on the Danish cultural community and its historic anchorage became from the second half of the 1800s a marker for the Danish road to the modern (Østergård 1984, 1992a:51-83, 1992b). The narrative emphasized a historical dimension which in itself implied a disassociation from elements, which in other nations became stars that guided contemporary development, as it was clear from the history of PH's film.

In the modern project, development is permanently associated with execution of power. The two sides are combined and at the same time they are each other's opposites. Marshall Berman identifies this contrast, the modern as a tragedy of development, with Faust, as the history of this modern, and intellectual, person was depicted by Goethe (Berman 1988:37-86). Goethe worked with the theme from 1770 until 1831 and, thusly, through the time when modern thoughts and practices started to be established as ideas of spatiality, as a modern maze. In Faust, development meant rupture, destruction, and it had no need for history:

> [Faust] won't be able to create anything unless he's prepared to let everything go, to accept the fact that all that has been created up to now – and, indeed, all that he may create in the future – must be destroyed to pave the way for more creation. (1988:48)

In Goethe's text, masculinity and male sexuality is connected with development. In contrast stands femininity, but also the small life-world, the family, and tradition: *Gemeinschaft*.[48] It is, thus, interesting not only that PH's pictures of manly women had consequences for the reception of his film in the 1930s, but also that Grundtvig, who understood the Danes as a nation marked by love, peacefulness, modesty, and lack of desire to conquer, much earlier precisely identified these deeds with the feminine.[49] While Grundtvig found the Danish to be associated with femininity, Sundbärg spirited claimed: "The main qualities of the Swedes are in an eminent

---

48   Berman refers directly to Ferdinand Tönnies' concepts from 1887 on the development from *Gemeinschaft* to *Gesellschaft* (1988:60). Tönnies' concepts are here employed to refer to imagined developments from a traditional to a modern society. This contrast is what Johan Asplund calls a "genuine frame of ideas" (*äkta tankefigur*) that is represented in many discussions and media (1991). As it was argued from the story about the snapcocks, persons earlier often tried to avoid unambiguous positions. That I employ this frame of ideas, thusly, does not mean that I imagine that individuals in pre-modern communities had firm identities and in modern societies fluent identities.

49   See Lundgreen-Nielsen's presentations and quotes (1992:28, 58, 78-80, 106, 133, 151f).

sense *masculine*" (1911:100). Pär Lagerkvist, later a well-known poet and the recipient of the Nobel Prize in literature, some years later pursued the same theme. He found, in contrast to Sundbärg, that his time's Swedish literature was overly psychological and dealt in excess with the human and sickly:

> There is no reason to object to that, if our time was sickly delicate and softened. However, the time is masculine sound. And so far away from apathy and sentimentality that it could deserve to be called brutal. (1913:20)[50]

As the modern's and masculine's associate, Goethe depicts the execution of power that Faust has to employ to be able to lead development forward:

> He knows he has made people suffer...But he is convinced that it is the common people, the mass of workers and sufferers, who will benefit the most from his great works. (Berman 1988:62)

The intentions with Faust's project, thusly, were identical with the grand prospects that the organizers of the Stockholm Exhibition invoked. The necessary execution of power, that made development possible, was in Sweden as in the rest of the West implemented by a growing body of public organizations. For the better of the whole community, they attended to construction projects, social welfare, and much else that benefited and were regarded as in the objective interest of common people. The stance of persons and groups towards the contrast, the tragedy of development, in the modern project has not necessarily anything to do with the division of the political specter – left and right – as Berman explicitly note. Different evaluations of the modern are present everywhere; also as contradictory parts of the same group's or the single person's worldview (1988:169f). The vocation of the free intellectuals as they have been known in the 1900s in continental Europe is exactly to dissociate themselves from this execution of power and thereby debate the essential qualities of life in their time. Such free intellectuals, as for instance the group of Danish cultural radicals in the 1930s, often were associated with one political direction, but they could be found everywhere in the political specter. Danish and other commentators' conception of the Swedish model foremost implied a stance on the modern and less on Social Democracy, which in this connection was more the servant than the master of the modern.

Berman's depiction of the two intertwined sides to Faust – development and execution of power – can be employed as an illuminating framework for the transformations of Danish and Swedish since the mid-1800s. The two nations together entered the first Faustian metamorphosis: *the dreamer*. In this phase, the isolated individual seeks inclusion in the community; in connection with the nation that means inclusion in the political community. The historical novel was an expression of this search.

---

50   Lagerkvist identifies the salvation for art in the visual arts, mainly among the cubists (1913).

In the next metamorphosis, Faust becomes *the lover*. Through his desire for the feminine, the familiar, and community, Faust at the same time crushes the pre-modern. Through his dynamic appearance, the lover himself changes and destructs the unchanging quality that he desires. In this phase the political community gradually becomes dominated by the cultural, which explicitly insists on the authenticity that the lover demands. In the national context, this was expressed through the invention and institutionalization of traditions (practices), monuments (symbols), museums (didactic collections) and university subjects (historical oriented investigations), all of which foremost were to substantiate and to celebrate the loved. In Denmark, the national project became fixated in this phase. Regardless of the tangible processes of modernization that indeed evolved, the cultural narrative maintained an understanding of the country as a pre-modern *Gemeinschaft*. However, as it was obvious from PH's film, it was a continuously more modern past that was desired.

Through this development, the Danish and the Swedish came remarkably out of step with each other as it became manifested in their bordering narratives. Until around 1930, the narrative about the peculiarity of the Swedish was commanded by a longing for development and modernity that should help it to get on in the world. To the Swedish dreamer and later the Swedish lover the – feminine – Denmark was a counterpart and to some extent an example due to its passionate character and its integrated social consideration. From 1930, the Swedish project entered the third Faustian metamorphosis. As a *nova stella* on the sky, the Stockholm Exhibition demonstrates that the Swedish project now had the character of *the developer*. The past was excluded from the narrative, which instead disseminated a utopian objective. To educate and execute the development, new nation-state subjects were needed and established. These subjects' gaze was not on the past but on the future and on planning for this future; in this period engineers, doctors, and sociologists became the leaders and heroes of the modern project. Even if the developers did not have any articulated ambitions on behalf of the imagined community, the political community, the state, became so much more tangible in their endeavors as the very object that had to be developed.

The Finnish historian Henrik Stenius has argued that differences between Danish and Swedish were founded in the 1800s. In Sweden, an evolutionary integration happened between the officials of the state and the popular movements. In Denmark, on the contrary, the popular movements with the followers of Grundtvig in their middle established themselves as an alternative public sphere, a civic society, independent of the state. The Danish liberal broad-mindedness, then, was not founded on a generous Grundtvigian philosophy. Rather it was a result of the failure of the nation to develop and establish itself with general norms for all citizens; regardless of this failure was due to a lack of ambition, will, or ability (Stenius 1993). In the 1900s in Sweden, the Social Democrats could establish themselves through the state and the popular movements as carriers of society, prescribes of the utopian goal, and builders of the nation. In Denmark, the equilibrium between popular movements, the state, and different political forces remained crucial. An integrative evolution between state and civil society therefore never occurred. While the urbanized civil servants of the Danish state formed one body in society, another was established around the popular movements who held Grundtvig as their chief inspiration. As a result, a group of leaders in society emerged from the Danish folk high schools and the farmers' cooperatives, which had its power base separate from the state. For instance they had at their disposal an alternative university (Askov Folk High School), an economy in the cooperatives, which was not under state supervision, and other institutions. The perpetual and necessary consideration of other parts within the whole halted the progress of the Danish lover into a Faustian developer. The fascination with the past and with *Gemeinschaft* was preserved and it was a built-in condition for the maintenance of the equilibrium between different parts of the whole. That Danish Social Democrats supplemented and later substituted for the Right as the carriers of the state is in this connection less significant (Christiansen 1984, Fonsmark 1990).[51] The understanding of the relation between Denmark and Sweden coincides with the contrast between development and execution of power. Swedes' desire for development made them in the narrative accept the execution of power. Danes remained so repulsed to the execution of power by their state onto their everyday life and so committed to the peaceful evolution of their community, that they in their narrative renounced to commit development onto themselves.

Danish geography books from the 1950s and on maintained the picture of Sweden as a country in tremendous development. Economy and industry bloomed, the people were now prosperous, and the earlier exodus had been stopped and turned around as new immigrants entered the country:

> At the same time as the work in the forest, the fields, and in the factories has become lighter it generates greater gain. Prosperity has come to everyone. Most people lived poorly and often had trouble acquiring the most necessary items. Today, everyone can afford to obtain a good apartment and to equip it with all kinds of technical aids. They can afford to dress well. Many acquire a car and go on vacation to the countries around the Mediterranean, the Canary Islands, or Africa. They only work five days a week, have a long weekend, and many buy a second home in the archipelago or in the forest. [...]
>
> The emigration stopped long ago; only few Swedes travel abroad to find work. On the contrary, they need many more people to work in the Swedish factories, and many now immigrate each year to Sweden. They come from Finland, Germany, Yugoslavia, Italy, Spain, Portugal, Greece, yes, even from Turkey. Also many Norwegians and Danes today work in Sweden. (Holm Joensen 1967:19, 17)

---

51 The close cooperation between the Danish Ministry of Agriculture and the organizations of agriculture through the 1900s must been seen as a promoting link for this perpetual balancing act. The film *The King Commanded (Kongen bød)* from 1938 is an expression of this fixation on the past, which I have analyzed in two articles (Linde-Laursen 1995b, 1999a).

The price for development was not specified in the geography books. However, as already pointed out, the execution of power was seen as a necessary expense and that influenced the debates, in which Danish distaste for the modern project clearly was expressed. This tendency left its stamp on the cultural bordering processes since the 1930s and reached its culmination 50 years later. In 1982, the Danish journalist and media person Mogens Berendt wrote a feature article in one of the big national dailies with the admonitory title *Close Sweden* (Luk Sverige, Berendt 1982). He obtained an overwhelming amount of reactions. Many wrote letters that supported his ideas and some Swedes in addition sent material that they imagined could promote additional critic of the Swedish state and its execution of power onto the people.[52] Berendt subsequently developed his points of view in a book *The Case of Sweden* (Tilfældet Sverige, Berendt 1983). In the book, he painted an atrocious image of a country in which "total-democrats" (obviously a word game with "social-democrats") have established such a thorough grasp of both the individual and of the state that basic human rights were disregarded. Mogens Berendt did not step onto virgin soil but continued a Danish tradition of depicting and despising Sweden. Moreover, the same image of Sweden was common also among other observers during the whole period in which Swedes contemplated their nation to be (the world's most?) modern. Berendt borrowed ideas and examples both from Roland Huntford's internationally much debated book on the totalitarian features in Swedish society (1972) and from Hans Magnus Enzensberger's elegant essay *Swedish Fall* from 1982 (1989). Thusly, Berendt did not renew the description of the bordering relation between Swedish and Danish. However, hardly any other Dane had assembled such a comprehensive account of the execution of power in the Swedish project; and with such an explicit purpose of advising Danes against the modern: "This book settle with *The Swedish Model* and is in itself a warning to other countries, for instance Denmark, against imitating it" (Berendt 1983: back cover text). Skeptical foreign debaters after 1930 found that Swedes' lack of a historic retrospect on the development of their society was a central problem: "They have forgotten something: the past. It makes the breathing of their explanations short and flat" (Enzensberger 1989:27). Berendt built his text around the metaphor *Prohibition Sweden* and its history, which he constructed as a unification of rationality and Swedish bureaucratic traditions since the days of king Gustav Vasa (1496-1523-1560) and chancellor Axel Oxenstierna (1583-1654). Berendt's Swedish history was propelled by the system's command of individuals' everyday life and the Swedes' acceptance of this. Any suggestion of Swedish society as a community of settlers as it was described around 1900 was completely absent. The Swedish Social Democrats and labor movement had conquered the political power and, according to Berendt, had employed it to repress deviations from their project without mercy. This development was legitimized with references to the formally existing democracy and the consideration for the common, and the System's, good:

---

52   Mogens Berendt has been so kind as to allow me to review his collection of clippings and letters with reactions to his article (1982) and his book (1983).

If one wants to understand Sweden of today, one has to know the system in society, which Gustav Vasa created during the 1500s. Its substance has not changed, only the nobility [the Social Democrats and the labor movement] is new. (1983:43). [...Gustav Vasa] founded the materialistic thinking that today sets its mark on Swedish society. It is at any time ready to breach principles that can be obstacles to the overall development. (1983:49). [...] From the outside it occur as if the Swedish nation has reached a kind of social Nirvana, in which the political changes and economic problems for a long time have appeared as fleecy clouds on a blue sky. A homogeneous, hard working population, which makes common cause with their shared society and have achieved an economic *Wirtschaftswunder*...The Swedish societal model is far more complicated than it appears at first: it takes a strong state, which is ready to use nearly all means, and an indulgent population, which has a certain, limited right to object. The wrapping is parliamentary democracy, the brand is totalitarian society, Made in Sweden. (1983:18) [...] They want to create The Good State inhabited by good people, and they do not recognize that their trust in common sense and good intentions have opened for an insidious development. Persistently more organization, control, and oversight are necessary to go forward. The result – at last – is despotism. (1983:172f)

Berendt's depiction of the Swedish tragedy of development was debated much in Denmark, where it during several weeks was number one of the national list of most bought books. It was also discussed in Sweden. The text soon was translated to Swedish, and to Finnish. From people on both sides of Øresund, Mogens Berendt again received many reactions. A few raged against him, but most supported his interpretations; some even drew parallels between Sweden and Nazi-Germany. In more sober Swedish reactions it was noted that something important had been pointed to, which needed to be discussed. Some urged Berendt to continue writing about Sweden: "Write more 'filth' about Sweden, please – it is only wholesome and entertaining" (letter from a Swede to Berendt from March 15, 1983).

The hegemonic narrative about the Danish cultural community from 1930 found strength from its relation to the Swedish. To Danes it was obvious that the Swedish nation had not realized the threat in the modern: the execution of power. This Danish realization was emphasized, when the Swedish model was depicted as corporativistic (Berendt 1983, Behrendt 1987, T. Knudsen 1994). Because the Danish and Swedish imagined communities after 1930 were out of step it appeared as if there were no similarities at all between the two countries, or as Mogens Berendt expressed it: "In reality Denmark has never had much more in common with Sweden than a jointed, un-peaceful history and two mutually intelligible languages" (1983:160).

The historic development, however, also illuminated other qualities, which contributed completely different bordering facets to the relationship across Øresund. Around 1900, Swedish nature was perceived as hostile and threatening. Writings about wild animals and the long winter were consistent elements in the geography books. If the people were less cultivated, the nature was directly ill

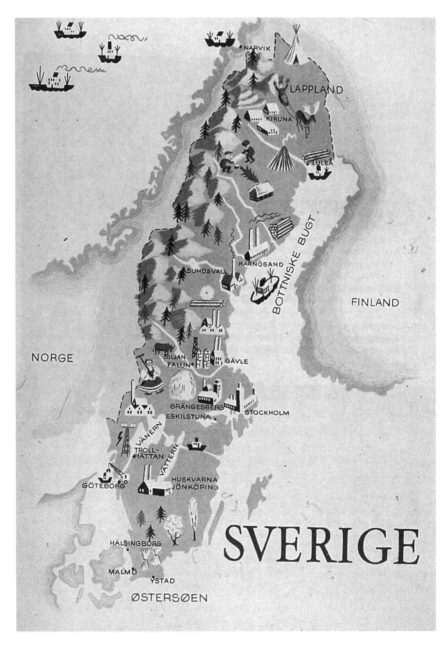

Figure 3.8   Sweden depicted in a Danish geography book, 1962

Notes to Figure 3.8: The natural state of the immense land, red cottages, forestry and timber floating on the rivers are customary elements in Danish depictions of Sweden from the 1880s until the 1960s; the locks on the Göta Canal, the hydroelectric plants at Trollhättan, the mines at Kiruna, and the iron ore railroad to Narvik are also commonly used elements.

To emphasize the distinctiveness of the Other, the exotic is often accentuated in geography books and travellogs. The exotic aspect of the Swedish population was illustrated by "Laps," who were nearly always part of the imagination, "These mongolish nomads are cared for by the Swedish state and benefit from many of the humanistic transformations, civilization has brought... they are bright and learn easily in their younger years when they without trouble acquire the ordinary school curriculum; however, their development stops early" (Spager 1935:17).

Source: Image from Holboe 1962:9; it has been impossible to determine who crafted the map for the publication. Photo courtesy of Björn Andersson, Historiska Media.

mannered, even if it also was regarded as a possible economic resource for the future. Writings about and pictures from the far north, often Kiruna, came to symbolize this in Danish geography books:

> In the northernmost part of the country, they have discovered such amounts of iron ore in the mountain *Gellivara* that it is enough to supply the world for many years. A considerable town with electric light and modern adaptations is now growing here, where a few years ago one met only Laps and their reindeer. (Andersen and Vahl 1910:134)[53]

This picture of Sweden's nature gradually changed. Other pictures, of small read cottages at the lakes and in the archipelago, were evoked; not any longer as examples of something lacking behind in economy and technique, but as tranquil, rustic, and idyllic places. Sweden became the land of the wide open spaces and "*allemansrätten*" (the Swedish legislated right for the public to cross privately owned land, to camp, to pick berries, etc.), which secured the people's free movements – in nature. This picture has since skillfully been employed in the marketing of Sweden to tourists.[54] The many Danes, who bought or rented deserted farms from the late 1960s worshipped this naturalness through descriptions of hikes in the forests, the picking of raspberry, blueberry, and cloudberry, or encounters with the moose; something that nearly reached mythical dimensions. In some deserted farms, dairies depicting such events were kept scrupulously. That the deserted farms mostly were found for instance in the southern part of the

---

53   In new editions of the same schoolbook, the same wording was maintained nearly unchanged at least until 1959. However, the text reads Kiruna instead of Gellivara, probably because the latter name was less known than the former.

54   A survey; *Image of Sweden in Europe* that the Board of the Image of Sweden (*Styrelsen för Sverigebilden*) made in ten European countries in the fall of 1992 showed that nature is the theme that is most frequently associated with Sweden.

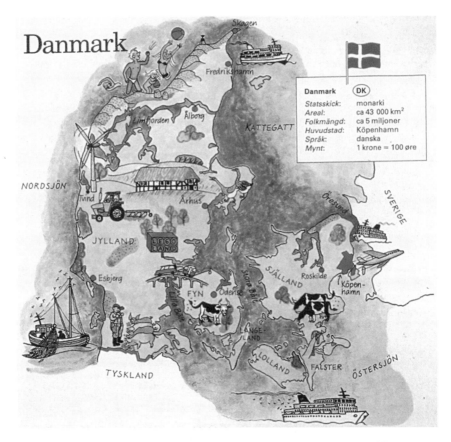

Figure 3.9   Denmark depicted in a Swedish geography book, 1982

Notes: In Swedish depictions of Denmark, a number of elements are customarily included: agriculture that dominates the tiny country's flat landscapes, Copenhagen – with the amusement park Tivoli, beech forests, bridges (Little Belt, Storstrømmen), sand dunes and bars on the west coast of Jutland, and half timbering. Denmark as it was depicted to Swedish pupils 1982; that Bornholm is omitted is not typical.

Source: Drawn map by Swedish illustrator Ylva Källström-Eklund (1933-1988). For additional information, see ylvakallstromeklund.blogspot.com. Reproduced with permission from Erika Eklund Wilson. Image from Nordling et al 1982:34; photo courtesy of Björn Andersson, Historiska Media.

landscape Halland, where the soil is very poor and where forests only recently had spread at the expense of bogs, did not change the perception of the naturalness of the landscape. Despite many visitors in reality moved around in landscapes of production, these were seen as more natural as they were of a different and more unregulated kind than similar areas in Denmark. The forests and lakes made Sweden appeared to Danes as much closer to nature than their own country, where trees in the plantations stood ruler-straight. Thus, within the well- and over-regulated Sweden, Danes found a desirable freedom from their modern everyday lives during the vacations they were ensured by their own modern society.

Narrating and Practicing

That the bordering narrative and the "actual" conditions or developments of the imagined community did not necessarily agree has been mentioned several times above. The hegemonic narrative for instance actively obscured the working women, whom PH filmed. There is consequently reason to emphasize that narratives as such are not different from other forms of reality. It is through imaginaries and practices in tangible situations, what we summarize as everyday life, that identity is bestowed upon individuals or obtained by them through bordering, supported by perceived differences within the modern maze. When experiencing, persons refer imaginaries and practices to one or more specific ideas of spatiality in the maze and turn narratives into tools with which they explain their complex surroundings. Narratives, or maybe rather elements of narratives, thusly, become explanations to life (Lakoff and Johnson 1980). Simple metaphors that references whole systems of narratives, overexpose some matters, while they obscure others and make them inaccessible to experience:

> [Metaphors] can have the power to define reality. They do this through a coherent network of entailments that highlight some features of reality and hide others. (Lakoff and Johnson 1980:157)

This cultural process is significant for the recognition and acknowledgement of difference. In such processes, a kind of deafness can prevail that makes a narrative self-affirming. For instance, Danes in general grumbled about Swedes' relation with alcohol in Sweden (referring to the *Systembolaget*) and observed them with an amount of indulgence when Swedes, as was expected, got drunk while visiting Denmark. Drunken Swedes in the streets of Copenhagen, thusly, were perceived nearly as a kind of alcohol-political refugees, who were prospecting on the other side of the ditch, where they were beyond the control of their modern System and could perform nearly as real humans.[55] It is of course impossible to estimate

---

55 Around 1990, the Danish brewery Carlsberg employed this understanding in a beer-commercial for movie theaters. A male Swede came into a bar in Copenhagen, but before being served "our" beer, he had to promise not to bother the Danish girls, not to throw up in Tivoli (a famous amusement park), not to urinate in the parks, not to sleep in Strøget (the central pedestrian merchant street in Copenhagen), and that he would get on the last ferry home to Sweden. In Danish theaters this ad was met with something as unusual as applauds from audiences (Erfelt 1991:284).

how many Swedish visitors returned drunk home from Copenhagen. Obviously, an estimation of ten percent (which probably in itself is an exaggeration) still indicates that nine out of ten Swedes did not act in accordance with the hegemonic bordering narrative. This, however, did not change the way Danes and Swedes alike told about Swedes' behavior when visiting Copenhagen. The vast majority, which did not get drunk, was either not recognized as Swedes, or the individuals were regarded as the exceptions that proved the rule. A selection of what constituted relevant impressions occurred. Contradictions and alternative experiences always exist; however, they do not necessarily affect the reigning comprehension of similarity and difference, culture and identity. If a new hegemonic narrative is established, it is often associated with an event that is bestowed symbolic significance, as it happened with Funkis. Metaphors refer to such central symbols that draw a definite difference between "a before and an after," or between a "on the one side and the other side" of a ditch. Cultural bordering narratives, which are out of step with actual performances and practices, therefore can obstruct reasonable communication. As expressions of recognition and acknowledgement, they establish expectations that it can be difficult to contradict. While it was probably quite easy in practice for a Swede not to get drunk in Copenhagen, it might be much more difficult to explain such a deviation from the narrative to other Swedes and Danes. Similarly, a Danish top civil servant, Erik Ib Smith, has revealed how difficult it in 1948 at international economic negotiations was to convince the other participating countries that Denmark was not to be seen as a farming society. Agriculture employed only one fourth of the workforce and the country concentrated on industrial development (1993:267f). His presentation was not in line with what the others told about the Danish; nor was it in line with what Danes knew about themselves.

Knowledge from bordering processes, thusly, was confirmed through narratives that through cultural deafness were able to resist many alternative realities and experiences. Narratives about the relation between the Danish and the Swedish were widely disseminated and were of course not only shared by authors of geography and debate books. In 1992, I arranged for essays to be written by students in high schools in Odder (south of Århus in Jutland, 23 students), Virum (a suburb north of Copenhagen, 19 students), in Malmö (47 students), and in Stockholm (78 students). The essays, thus, represented both capital and provincial (center and periphery) settings, as well as both closeness and distance to the ditch in Øresund. The students were asked to describe their knowledge about and impression of the people on the other side of the border. As an inspiration, the students were given an article on Danish-Swedish relations (Löfgren 1986). It was obvious from the essays that they were products from very different school systems. The Swedes for instance wrote their essays during a class, while the Danes had the assignment with them home and worked with it. While the Danish essays, thus, were longer and often more thorough, this difference did not make the material less relevant as comments on Danish-Swedish cultural bordering.

Figure 3.10 *He who owned the truck sometimes brought presents*, **diorama
by Ola Terje**

Notes: Photographer and diorama builder Ola Terje catches something essentially Swedish
as he depicts Småland (in southeastern Sweden) as remembered from his childhood. He
depicts not only a Sweden that existed in everyday life, but also a Sweden that existed
in narrative: the red cottages of Småland with white corners and windows, situated on a
lake. This imagination is recognized as Swedish and appears in many different contexts:
Danish geography books, advertisements for Sweden to tourists, and the cover of the 1989
annual report from the Swedish National Audit Office. Detail from: *He who owned the truck
sometimes brought presents.*

Source: Photo by Ola Terje; reproduced with permission from Ola Terje.

Without hesitation, the students applied metaphors that refer to the bordering
narrative as it has been drawn above (see also Gertten and König 1985). To Danish
students, Sweden represented magnificent nature, slightly boring people, strange
alcohol habits, industry, and a frightening modernity (as an intolerable execution

of power): "In short, Sweden's richness is the nature and not their culture and people" (student from Virum). Only a few students turned the dystopia upside down and described also a utopian element in the Swedish project.

The Swedish students regarded Danes as more relaxed and humoristic, and they described a somewhat provincial, slightly old-fashioned, cozy, petit bourgeois condition: "The Danish culture is actually enormously warm and natural…The merits of life appear in the light of day in some enchanting way" (Stockholm). Those, who mentioned trades, all referred to agriculture; and in that case often in connection with a description of the smallness of Denmark and its lack of real hills. The Tivoli amusement park in Copenhagen, the hot dogs stands, the Danish pastry, and beer drinking were central elements in the students' image of Denmark as was the country's closeness to Europe.[56] "Danishness is a unique and delightful mixture of Swedish settled trustworthiness and southern European festivity!" (Stockholm).

To most high school students foreign countries are associated with vacation. Some of the Danish students had been on deserted farms in Sweden, while some of the Swedes had visited Copenhagen or been in Denmark on seaside vacations, or had just passed the country on their way south. However, most had never visited the other side of Øresund. Even if many students comment on their lack of knowledge of the relation and personal experiences with bordering, and although some expressed that Danes and Swedes were mostly very similar, they all reproduced the hegemonic narrative about the differences. As one student wrote: "One does not always need knowledge about something to have an opinion" (Stockholm). The personal experiences, which some of the students described, were inserted into this already known framework of cultural bordering.

Regardless of the overall harmony that existed between the descriptions of the Danish-Swedish relation and the bordering narratives, it was obvious that the students from Malmö stood out in one respect. They regarded crossing the border as if not a daily event, at least as something quite un-dramatic and also highly ritualized. They could substantiate their narratives with personal experiences from many trips with their parents, the school, or their friends. While they were kids, the trip to Copenhagen meant Tivoli, candy, and buying Christmas presents for the family. When they grew older, Copenhagen to them had new and fascinating qualities: to travel on the hydrofoil from Malmö, to drink beer, to go in Tivoli or on a disco, and to consume – to walk up and down Strøget (Copenhagen's pedestrian, main shopping street) and use a little more money than intended. To them, then, a trip to Copenhagen constituted a pleasurable pastime, a hobby, an avocation,

---

56   The Tivoli Park is internationally well known, the hot dog stands are typical elements in the streetscapes of Copenhagen, the Danish pastry (which Danes by the way call "*wienerbrød*," literally "Viennese bread") explains itself, as does the beer. Swedes, at least in the south, often refer to Copenhagen as "the continent"; disregarding the fact that Copenhagen is on an island and that Sweden is part of a peninsula connected to the European mainland.

and/or a devotional rite of consumption (Miller 1998:9): "To me, Denmark is freedom, a moment's possibility of escaping from reality, and relaxation" (Malmö). Their experiences, however, still were described within the framework of the hegemonic bordering narrative. Contrary to the situation in Malmö, the closeness to the border meant nothing to the students from Virum, whom not more than the more distant groups in Odder and Stockholm had experiences with transgressing the ditch. In the 1990s, then, local traffic across Øresund consisted largely of Swedes going west.

Some of the students explained that they were aware, that they reproduced a stereotypical narrative, and a few of them directly dissociated themselves from such a practice. "Danes are as any other humans" as a student in Stockholm very shortly answered the assignment. However, their explanation of the relation between Danish and Swedish existed on the whole only in one form, which corresponded with the hegemonic bordering narrative. Only a few tried to establish a counter-narrative and turn the relation upside-down. When that happened, it had the form of a reflection from the hegemonic narrative. Again cultural deafness played a role; to be able to disassociate oneself from the narrative, one has to take as a point of departure its very existence. Consequently, even if the narrative is moderated, its hegemony is reaffirmed. While the narrative was stable, the interpretation of its elements could be negotiated. The students, thus, had an understanding that corresponded meticulously with the narrative about Danish-Swedish bordering. Compared to the debate literature and especially compared to the newer geography schoolbooks the center of gravity in their picture, however, was shifted. Often inserting themselves as the center of their narrative, they put less weight on society, economy, and politics and more emphasis on people and culture.

Swede among Swedes

By analyzing imagined communities through the relations between them, differences and transformations in the narratives about the societies become clear. It is exposed, how cultural bordering processes provide nations with identities and how this provision has changed through the times. However, as it has been mentioned, especially in connection with the older geography books, the debate around Sundbärg's aphorisms, and the student essays, the narratives also contain understandings of how Danes and Swedes are different and behave as individuals. The anthropological *culture and personality*-school[57] sought similar understandings of citizens of nations and in the early 1980s inspired a discussion on "national mentality" introduced by Swede Åke Daun in Nordic ethnology (see 1983, 1989a, 1992, 1996). The concept "mentality," which was the central methodological tool in Daun's work, was not always received well among other scholars. They pointed

---

57 See the discussions in Daun (1989a:14, 40ff, 1996:20) and Østergård (1991a:153ff, 1992c:20ff).

to ethnologists' and folklorists' involvement with nationalist movements in Europe between the two world wars who held similar understandings of nationhood as something founded in personalities.[58]

The aim of Daun's research was to analyze the incidence of a Swedish type of personality that was different from the type of personality found in other nations. Consequently, Daun employed a comparative perspective in which the Swedish was found through being contrasted against the non-Swedish (Daun 1989a:12, 1996:4). In the end, Daun summarized his findings:

> If now indeed many Swedes are shy and socially insecure, if they strive for autonomy, feel extremely uneasy confronting personal conflicts, have difficulties expressing and receiving strong emotions, feel all the more secure with matter-of-factness and rationality – it comes from somewhere. If they truly conceive of themselves as individualistic but are not all that happy without the support of a collective, it is not just by chance. If many Swedes are comparatively quite, slow, serious, possibly even gloomy – in any case hardly light-hearted even though theirs is a self-image of being nice, cheerful, and happy – then all these tendencies are consequences of Sweden as a social organization and historical product, of those institutions and natural preconditions that apply to our part of the world. (Daun 1996:195f)

He stated, that "seriousness and rationality" probably was at the core or the point of gravitation of Swedish mentality and related it to melancholy (1996:151ff). This core was not uniquely Swedish, but summed up a for Sweden peculiar, "modal" statistical distribution of personality traits:

> the modal personality in a population refers to the most frequent type, which is not the same as the "average person" and moreover need not be represented by a majority. (Daun 1996:25)

The basis for Daun's statistical investigations was a survey (CMPS) that was developed to study personality traits. In 1970, a thousand persons around Stockholm were asked to answer the 165 questions. Instead of analyzing the answers to the whole questionnaire, Daun focused on the information provided through the answers to some of the 165 individual questions, which were designed to address eleven different "needs variable" (Daun 1996:27f). The results from the tabulations were in the book supported by Daun's own experiences and other authors' descriptions of Swedishness (Daun 1989b). Daun stated that personality traits were both something collective, formed by culture, and something individual, formed by the each person's upbringing and genetic inheritance (Daun 1996:19, 26f).

---

58   For a discussion of the reception of Daun's first paper, see Skarin Frykman (1983:163). The author of the foreword to the English edition of Daun's book from 1989, David Cooperman, mentions similar, possible objections (Daun 1996).

He reported that "more than half of all personality variations are due to biologically inherited factors" (Daun 1996:114).

Methodological reservations against Daun's investigations have been raised. The percentages derived from the statistical material had to be interpreted and it was sometimes difficult to know, what questions they answered. In some cases, Daun's arguments seemed more shaped to prove that this or that number met the overall picture than to take the number at face value.[59] That the survey was developed through fifteen supplementary questions to measure one need variable might make it questionable to pick out the answers to single questions for individual interpretation. It was, for instance, obvious, that there were no statistical significant national differences on many of the questions.[60] Daun avoided discussing linguistic problems connected with his use of the questionnaires (1989a:227, note 18); however, as it was emphasized above in connection with the Swedish and Danish understandings of the term modern, different meanings of similar concepts might very well be a problem to comparative investigations. It can hardly be taken for granted that words like: friend, bad conscience, independence, progress, and disappointment, have identical semantics in Swedish, Finnish, Italian, and Korean. However, it was the possibility of making such comparisons that were at the center of Daun's method.

Nordic ethnologists have not copied Åke Daun's investigations. However, statistical comparisons between nation-states are not foreign to other subjects. The aim might not explicitly be to study national mentalities, but to examine differences and similarities, conditions and developments, and to connect that to the character of the countries (see for instance Riis and Gundelach 1992, Sundback 1993). In Daun's studies, national mentality was described as a product of history, climate, and the institutions of society, moderated by the available gene pool. Mentality constituted a specific distribution of personality traits and, consequently, national

---

59   As an illustration, the arguments around a survey among school children can be mentioned. Daun concluded, that when 69 percent thought of Swedes as stressed, they thought of their parents, while the 61 percent, who thought of Swedes as interested in sports, thought of themselves. It was equally difficult to know, how Daun knew, that the 25 percent that answered "no" to the question: "Do you possess self-esteem?" right away could be combined with the 6 percent who answered "very difficult" or "rather difficult" to the question: "Is it easy or hard for you to find friends?," and, furthermore, how was it self-evident that these percentages had anything to do with "shyness" (Daun 1996:43). More examples could be found (see also Montgomery 1989).

60   See the presentation of the answers to CMPS in Daun, Mattlar, and Alanen (1988:288-294). It was not certain that the questionnaires were answered with the same eagerness in different countries. It was for instance remarkable how many results around 50 percent the statistics from Italy show. No distribution was outside the spectrum of 40-60 percent on no/yes questions (Daun 1992:25-95). However, it cannot immediately be concluded that the Italians did not read the questions. On a question on individual's ability to feel hatred there was a significant difference in the answers between Italian women and men (Daun 1992:38).

mentality was something that was anchored in the individuals that constitute a people. Even if it was probable that the statistical distributions would change with time, Daun's held that national mentalities have always existed and will forever be a crucial aspect of individuals. Thus, according to Daun, national differences existed in humans as something else than narratives. From this conclusion, it seemed peculiar that employing his different method, Daun reproduced the description of Swedish mentality that others and he himself (Daun 1983) had made on the basis of qualitative investigations. A different method, thus, did not change the result. This might be taken as proof that the findings were valid. But it might as well be taken as an indication that in many cases, people answered the questionnaires from within the national narrative; that each person answering was practicing cultural bordering against an imagined or in the survey stated other. Thus, they already knew, what was expected. Alternatively, the researcher was predisposed to pick out such information that fitted nicely into the already hegemonic narrative. If this was the case, it explains the immediate recognition of the Swedish in Daun's studies, and it enables us to regard his results also as notable expressions of the narrative itself. *Swedish Mentality*, thus, can be, and was in the public when it appeared, read as scientific proof that Swedes were exactly as the narrative predicted. Thus, Swedes were through the statistical arguments seduced to be recognized for what they always were: Swedes (see the discussion of Harbsmeier's ideas above). Another method, like Daun's, was, thus, likely not to produce alternative results. His investigations of national personalities followed in the footsteps of the hegemonic cultural bordering narratives.[61]

There is in general no reason to expect an agreement between an approach that regards bordering processes between nations to be constructed as identity providing narratives and another that seeks the national as a trait in each individual. However, this does not mean that the two different understandings of the scientific object must contradict each other. As illustrated above, they might very well support each other's interpretations of the national characteristics. Swedes have been described very similarly from Gustav Sundbärg to ethnological investigations in the 1990s; even if the shy, conflict avoiding, serious, and rational Swedes of course must be understood in relation to their possible individual success in a gradually much transformed modern project (Frykman 1989). Åke Daun wrote about his object, that "it should be recalled that what are measured in the CMPS are personality traits, not how people behave in real situations" (1996:74). Likewise, it has been stressed above that the hegemonic cultural bordering narratives can have very limited significance as a measure of what constitute the actual practices of a nation. However, this does not necessarily imply that it is impossible to regard the

---

61   On many issues, Daun's description was similar to earlier authors'; see for instance Susan Sontag (1969) and Hans Magnus Enzensberger (1989). See also Daun's mentioning of Phillips-Martinsson's description of the Swedish (Daun 1996:22). In a discussion of how the national can be understood as a narrative, Jonas Frykman criticized Åke Daun's investigation (Ehn, Frykman, and Löfgren 1993:127ff).

single person's life as significantly formed from bordering processes. The border between two nations is not only the ditch that represents the identity-providing narratives about self and Other. We do not only live our lives as narrators, but also as individual bodies, which are born and get older, eat and drink, walk through doors and across streets, drive tricycles and cars, go to school and to work, cook and dress the table, and many, many other activities. The political border is also a separation between two bureaucratic systems that on many levels embrace such activities, which taken together constitute our everyday lives. Consequently, our everyday life is managed and formed within frameworks that are created in particular historical conjunctures and are shaped by particular agencies. Without considering it, each of us in specific situations materialize and perform aspects of these national bureaucracies and, thus, we engage in bordering practices.

Chapter 4
Personal Bordering

"Goodbye," said the fox. "And now here is my secret, a very simple secret: It is only with the heart that one can see rightly; what is essential is invisible to the eye."
"What is essential is invisible to the eye," the little prince repeated, so that he would be sure to remember.

(Antoine de Saint-Exupéry 1943:70)

Memory and Recollection

As argued above from Åke Daun's investigations, comprehensions of an individual's identity are connected with cultural bordering narratives. Individuals speak of themselves and are recognized within certain hegemonic understandings of cultural communities. In this sense, people not only talk in terms of nationality, but also with reference to other markers such as gender and class that identify differences. Identity remains and must be recognized as an open-ended process.

The narrative of the cultural community represents one edition of a society's imaginary. As it was illuminated in the previous chapter with a multi-sited ethnography (Marcus 1998), this imaginary – its origination, condition, and/or future emerges in geography books, novels, public debate literature, films, exhibitions, school essays, and other ethnographic arenas. In relation to different other imagined communities or relating the present to different other epochs of the self, the narrative helps a people remember its peculiarity, its past, and what is to come. All individuals in the nation ideally share one hegemonic national narrative as the foundation for this aspect of their identity. From the 1800s, it has been debated within modern sciences whether this or that element from the narrative is correct or erroneous; if a certain event deserved its position within the narrative was seriously discussed and evaluated through critical investigations of historical sources. Still, all participants in a cultural community are qualified to call upon and discuss this collective memory. "In nations the truth has been scattered to the four winds: in principle we all hold the truth both about ourselves and the others" (Harbsmeier 1986:53).

The question remains though, whether national culture exists only as such culturally invested narratives about others that are suitable for demarcation at Febvre's ditches (1973). Maybe an individual can experience and practice bordering in completely different ways and in different arenas than what the emotionally and morally charged narrative suggests? This will be discussed below without turning, as Daun does, the investigation towards psychological perspectives; bordering will be maintained as a *cultural* process.

Philosopher Søren Kierkegaard, in the introduction to *In Vino Veritas* from 1845, suggests that recollection is different from memory. In the introduction to William Afham's recollection he writes:[1]

> To recollect is by no means the same as to remember. For example, one can remember very well every single detail of an event without thereby recollecting it (Kierkegaard 1988:9).

Kierkegaard argues that a *recollection,* contrary to a *memory,* cannot be shared with others: "Strictly speaking, a fellowship of recollection does not exist" (1988:14). The recollection is personal, essential, and related to specific situations: "Actually, only the essential can be recollected [...] To bring about a recollection for oneself takes an acquaintance with *contrasting moods, situations, and surroundings*" (1988:12f, my italics). Consequently, the recollected is beyond the character of a memory as right or wrong, as a reflection that can be different from one moment to the next (1988:13); however, it is always "accurate" regardless of its relation to the truth (1988:9). Importantly, a recollection is a movement back in time to a similar or contrasting situation that allows the person to relive the earlier occasion. The individual, though, is not in control of this movement in time: "What is recollected can be thrown away, but just like Thor's hammer, it returns" (1988:12). Remembering is the opposite – what happened is brought from the past to the present, where it is remembered and contemplated as in the modern historian's thorough critic of her/his sources (Koskinen 1980:93).

If bordering is enacted *outside* the culturally invested national narrative, such a practice might be found in the individual's recollection of situations where contrasts were revealed. Furthermore, such a recollection is beyond the memory that is shared and shareable with everyone. A persons' recollection, then, must refer to specific situations that are filled with feelings and meanings.[2] From this, the methodological problems in studying individuals' recollections are obvious. Not only are recollections connected with specific situations as sets of texts and contexts, but also they are essential and personal. Consequently, it can be expected that persons are both reluctant and unable to share their recollections with others. When asked about cultural bordering and their relation with their nation it is not surprising that individuals answer from within well-established narratives that are marked out in books, mass media, jokes, and many other vehicles for information

---

1   Søren Kierkegaard always wrote under different pseudonyms. William Afham literally means "William by himself." *In Vino Veritas* essentially is about men's fascination with women.

2   Roland Barthes distinguishes similarly between *punctum*, which is the carrier of meanings and feelings, and *studium*, which is a result of the systematic investigation (Barthes 1981). Hans Fink's reflections over *arenas* as wholes of scenes, actors, performances, and spectators are illuminating for how we can grasp what constitutes a situation in this context (1989:7ff).

about self and others. As it was shown in the previous chapter, these narratives form safe paths that additionally can be expected to satisfy the inquiring person – someone who in this case could be understood as being the ethnographer. As the narratives are both safe and fulfilling there is thus, no reason to share with others recollections outside the narratives, which might be minefields of personal experiences and feelings. And in particular, there may be no reason to share such recollections with official researchers. In such a context, the narratives constitute a refuge for the ethnographic object, or person asked. Interviewing about national identity and bordering becomes soon enervating. Together, the interviewer and the informant follow safe paths, noting the Swedish nature, the Swedish Model as a modern System, the old-fashioned, cozy, popular Danishness, and so forth. Of course, every interview can contain more or less developed reflections on the bordering relation. The perception of this or that can be turned upside-down, the significance and development of the relation, and much more can be discussed. Nevertheless, it is very difficult to move beyond the safe paths of hegemonic narratives and out to the grounds of feelings, situations, and contrasts of the recollected.[3] The weaknesses of interview materials might be compensated through long periods of alternating intensive and extensive fieldwork (Gerholm 1993). Personally, my experience of living on both sides of Øresund is in this context, a valuable source of knowledge about Danish and Swedish. However, time alone does not solve the problem. The essential, the recollection, is personal – or "a secret" as Kierkegaard writes (1988:14).

While persons have feet and not roots, many individuals have through the years moved across Øresund for different reasons: love, work, dwelling, tax reductions, and adventure.[4] These individuals can be expected to experience bordering, or as Kirkegaard wrote, contrasting situations in which their cultural community is exposed through confrontations with related, yet different ways of doing things on the other side of Øresund. Furthermore, they can be expected in specific situations to have bestowed or to obtain identity through recollections of difference.

The topic – bordering between Danish and Swedish – has not been seen as problematic in the many interviews others and I have made with people who have

---

3   I have made series of interviews, and in connection with coursework or externally funded projects, my students at Lund University have conducted many more. Most of this material, gathered by Sara Berglund, Lars G. Gustafsson, Malena Gustavsson, Mette Kirk, Fredrik Nilsson, Elisabeth Rysén, Lotte Mortensen Vandel, and Astrid-Louise Walther, is now in the Folklife Archive at Lund University (see for instance LUF M: 21245, 21246, 21247). See also Berglund et al. (1999).

4   On December 31, 1988, 41,533 persons who were born in Denmark lived in Sweden. Persons born in Denmark formed the third biggest category among immigrants (after people from Finland and Norway). In Sweden lived 25,664 Danish citizens. Thus, many Danes moving to Sweden change their citizenship (*Folkmängd 31 Dec 1988* 1989). At the same time 8,178 Swedish citizens lived in Denmark (Ploug 1990). As will be discussed below, a bridge that opened in 2000 across Øresund later affected moving patterns in the area (see Linde-Laursen 2005).

such mixed Danish-Swedish lives. Everyone spoke happily, freely, and often for a long time. The problem of these interviews is that as soon as the questions become more specific, the answers become more elusive. That certain topics are regarded culturally as tangible and less problematic does not indicate that it is easier to gather information about them (Gerholm 1993:17); in some cases, the opposite is true. When questions get particular, the desired information is anchored in experiences for which one can lack a language to describe (Frykman 1990). Or, the questions transgress the border between the remembered and the recollected, while the informant declines to follow often by recoiling to the very well known and seemingly satisfying bordering narrative. The reaction to such, in a double sense, border-crossing inquiries might be dismissal, disturbance, silence, or a short laughter. To be able to conceptualize such interruptions demands more than what Clifford Geertz terms *thick description*, which presuppose that persons are rational writers and readers of cultures as texts (Christensson 1991, Geertz 1975). Thorough contextualizations enable analysis of more layers of memory and additional information can be attracted in this way by moving the past to the present. Nevertheless, contextualizations alone do not afford the recollection of a contrast in a specific situation that passes the present to the past. As the fieldworker is left out of the informant's recollection, the researcher cannot gain an insight into the identity-producing experiences of difference that – irrationally – can be provoked repeatedly in similar or contrasting situations. It ought to be underlined that Geertz's methodological problem seeks, through ethnography, to establish an understanding of a culture from which the researcher has no problem maintaining separation.[5] The meeting of the anthropologist with the informant(s) thus often can provide a recollection of contrast that is later written as experience (Tyler 1986:138). This in itself is not a problem but can obviously easily lead to the exoticization of "others."[6] My methodological problem is the reverse, however. How does one understand the ways in which bordering on different arenas is established across a cultural divide when on both sides I am an insider.[7] For example, I do not think that Geertz, when running from the authorities during a raid on a cockfight,

---

5   See for instance the explanatory note "anthropology *can* be trained on the culture of which it is itself a part, and it increasingly is; a fact of profound importance, but which, as it raises a few tricky and rather special second order problems, I shall put to the side for the moment" (Geertz 1975:14).

6   This, of course, is not a new realization. Paraphrasing Edward Sapir, Richard Handler stresses the point, "Writers tend to concentrate on the surface of an alien way of life because immediately palpable exotic externality commands their attention as the proper object of discourse" (Handler 1983:224).

7   For a parallel discussion of the ethnographer's self in writing and of the remembered and recollected, see Fischer (1986). While Fischer identifies this with postmodernism, I suggest that this understanding is very much a continuation of modern thoughts on how ethnography is constructed. I do not recognize any discontinuity that suggests a "post."

was suddenly recollecting that he was always Balinese.[8] Of course, this does not preclude that Geertz, when writing his ethnography, recollected this as the specific situation in which he first imagined himself to be included as Balinese in the specific situation; and this is exactly what he wants us to understand (Crapanzano 1986:68ff). Anthropologists have many ways of reacting to such moments and situations of experienced inclusion; whether the anthropologist shares this experience of inclusion with her/his objects is, in this context, inconsequential as the words remain ours and not theirs (Hastrup 1988:133). Geertz obviously cherishes his moment of imagined inclusion. Contrary to this, Japanese-American anthropologist Dorinne Kondo describes the continuous erosion of insider/outsider distinctions in her everyday life doing fieldwork in Tokyo as a highly problematic "collapse of identity" (1990:17).

The transferral of experience from recollection to memory is not facilitated through questions. It demands going beyond interviews and asking, and instead to participating in and witnessing all forms of histories – histories narrated and performed on different arenas (Klein 1989). This, however, is connected with obvious ethical problems (Caplan 2003). For instance, ought we to ask whether "the Balinese" appreciated Geertz's writings about cockfights? Many illustrative situations must (or, ought to) be left out of any text from such considerations. A funny or interesting situation or story that was not intended to be stored in the memory of the participants or listeners can later become a problematic issue. Let me provide an illustration. I once asked a friend who is a cultural researcher to contribute an article on a topic closely related to her own recollection of national and ethnic bordering; it was something with which I was familiar due to a story she had told me earlier while at a party. She declined: "That, however, is something one cannot talk about, at least not loud enough for people to hear" (letter from NN, November 22, 1993). Martin Stokes found the same silence, when he was studying the role of Arabesque music on the Turkish-Syrian border. "It is in the non-verbal domain that people are often able to embrace notions of hybridity and plurality which are often unsayable; this domain is consequently a vital cultural resource in the management of border lives" (1998:264).

Above, cultural bordering was discussed from invested narratives as *national cultural processes* (Gullestad 1989:84). These narratives explicate the national idea of spatiality particularly when the nation in its different forms is debated in public arenas and/or when one nation demarcates itself from an Other. An alternative arena for studying bordering, as a situational encounter between the national and the personal, has been established here through the separation of the remembered from the recollected. Instead of focusing on culture in terms of what people say, it is what is left out of discourse that has to be characterized. The problem identified is to distinguish arenas on which people experience bordering

---

8 See the essay "Deep Play. Notes on the Balinese Cockfight" (in Geertz 1975), which often is the focal point of others' discussions on Geertz's understanding of culture and ethnography.

as national belonging and from where they carry this experience as a recollection. Thus, the perspective shifts to *cultural processes in nations* (Gullestad 1989:84, Löfgren 1989:15) that focus on a person's everyday life as shaped within the nation-state as a cultural framework. Those who look for bordering in the culturally invested narratives disregard that the modern project not only provided people with bordering narratives and identities, but as much was an attempt to transform people's everyday lives, their taste, and habits. This was, for instance, the explicit intention of the organizers of the Stockholm Exhibition. The planning of dwellings and towns were matters that the states (through the 1900s) intensively tried to regulate. They are, thus, obvious arenas to investigate if one wants to research alternative understandings of associations between the everyday lives of people and the bordering, national projects. Or, as Michel Foucault expressed: "What we have to do with banal facts is to discover – or try to discover – which specific and perhaps original problem is connected with them" (from Abu-Lughod 1999:114). What we should expect is that we, in the form of recollections, live with what Marcus calls an "unseen world" (1998: 152ff),[9] albeit also a personal and "secret" such.

I am here interested in the same quality of banality ("repeated daily actions and gestures," Mbembe 1992:1) in one out of many possible arenas that is discussed by Mbembe (1992) and Richman (1992). Mbembe (1992:11) finds that this plurality, in a postcolonial situation, in effect shreds associations between state and people. An implied argument here is that within the modern Nordic Welfare State, different ideas of spatiality happened to be synchronized in the modern maze. Cleanliness, economy/capitalism, political representation/democracy, urbanization, religious secularization,[10] gender roles, nationhood, etc., evolved in a relatively synchronized manner, resulting in the maintenance of a mutually beneficial equilibrium between on the one hand, the interests of the state as an instrument of power execution and on the other hand, the population's appreciation of the collective pursuit of advantages from modern developments. In a general perspective the vast majority of Danes and Swedes experienced quite similar effects of modernity and approximately in the same periods, which is exactly where discussions of the Scandinavian countries

---

9   It is difficult to make this shift in focus from the culturally invested narratives to everyday life, as it is obvious from attempts to focus "habits." See for instance the contributions in Frykman and Löfgren (1991). Some of the articles from that volume are also published in English (Frykman and Löfgren 1996) although in general the English articles focus more on habits than the Swedish original.

10   "Religious secularization" might look like an oxymoron but stresses that moral and institutional elements of the Lutheran churches continued to affect Danish and Swedish society, profoundly shaping legislation and other practices after formally religious habits (such as attending religious services) were in decline or had more or less disappeared. An investigation into how Lutheranism has formed Scandinavia societies is, however, outside the scope of this book. A standing joke among Scandinavian social scientists in the 1980s was that "Scandinavians are practicing Lutherans who do not go to church, while Italians are atheists who regularly attend Catholic mass."

as extraordinarily socially homogeneous is located discursively. This equilibrium cannot be compared directly with the mutual obligations towards and protections from state patriotic society, but constitutes an understanding of self and society within the modern episteme. It probably should be emphasized that it is exactly the condition where the individual and the "us" mutually are identified and find benefits with each other that conceals the execution of power in the Welfare State system as such (Rabinow 1986:260). It might in fact be specifically the lack of similar synchronization of modern processes that characterizes the postcolonial condition and allows for such erratic use and abuse of power, as described by Mbembe. Such inconsistencies will also seriously affect the possibilities such societies would have for establishing and employing a window-of-opportunity for the development of what could be understood as relatively ideal bordering processes.

In the following, I discuss what I know (my recollection) as different from what we narrate (the remembered), taking my point of departure in my personal recollection of a situation in which bordering occurred. I recognize, and I hope that the reader will appreciate, that the presentation suffers greatly from limitations in my ability to express myself. The recollected does not have a language of its own, apart from the language that we employ to share and debate the remembered. Thus, regardless of how something is communicated through a language, the feelings emanating from the recollected cannot be passed on; they remain with the person in the specific situation, in the past, as "a secret." What further happens to this secret, when written and shared, will also be discussed below.

A Foundation for a Bordering Recollection

In October 1988, I had lived in Lund for ten months and I was visiting some of my Swedish friends. Besides the host and hostess, there was one other guest – a German who earlier had lived in Lund and now was visiting.

Late in the evening, after we had dined and sat talking, the host went to the kitchen to clean up. I followed and started washing the dishes. While I put the first washed plates in the dish drainer, the host turned to me and said: "You are not doing the dishes properly; you are doing them in *Danish*. Here, let me show you…"

There was no doubt about it; the host was unpleasantly affected by my method of washing the dishes. Thus, I let myself be taught how the detergent had to be rinsed off before the dishes were placed in the dish drainer; strangely, however, I was not allowed to dry the dishes by hand.

While I continued washing the dishes, I felt pointed to as Danish. It was obvious that I had violated the acceptable norms for proper dishwashing, and that my host identified my lack of compliance with my national background.

This recollection refers to a situation in which dishwashing became a marker of difference between Danes and Swedes, even if the difference revealed in the situation also could be investigated as a matter of cleanliness. To be able to comprehend this situation it has to be considered which space cleanliness occupies in the modern maze, in what way dishwashing became associated with the national project, and how the interface between cleanliness and the national can become a foundation for recollection.[11]

What is (un)clean is an often-discussed theme in cultural research. All classifications of the world assume that there are things that are left outside. With reference to Mary Douglas' description of this by-product of humans' classification as the unclean, the division between clean and unclean has been used to analyze many different groups and phenomena (Douglas 1966:35f). A culturally invested border is drawn between what is clean and unclean. As a marker of obvious difference, such a border constitutes a strong foundation for processes of identity formation (Schmidt and Kristensen 1986:21). Douglas finds that people who are confronted with transgressions of such cultural categories try to re-establish order (Douglas 1966:39f). They label disorder as, for instance, "Danish" in the example above, or they terminate disorder physically, for instance through upbringing or education, as it also happened that late night in Lund.

In an historical investigation of the modern battle against the unclean, Lars-Henrik Schmidt and Jens Kristensen have shown how cleanliness from the end of the 1800s became associated with physical dirt (1986). To promote healthiness, cleanliness was achieved or maintained through the development of medicine, through the promotion of outdoor life and gymnastics, and through the creation of a functional division of everyday life and the substances (water, sewage, and so forth). These measures should prevent "the disorder of contamination" (1986:86). This development resulted in a general attack for cleanliness, which transformed the sterile to an aesthetic as well as a moral strategy. "The road towards a higher moral goes through the bathtub" as it was expressed in 1909 in *Sundhedsbladet* ("Journal of Health"; from Jørgensen 1994:104). It is also in this context we must understand the appeal of milk and not coffee, tea, or chocolate (Schivelbusch 1993), which after around 1900 became *the* modern drink. Milk's whiteness also reflected the sleek, white surfaces of functionalist architecture (for instance, the architecture of the Stockholm Exhibition).

The housewife became, as was mentioned earlier, responsible for the upbringing and integration of future generations into the imagined community. It was also she who became responsible for maintaining cleanliness (Smith and Kristensen 1986). Home economics schools and journals, many of them created by housewives, informed the housewife how to conduct the battle against dirt. Moreover, her efforts became crucial to her own and her family's identity as she was the guardian of the crucial line between what was clean and what was not. As Jonas Frykman

---

11 I have earlier written more in technical detail about Danish-Swedish bordering through the nationalization of dishwashing; see Linde-Laursen (1993).

has pointed out, paradoxically women's domestication became associated with an intensification of cleaning that provided identity at the same time when blemishes originating from much practical work disappeared from the homes (Frykman and Löfgren 1987:244ff). Thus, one kind of task substituted for another. This feminine role was not unambiguous as it became obvious from the reactions to PH's film. Comprehensions of the woman and her work and world could be turned upside down again and again. Consequently, the development of women's everyday life has been perceived either as an emancipation from something obsolete and restrictive or as (male) society's execution of power directly onto women's lives. The interpretations change as the pendulum swings. Some have disassociated themselves from the historical, ideological foundations for the actions executed by male society and the state onto the housewife and her activities (Hirdman 1989). Others have regarded the development of the everyday as a relief for the women, as earlier many were isolated in their homes. They emphasize that wives and mothers also often participated in the production and through that, the provision for their families (Hagberg 1986). Moreover, domestic work and discussions about it sometimes formed a platform for a critique of society and became a vehicle for efforts to gain influence (Lövgren 1993). Hirdman (1989) and Nilsson (1994) paint contrasting evaluations of the life and ambition of the Swedish Social Democrat, feminist, and public educator Alva Myrdal, who was also an international expert on disarmament; these depictions provide an illustration of how differently these processes later were perceived. Where the former sees execution of power, the latter emphasizes development. The Danish cultural community does not include a similar female figure that in a comparable way embodies the discussion about modern everyday life. This is an illustrative consequence of the hegemonic Danish narrative's backward-focused scope onto the feminine *Gemeinschaft*. However, this does not imply that Danish modern women, who can be compared with Alva Myrdal, do not exist. Examples could be Karen Braae, who among other things was the chairman of *The National Consumer Agency of Denmark* and a member of several commissions, or the somewhat younger Lis Groes, who was active in *Danish Women's Society*, chairman of *The National Consumer Agency of Denmark*, and Secretary of Trade.

In the tragedy of development, as painted by Goethe, the community that is trashed by modern development is identified as feminine. Gretchen, the woman Faust loves so dearly, becomes the victim. Her fate exposes the lovable Gemeinschaft as brutal and crude (Berman 1988:51-60). Gretchen's longing for the genuine and authentic was not necessarily shared by many who lived and developed the frameworks for modern everyday life. Women, thus, were not only taught to clean as in the perspective of Schmidt and Kristensen (1986). Many women were agents for and driving forces in the development of cleanliness as an idea of spatiality in the modern maze. This drive towards new, cleaner practices happened approximately at the same time in Denmark, Sweden, and other northwestern European countries. The same thoughts were promoted in Sweden and Denmark around the same time – and by persons (among whom many were women) – from the same social environments. From the end of the 1800s, there is a remarkable coincidence in time and content in the development of organizations and institutions created to execute cleanliness.[12] Everywhere, national associations of housewives, educational facilities for teachers, and bodies of advisors were established; as these organizations were demarcated by the nation-states and provided career-paths for women, they came to provide for women's "pilgrimages" within the emerging nation-states in ways similar to the bureaucrats of late colonial, early national Americas; this also illustrates how later nationalisms replicate earlier processes (B. Anderson 1991:57). In Denmark and Sweden, the same national organizations, with financial support from the states, were founded approximately simultaneously and as a result of the practices by very similar agents. Thus, the national association of women in Sweden was actively engaged in establishing the Domestic Research Institute in 1944 (*Hemmens forskningsinstitut*), and the Danish organization was similarly central in the founding of *The National Consumer Agency of Denmark* in 1935. Moreover, institutionalization on a Nordic level also occurred. In 1919, the Nordic association of women was established simultaneously with the national associations in both Sweden and Denmark. It was no coincidence therefore, that the same ideas of testing different supplies and methods and the dissemination of the findings came about in all of the Nordic countries at the same time:

> The Nordic Association of Housewives [*Nordens Husmodersförbund*] (which still exists) meant a great deal for the exchange of ideas and experiences. The threads went from country to country: "we have tried this, so perhaps it would be the thing for you too." One can find a touch of stimulating rivalry, for example, in the occasional discussions of who was first with an idea. Mostly, however, they were good friends and admired each other at the large Nordic meetings that were held regularly. (Åkerman 1983:135)

Emerging practices related to this clean idea of spatiality were consequently shared. However, because they were developed within and disseminated from organizations identifying themselves as *national*, they also became directly associated with that idea of spatiality. The organizations were not only interested in promoting cleanliness, but in making Danes, Swedes, etc. clean.[13] Discussions in the organizations were directed towards clarifying the role of the housewife in both society and family as well as investigating how she could do her homework as reasonably as possible with regard to economy and time. Assembly-line rationalities were an integrated part of the shared ideas (Jeansson 1936); saved

---

12   The Swedish debate and institutionalization of hygiene and the role of housewives is well illuminated, see Lövgren (1993) and Åkerman et al. (1983a, 1983b, 1984). On the Danish institutions, see Dam (1960) and *Statens Husholdningsråd* 1935-1985 (1985).

13   For an illustration of how the same process developed in the US, which fails to recognize the ordinariness of this process, see Hoy (1995:140-149).

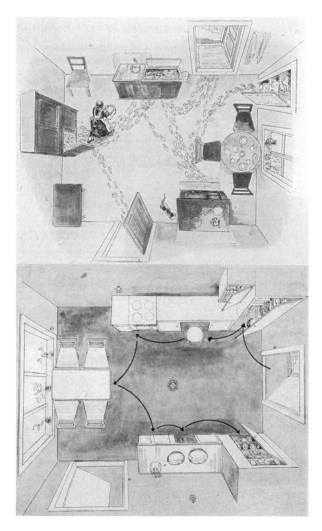

Figure 4.1 *"Ms Samuelson studied these pictures and realized with a sigh, how things were in her own kitchen,"* **1936**

Notes: *To organize* (predict, plan, coordinate, instruct, exercise control), and to *maintain order* were the essential measures for women who aspired to save time on their housework. Time studies were a crucial element in this rationalization and included all aspects of the mother's/wife's daily tasks:

Routine tasks have to be practiced so that they can be executed mechanically, semi-consciously, as when one is dressing or undressing. Improve your performance by measuring the time involved in, for instance, the daily making of the beds or dishwashing and try to beat your own record. Practice improves performance…The elasticity of the mind returns, as work no longer demands all hours of the day and all of one's energy too. But most preciously is the possibility for an improved family life provided by the decreasing tiredness and the increasing timesaving. (Jeansson 1936:36)

Source: Image from Jeansson 1936:24-25; photo courtesy of Björn Andersson, Historiska Media.

time should be transformed into a more modern (family) life with new qualities (Hirdman 1989, Lövgren 1993). Through many different media – schools, information materials, mass media, and new national radio – housewives were alerted to expectations of who she ought to be and how she ought to act. This execution of power was accompanied by the other side to modernity; women through the rising standard of living and housing actively formed their everyday lives in new ways and consequently, shaped modern society. Both development as emancipation from earlier unhealthy and filthy conditions and modern executions of power were simultaneously present in thoughts and practices. It is in this context that the enthusiasm, which functionalism provoked in the 1930s (at least in some circles), must be perceived. Thoughts about development and the transformation of everyday life could be translated into instantaneous practices, as Sweden's Social Democratic leader Per Albin Hansson also pointed out (Kristensen 1938:290). As a consequence, there is an intimate association between consumption and the personal conquest of new physical spaces in which to act as a modern being. As Orvar Löfgren has noted, spaces and things often became the hooks onto which individuals hung their autobiographies (1992; see also Hoy 1995:117). The first time I had a room to myself, when I moved into a new and better dwelling, or when I got my first car became emancipations that facilitated new and most often modern conquests as additions to the life story.

The Nationalization of Dishwashing

In the beginning of the 1900s, people still ate from a common dish, each with her/his separate spoon that they then put away until the next meal. Soon though, dishwashing became a daily chore for the overwhelming majority of families. Dishwashing became one issue about which the new, similar organizations on household economy and family life in Sweden and Denmark had to investigate, enact decisions, and disseminate information.

In the same year that it was established, the Swedish Domestic Research Institute embarked on a large-scale study of all the problems of dishwashing – hygiene, time, and economy. The study was published two years later and addressed all aspects related with dishwashing including the calcium content of the water, different detergents, brushes and dish drainers, as well as the placement and shape of the kitchen sink (*Diskning* 1946). The study arrived at clear conclusions and consequently it was possible to establish a recommendation for how dishwashing should be carried out to ensure economy of time as well as proper hygiene:

> The following general guidelines for the organization of dishwashing can be presented. [The preliminary treatment requires] rinsing under the tap with the aid of a washing-up brush...When washing up one should ensure that the water temperature is kept high, preferably around 45 degrees Centigrade [113 F]...A washing-up brush or a paper dishcloth and detergent of a suitable type should

be used. The dishes should always be rinsed, preferably in hot water (over 50 degrees Centigrade [122 F]). [...] The dishes should be allowed to dry in an appropriate drainer...Air-drying is preferable to hand drying with regard both to the time consumed and to hygienic and economic considerations. (*Diskning* 1946:74)

The ideal grammar for dishwashing is thus: rinsing > washing > rinsing > air-drying in a dish drainer. Moreover, the kitchen sink must be situated so that the washing up can be done from right to left (*Diskning* 1946:69). To optimize the process, a kitchen counter out of stainless steel with two sinks is required, one for washing, one for the second rinsing, so that the process can be carried out continuously. Information about these findings was in the following years disseminated through pamphlets and journals.[14]

The study also included a survey, which clearly demonstrated that this correct grammar of dishwashing was not employed in most households. Many kitchens were not equipped with suitable counters or had no running water or drain; even fewer had hot tap water. Consequently, the lack of improvements of housing standards left many households incapable of performing the desired practice of dishwashing. From statistical information, it is clear that 60 percent rinsed their dishes before washing them and 80 percent did so after. However, 70 percent dried their dishes by hand (*Diskning* 1946:91). Obviously there were good reasons to make an effort to improve information and eventually, the equipment in many kitchens:

> The study of dishwashing shows that by introducing improved dishwashing facilities and methods in the household, one could gain a great deal from both the hygienic and the economic points of view, while also ensuring that dishwashing became less of a physical and mental strain. (*Diskning* 1946:96)

However, one could as well be surprised that dishwashing in many households was carried out according to the recommended grammar. The women in the Swedish Domestic Research Institute were not spearheading a radical novelty; others had earlier promoted the same ideas. An analysis of Swedish housekeeping books from the 1920s to the 1990s, some of them aimed at schools while others at young housewives, shows that the general pattern of recommended dishwashing is: rinsing > washing > rinsing > air-drying.

An investigation of equivalent Danish material clearly demonstrates that the recommended grammar of dishwashing is different. Whereas few Swedish texts suggest drying with a tea towel, this is the method recommended in almost all the Danish texts; only a few Danish housekeeping books mention that tableware could

---

14   See for instance *HFI-meddelanden* (1948-49 nr. 5: Bra hushållsredskap and 1954: *Matlagning och diskning*) and *Vi husmödrar* (1953/6: special issue on the Swedish Domestic Research Institute, its efforts and results).

Figure 4.2   *Kick* saves time, is hygienic, Americanizes everyday life
                and indeed makes "housework easier and more fun,"
                advertisement 1949

Source: Detergent ad in the Swedish weekly magazine *The Housewife* (*Husmodern*, N:r 39,
1949); photo courtesy of Björn Andersson, Historiska Media.

be left to dry in the air. A consistent recommendation of air-drying was found in only one book, which explicitly referenced the saved time of four minutes or 30 percent. "That one, in this way, also is forced to wash the dishes hygienically is indeed only one additional advantage" (Begtrup and Svensson 1949:39). The Swedish texts without exception prescribed rinsing the dishes after washing and before drying. This agrees with the Danish recommendations until the mid-1940s. However, after the Second World War, about half of the Danish texts omit mentioning this rinsing or else make it optional. After the Second World War, the second rinsing increasingly disappeared in the Danish grammar for dishwashing: rinsing > washing > hand drying. It is impossible to determine whether there is any rationality behind these different grammars of dishwashing in Sweden and Denmark. In this context, however, the reason – if there indeed is one – is not so important; it is the difference that is interesting.

It is not only in the work and in the bodily practices themselves that the difference between Danish and Swedish dishwashing leaves its trace. The design of the kitchen counter itself is intimately associated with dishwashing. From the 1920s, test kitchens were established in both Sweden and Denmark; architects and planners discussed the optimal layout for a practical kitchen. The more radical functionalists wanted to reduce the kitchen to a workroom, which was in agreement with the basic idea of differentiating between the functions of each of the dwelling's rooms such as socialize, work, and rest (Asplund et al. 1980:58-71). The smaller dwellings exhibited in Stockholm in 1930 had thus, very small kitchens lacking direct light from the outside. In general, female reviewers did not applaud these "laboratories" as they were called. This exemplifies that people often did not want to *practice* as planners and architects *thought* (Hagberg 1986:117-128). People wanted space to be able to eat in the kitchen; the reformers soon found a reasonable, rational explanation that was in agreement with what people wanted. Kitchens of course had to be equipped with a table so that the housewife, while taking care of her own duties, could keep an eye on the kids at play and at their assigned schoolwork.

In Sweden, state subsidies for the building of dwellings were introduced during the Second World War, requiring that houses adopt certain standards. At first, this applied to standards for the size of rooms (Westholm 1942) but then soon came norms that prescribed in detail the particular dimensions and elements that had to be used. One reason for this standardization was that it made houses cheaper to build. In 1954 came the publication "Good Houses" (*God bostad*), which became normative as it shaped the requirements for obtaining state subsidies; however, these norms were never established by law. "Good Houses" was based in part on the studies on kitchens by the Domestic Research Institute, which were collected and published at the same time (*Kök. Planering. Inredning* 1954). With these publications, the Swedish standard became a kitchen counter made entirely of stainless steel, and with a dual sink, one for washing up and one for rinsing (that was also to be used for slops). It is specified that the rinsing sink must always be to the *left* of the washing sink, "The reason is that one can wash the dishes with the least possible crossing of the hands if one always has the rinsing bowl to the left of the washing-up bowl" (*Kök. Planering. Inredning* 1954:49). Thus, in general such norms did not allow for the specific problems of a left-handed minority to be considered when outlining the regulations by which modern everyday life could and ought to be practiced. The idea was not that all kitchens had to look alike. Nevertheless, the rules and directions had a great and immediate homogenizing effect on the building and organization of dwellings:

> Unfortunately, kitchens came to be planned much in a similar way and this resulted in something that can be called "standard kitchens." The intention was, however, that the parts could be combined more freely. The norms for kitchen equipment established themselves very fast. The producers were interested; the mortgage requirements promoted the enforcement of the standard, and within a couple of years all kitchens in rental flats had become bigger, had spaces for eating, and had suitable equipment. (Boalt and Lindegren 1987:66)

The development in Denmark took a somewhat different course from that in Sweden. This was not least due to the fact that private credit associations financed a very large proportion of Danish home construction so the state had no direct instrument with which to dictate standards. In Denmark, standards are therefore seen more as guides than as norms. However, the Danish guide for the kitchen sink is still different from what emerged as the standard in Sweden. In 1959, the Danish State Building Research Institute published guidelines, which directly referred to the study by the Swedish Domestic Research Institute. On the subject of the kitchen sink the guidelines say: "this should be a dual sink, preferably in a steel counter…The rinsing sink in a dual sink should be placed to the *right*, which means that the total amount of reaching movement over the sink is lowest, and that it is closest to the worktop which is used for cleaning and preparing food since this is best located to the right of the washing-up place" (Arctander 1959:8, my italics). Otherwise the Danish Government Home Economics Council has always recommended that: "Higher priority should be given to counter space; in other words, if there is a shortage of space, then a single sink is better" (*Nyt køkken* 1986:4). Although a complete steel counter was recommended, many people in Denmark chose to buy a separate sink that is set in a hole in the counter, which can be of wood or some other material. In Sweden by contrast, they insisted on the need for a dual sink. Even when space was limited, in Sweden one sink was placed at the front edge and the other along the wall behind it. In interviews in the summer of 1992, the marketing department of *Ifö Sanitär*, one of the largest Scandinavian producers of kitchen sinks and counters, illuminated that Sweden and Denmark constituted very different markets. In Sweden, 85 percent of sales consisted of whole counters with sinks of stainless steel; 98 percent of the units they sold have two sinks or a washing sink and a slops sink. In Denmark, separate sinks for insertion into a tabletop of a different material accounted for 90 percent of sales, and 90 percent of the customers had only one sink.

In both Sweden and Denmark, it is recommended that people wash up from right to left but both the bodily practices of dishwashing and the material form of the dishwashing arena are very different. Thus, a nationalization of details of everyday life and its spaces occur and through this, a foundation for experiencing bordering differences in specific situations is established. The Swedish grammar requires two sinks and a counter that is resistant to moisture from the dripping of the dishes that are left to dry. In addition, the slops sink has to be located to the left of the washing sink so that it can be used to rinse the dishes after they are washed. The Danish grammar requires only one sink since there is no second rinsing. As the dishes immediately are dried with a tea towel, there is no need to worry about moisture. The counter can be protected with a freestanding draining tray that lets water drain back into the sink before the dishes are dried and can be put away when dishwashing is completed. Afterwards, the counter is wiped and consequently, there is no problem with moisture. If there is sufficient counter space, a slops sink should be placed to the right of the washing sink so that dishes can be rinsed off there before they are washed. In the guidelines from the Danish State Building Research Institute a double sink is directly called, "model: Swedish standard" (Arctander 1959:7). Those who drew up the Danish recommendations directly referred to the Swedish dishwashing investigation (see for instance Heiberg, Kristensen and Salicath 1950). Furthermore, Danes who during the Second World War had sought refuge in Sweden from the Nazi occupation of Denmark had carried out parts of the Swedish dishwashing research. Thus, information about this groundbreaking research was readily available on both sides of Øresund and the arguments employed on both sides of the border are identical. Both sides seek the most economic, rational, hygienic, and time saving practices that cause the least strain. Although they shared a basis of ideas, the results of their recommendations turned out many differences in the details. Additionally, it can be noted that people in Denmark often wash up in a separate plastic basin inserted in the sink (Møller et al. 1949:58). This is necessary as the cooled water from the draining tray otherwise would cool the dishing water. In Sweden, the water from the dish drainer runs down into the slops sink and people therefore, wash up in the washing sink itself. Consequently, Swedish recommendations for heights and depths of kitchen counters and sinks cannot be transferred to Denmark.

The association between the space of cleanliness in the modern maze and the specific ways that dishwashing was manifest both as bodily practices and as materialized kitchen space is an example of how the border between nations can make itself noticeable as real difference. Supported by observations in many homes it can be established that dishwashing is *one* example of how the nationalization of modern everyday life after the Second World War resulted in an objective materialization of the border between Denmark and Sweden. Even if not all Danes and Swedes perform as recommended, dishwashing has been shaped in such ways that many people who have moved across Øresund will have recollections of situations in which they practiced and experienced bordering based on this specific interface between these two modern ideas of spatiality: cleanliness and nationhood.

Sweden — Dishwashing process from right to left :

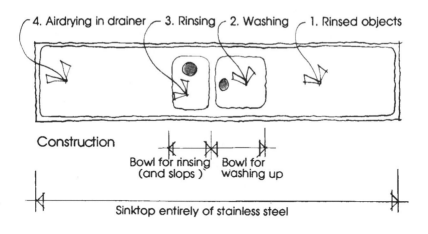

4. Airdrying in drainer — 3. Rinsing — 2. Washing — 1. Rinsed objects

Construction

Bowl for rinsing   Bowl for
(and slops)   washing up

Sinktop entirely of stainless steel

Grammar of dishwashing

Denmark — Dishwashing process from right to left :

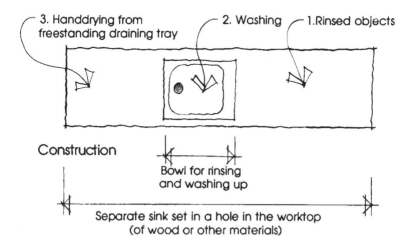

3. Handdrying from freestanding draining tray — 2. Washing — 1.Rinsed objects

Construction

Bowl for rinsing
and washing up

Separate sink set in a hole in the worktop
(of wood or other materials)

Figure 4.3   Danish and Swedish grammars of dishwashing

Culture – Nature – Culture

Because of the modern, technical improvements of dwellings during the time following the First World War, it was necessary to have a public discussion on how proper dishwashing was to be conducted. This debate lasted decades and was of interest to many at the time. "Moreover, many of the housewives express complaints that there is sometimes a lack of information about the correct working method," as this desire is expressed in a study of kitchens in Denmark from 1949 (Møller et al. 1949:61). Since then, the nationalized kitchens and bodily practices have become "naturalized," which makes it highly provocative to experience how other people wash dishes in their, different, way. That the agents of this whole development were aware of this nationalization is obvious from interviews conducted with kitchen consultants and producers; it is also patently evident from Swedish consumer information material, which expressed a concern with maintaining the achieved rationality and hygiene in everyday life.

> There also exist kitchen counters modelled on foreign types, where stainless steel is not so common. A washing bowl and slops sink of stainless steel or enamelled sheet steel is inserted in a top of wood or plastic laminate. These sinks have obvious disadvantages. They generally lack an upturned front edge, with the result that water can run over the front and down the cupboard doors to the floor. Another problem is the joint between the sinks and the counter, and between the counter and the wall/splash-guard. (*Kök – planering, inredning* 1981:36)

It is however, difficult to obtain information about such nationalizations of everyday life. When the public discussions are concluded and the recommendations become natural instead of cultural, the details are put beyond discourse. The differences remain in our bodies and spatial surroundings, revealing themselves only when one's practice is not recognized as proper by others with a similarly naturalized, but somewhat different cultural organization of their bodies and practices. Such a contrasting and bordering experience can become a recollection, which as Thor's hammer keeps coming back in related situations. Since that night in 1988, I have washed dishes in Swedish homes many times and each time, the same feeling and consciousness of being Danish turned up – even if I adopted the Swedish practice also in my own home.

During the summer of 1992, when I was gathering information about kitchens and dishwashing, I conducted and taped a several-hour interview with a Danish-Swedish couple, Dagmar and Sven. They met in Stockholm where Dagmar lived for many years but had recently moved to Copenhagen. Throughout our conversation, we examined the culturally invested bordering narrative about the Danish and Swedish. We all agreed that experiences from everyday life are crucial if one wants to develop a more thorough understanding of bordering. Nevertheless, it turned out to be difficult to pinpoint such examples and experiences. To guide the conversation towards something that I knew for a fact, I tried several times to raise dishwashing as a possible topic. In the following excerpt from the interview, we are discussing when Dagmar and Sven talk about national difference in their everyday interaction:

> *Anders:*   What kind of situations are you thinking about?
> *Dagmar:*   Oh…Well, that's how Swedes always do, I can say, how typically Swedish! – Do you really have to behave in that way!
> *Anders:*   But, what is topical? Television-habits, dishwashing – or what?
> *Dagmar:*   Not dishwashing, it's more things like…Gosh, it is – suddenly difficult to find…because it is small things.

The conversation then approaches other matters. However, I maintain dishwashing as an issue and later return to it. Eventually I am pressing hard to learn if Dagmar or Sven have experiences with dishwashing.

> *Sven:*   It can be dishwashing, it can be traffic…
> *Anders:*   Dishwashing?
> *Sven:*   Different systems.
> *Anders:*   Which systems?
> *Sven:*   Here [in Denmark] many save water –we don't have to think about that in Sweden – well, maybe in Skåne. Traffic…[the conversation continues onto traffic and language, until I persistently return to dishwashing].
> *Anders:*   Are there such different ways of organizing things? Traffic, dishwashing – Swedes always rinse the detergent off afterwards… (Dagmar is laughing)…have you reasoned over that?
> *Dagmar:*   Not a lot – Sven doesn't do the dishes much! Thus, it is impossible. However, it could be one of them. On the other hand, Sven is running around with the water canister and waters on and on – and I say that he shouldn't do that. First, water is more expensive here, and second, we'll have water-shortage – it is our drinking water – and so forth.

Again, our conversation takes other paths. I realize that questions on dishwashing will not be answered with stories similar to the one I recollected. However, when I am reading the transcript from the tape over and over again, and subsequently when I go back and listen to the tape repeatedly, I realize that all three of us assume that there are differences between dishwashing and dishwashing. I am furthermore convinced that when I press hard by shortly sketching my recollection of a bordering situation to obtain a reaction, Dagmar in a split second disappears back to an episode in the past – into her recollection, and she laughs. Nevertheless, the recollection about a bordering, personal experience of nationality remains unshared.

As it can be expected from this, my interviews are not filled with references to nationalized thoughts and practices that have been turned into nature where they can be experienced in bordering situations. At least such experiences are not distinguishable in ways that make them recognizable as recollections. This does not mean that bordering recollections cannot hide in the material even if nobody is able to see them. As pointed out, recollections share language with memories, and they are therefore not only, as Kierkegaard pointed out, impossible to share, but also they are impossible to recognize through our ethnographic senses. From my many interviews, only one person told a story that she herself pointed to as "funny" about differences between Danish and Swedish dishwashing. Maybe for Dagmar and Sven, it is the question about watering plants that really constituted a bordering issue in their everyday life.

The cultural history of dishwashing reveals that national culture does not only exist as culturally invested narratives about the peculiarity of nations and their differences from Others. I am not arguing that national culture exists as separate social systems. However, I hold that within the borders of the nation-state, homogenizations of practices, bodily movements, and spaces have occurred and for that matter, still occur. Usually such practices are through time turned into nature and nobody is able to talk about them furthermore. Modernity is described often as cultural processes across borders that have great homogenizing influences. It is, nevertheless, not only the nation as an idea of spatiality within the modern maze that is filled with comprehensions of difference. From the 1800s, cleanliness also marked itself as another arena on which identical efforts and institutions were shaping up all over northwest Europe. Everywhere people faced identical problems. Debates and information developed in quite similar ways about how hygiene, bodily practices, and economic rationalities should enable housewives to attend to the cleanliness of themselves and their families. However, through these identical practices and executions of the idea of cleanliness, new differences were created that could potentially supply identity providing bordering experiences. Thus, modernity has a homogenizing influence and border-crossing side, but it also, at the same time, has a differentiating side where (possibilities for) bordering are constantly being created. "The very ways societies change have their own authenticity, so that global modernity is often reproduced as local diversity" (M. Sahlins 1993:2). While the solutions to the problems of modern everyday life were discussed and established within the framework of the states, many of these new differences coincided with the political borders. In bordering situations, such differences can be experienced and be established as recollections through which the personal is directly associated with the national; most often though, such recollections are formed outside the narratives about cultural communities.

When naturalized differences between nations are pulled out of the collective's shared oblivion through research and thereby (again) turned into issues for public debate, the differences become marked as cultural. When I was writing about dishwashing, it made it again possible to talk about it. Dishwashing was restored as part of culture and was, thus, regained from nature. While dishwashing in the

1930s and 1940s was debated within the idea of cleanliness, in the 1990s it was possible to speak about it within the idea of spatiality of the nation – as a sign of bordering difference between the Danish and Swedish imagined communities. Again, it became possible to think about and express the problems of dishwashing. After having read an article I wrote, a woman from Malmö sent me a letter:

> I read it with great interest as I have had a similar experience. It happened the first time I visited my pen pal in Copenhagen. I wanted to make a good first impression and offered to wash the dishes after dinner. They found that I was both wasteful and careless as I rinsed the dishes in warm water and did not dry it by hand! (Letter from BB, July 14, 1993)

At the same time as dishwashing through research anew is turned into culture, the relationship between memory and recollection is changed. Bordering experiences in dishwashing situations lose the essentiality with which the recollections were endowed. It is again possible to discuss which is better or good enough while at the same time discussing how many Danes and Swedes really perform as described. Is it right or wrong to suggest that dishwashing provides bordering experiences between Danes and Swedes? Does it mean anything? Dishwashing is again an element of memory that everyone can discuss and share. Thereby it is also again accessible to modern cultural research that can try to answer the many questions. References to dishwashing as part of national narratives started to appear in interviews after my research had been referenced in national newspapers and radio. In this way, I was able to invent elements for a narrative that I afterwards could discuss as something of interest to science. However, the expense was the recollection; I no longer felt anything special about doing dishes in Swedish homes. With Kierkegaard's word, the essential has disappeared (1988). In a bordering situation, a seduction of the involved had happened. They got an experience of becoming "the one which one in reality always has been," as Michael Harbsmeier expressed it (1986:53). Through the re-culturalization of dishwashing, this seduction, which of course depended on me crossing the border, was blocked. While washing the dishes in Swedish homes, I am no longer seduced to become the one I had always been.

The Little Difference

The extension of the scope of bordering to include everyday life and its association with the state as a framework for cultural processes illustrates features of the complexity of the modern maze. The example of dishwashing throws light on the relation between nation and gender. National narratives were and are to a large extent provided by male scientists who have not felt a need to investigate how the female gender actively participates in the formation and development of the nation. By shifting the interest towards everyday lives and repetitive situations, much more attention to women is needed as well as to their active involvement

Det är inte tråkigt att diska, om det görs rationellt.

Figure 4.4   Washing the dishes, 1959

Notes: Discussions of cleanliness are clearly associated with understandings of the work of the wife/mother in the home and consequently, associated with the cultural construction of gender. As conceptions of roles and responsibilities in the home evolved over the years, maleness also became associated with aspects of housework. Dishwashing was one of the earliest tasks, which seriously was discussed as a possible husband/father responsibility. Even if dishwashing became a task in which males could or should partake – they could (in 1959) relax with a pipe of tobacco while such engaged – it was in 1975 only in 8 percent of households (in Sweden) that males were principally responsible for this task; in 14 percent of households it was a shared responsibility, while in the rest it was regarded mostly as a task for females (see Boalt 1983:55). Men seem later, in many cases, to have been the driving forces when households acquired dishwashing machines (see Kjellman 1993:55-59).

Source: Illustration by Sonja Ceder, 1917-2001 printed in Corall and Calminder 1959:32; photo courtesy of Björn Andersson, Historiska Media.

in debates revolving around the implementations of programs associated with dwellings and consumption. Through such a lens, it becomes clear how the two genders contribute differently to the national project. Thoughts about cleanliness, the differentiation of the rooms of the dwelling, the role of the housewife, and the rationality of everyday life were parts of transnational cultural processes acting through the market and social movements since the 1800s. However, the elaboration of details happened within states in very tangible ways so that differences in forms of life were established (Hannerz 1992:40-61).[15] These differences can, in

---

15   There were differences of course with regards to the importance of these different frameworks in different places. Even if a measure for this might be very difficult to establish, it appears that the market, business, and advertising (also very important in the

bordering situations, become markers of cultural difference and thus be turned into consciousness of identity.

On his renowned journey through "Filth-Sweden" (*Lort-Sverige*) in 1938, Lubbe Nordström visited provincial doctors who he saw as the real engineers of the good and hygienic society. "They are probably the foremost reformers and revolutionaries in the country. They have continued in practice, where the social and political reformers and revolutionaries have become petrified in theory" (1938:23). They were not only the engineers of the future, as Nordström saw them, but also the designers of a Swedish everyday life, and consequently of Sweden. The unambiguousness of such processes must not be overstated – or rather, it must be seen within its proper context. In the essays from high school students from 1992, quite a few students mentioned the difference between Danish and Swedish hairstyles and colors. Some Swedish students found that the typical Dane had brown-reddish hair. Similarly, many Danish students mentioned the hairstyles of Swedish men and boys. "The guys have Swede-hair (short at the top and long in the back of the head) [like the American 'mullet']" (Virum).[16] When I inquired among Swedes, I got a completely different explanation for this hairstyle. To them it had nothing to do with nationality; all agreed that it was a signifier of the working class. The hairstyle, thus, was on the two sides of Øresund referred to by two different ideas of spatiality in the modern maze. From both spaces bordering experiences and identity can be established. However, the identities are in the different circumstances built on different foundations – nation or class.

As with dishwashing and hair, references to the different spaces of the modern maze are turned into metaphors that explain what people experience in situations. Danes, Swedes, and all others who travel across borders for shorter or longer time can explain through references to nations why they are bypassed in certain situations, not understanding or misunderstanding what is going on. This exclusion from situations in everyday life is in principle alike for all who transgress Febvre's ditches of cultural difference (1973). Even so, individuals can have more or less of such bordering-situation experiences as a direct consequence of the relation within their autobiography between their place of origin and the place of their everyday life. Public debates about people who have moved across borders most often focus on groups that can be recognized as "immigrants," who look different, have limited economic resources, live or work in socially tainted urban arenas, and so forth. Consequently, it is assumed often that no differences exist between Danes and

---

Nordic countries) played a relatively more prominent and aggressive role in the US (Hoy 1995:140-149).

16   For inspiration, students were given an article by Orvar Löfgren in which hair is mentioned (1986:49). However, it is mentioned in the same way as other elements that very few or none of the students mentioned.

Swedes. Or, "Danes are not immigrants"[17] as directly stated by a cultural worker in Halmstad, a town in southern Sweden of 55,000 inhabitants in 1997.

Geographical and political borders are translated into cultural borders as everyone who moves across borders can be expected to have the same foundations for bordering experiences. Danes and Swedes who have not moved across the border not only have difficulties with recognizing how their nations as cultures have been established in and through them as persons (Arnstberg 1989) but also have similar problems in recognizing each other as culturally different.[18] A Danish woman who moved to Malmö in 1991 in order to get a job explained in an interview one year later:

Anders: But you told me on the phone that you are tired of being recognized as a Turk or a Pakistani?

Maren: …At work, it is good; everything else [meaning: in all other social situations] is without social contacts – as for other immigrants. The only people where I have been able to get across the threshold and visit are Iranians or Italians. And I feel more comfortable talking with them. We're equal – by virtue of coming from the outside; that one has come from where one is raised.

Anders: You had never considered, that it could be a problem suddenly to be classified as an immigrant?

Maren: No. I didn't imagine that to happen. Maybe we Danes are also too poor at it – I didn't know anything about Sweden. They are Nordic people, as the distances all over are not so great. I hadn't thought of it as a problem. If I had to go and work in Africa, I would most certainly have prepared myself better, investigated the conditions, the mentality, and how one ought to behave.

Anders: Is it hard suddenly to be recognized as an immigrant, without really being aware of it happening?

Maren: Yes…

---

17   The concept *immigrant* is problematic in many ways, which there is no reason to discuss at length here. It accentuates certain groups as more "immigrant" than other groups. It also hides the fact that to become an immigrant an individual first has to be an emigrant. The concept is the framework for the execution of power and counter-power (B. Svensson 1993), as people who are identified with the word can use it to manipulate society and for instance, gain access to resources for cultural events.

18   Because of this, there is an insignificant amount of Danish associations in Sweden and vice versa. Activities are started yet often disappear with the people who began then, obviously making it difficult to maintain some kind of continuity. The only "old" Danish association in Sweden is more than one hundred years old. It is in Stockholm and limits its activities to gathering Danes at Christmas, celebrating the Queen's birthday, and a few other occasions.

In bordering situations, Danes and Swedes meet each other as different. When such differences are threatening to the order of things, two solutions are at hand. The difference can be explained as a result of the single person's inability to act in agreement with nature for pathological reasons. Yet, this explanation leaves the person solitary and perverted (Arrhenius and Sjöholm 1995), which might result in the deterioration of social relations. If the involved do not wish to confront each other they can instead refer to the only *natural* reason for their difference, "You are not doing the dishes properly; you are doing them in *Danish.*" A direct reference is made to what everyone knows and recognizes from the narratives – the natural organization of the world as divided into nation-states. Imagined communities are thus, not only reasons for conflicts, which might be how they are most often depicted (see for instance Schlee 2002) but also, they can be employed to avoid conflicts in everyday life; this is indeed likely to be the way bordering invocations work in most everyday situations. This, of course, under the condition that minute details inscribed into bodies, as the movements of dishwashing, are not turned into signifiers for a perceived difference that demands that the Other be treated poorly or be obliterated. Differences inscribed onto bodies can in specific situations become the sign of a real or imagined threat and thus, can have deadly consequences. The haunting aspect of this is that while no one (at least thus far) has been killed as a consequence of doing their dishes differently across the Danish-Swedish border, what would in most situations be regarded as similarly insignificant, have in other cases turned into reasons for killing.

Even if Danes and Swedes in this way are able to socialize, they still express surprise when persons who have moved across Øresund do not wish to return to their natural surroundings. "Many stayed behind and died here and everything," as an informant expressed it with astonishment when writing about the not very many Danish immigrants to Sweden around 1900 (LUF M:14479). On the whole, death – as a reunification with the parental dirt through the funeral – is a situation that makes bordering pivotal (Kürti 2001:158). In his autobiographical memoir, *Out of Place*, Edward Said makes this connection explicit as he describes how writing a book about his own feelings and experiences of displacement is intimately associated with him being diagnosed with a leukemia that will eventually kill him (2000:133, 216). Likewise, Per Olov Enquist recollects a situation in which he discussed death with his mother. She found, everyone had to be buried there, where a person through his/her birth naturally belonged and consequently had as a natural point of reference and vantage:

> Where should I otherwise rest. Under no circumstance abroad. That would be shameful… One should not be buried abroad but in the center. (1992:11)

This sentiment and insistence on the material reality of the border, again associated with the end of life, is similarly reflected in a novel by Amitav Ghosh as summarized by James Clifford (1997: 332f). In the novel, the narrator's ailing grandmother is going from Calcutta to Dhaka by plane and is puzzled that the border cannot be seen from above. The old woman reflects on the separation of India and East Pakistan (today Bangladesh). "But if there aren't any trenches or anything, how are people to know? I mean, where's the difference then? And if there's no difference, both sides will be the same; it'll be just like it used to be before, when we used to catch the train in Dhaka and get off in Calcutta the next day without anybody stopping us. What was it all for then – Partition and all the killing and everything – if there isn't something in between."

Memorizing Space

> Even if the journey across Øresund does not last more than barely an hour, one anyway feels that one is abroad, where one meets an abundance of things that one does not have in one's everyday life. (Student essay from Malmö, 1992)

To pass borders in space means that persons do things and recognize their surroundings in different ways (Löfgren 1999b). All kinds of experiences of difference are through metaphors referring to the border itself. What is it that endows borders in space with such cultural power that they can change our way of acting and seeing?

Where and when persons could or would be located in pre-modern society very much depended on and varied with their position in the state patriotic system; this was reflected in their dress, name, furniture, and field of activity (Balle-Petersen 1978, G. Hansen 1944, 1947, Hanssen 1973b, 1977). As it was discussed above, the modern nation is a historicization of territory and a territorialization of history. Persons are more likely identified with where they are from and when they are in a certain place than from which social and political position they occupy. The modern maze can be perceived also as a specific association between time and space.[19]

Modern time and space are of different qualities in regard to their bordering potentials and functionality as foundations for processes of identity formation. *Time* is understood as an absolute. Its only real border is the "now" that in the next second has become a part of the past. It is impossible to turn back to a previous now or go ahead to a future; many categorically modern, fantasies about time machines illustrate this problem. Time is thus, an irreversible process. Presumably, time is relative – how long ago is long ago? Nevertheless, nothing can be done to the continuous and transient character of the now.

The character of *space* is different in that all border crossings are in theory reversible. While it can be debated how far away, far away is, it is beyond

---

19   In his sociological biography of Alva Myrdal, Jan Olof Nilsson has argued for an understanding of the modern as an organization of time and space. He does that through joining Arthur Schopenhauer's statements on time and space with the time-geography of Torsten Hägerstrand (1994).

discussion that one can go there – and come back to the point of departure. Of course this is the whole rationale of another modern phenomenon – tourism (Urry 1990, Löfgren 1999a). At the return, the space will have changed; the now of the departure will be in the past. But in the spatial dimension it is still the very same space. This does not mean that space in any way is objective (Fink 1992). It is to the highest degree a cultural construct (Gregory 1994, Lefebvre 1991, Linde-Laursen and Nilsson 1994). However, due to its tangibility, it appears as real and unambiguous.

Time forms an absolute border that continuously shifts, and in its character is unchangeable. Space on the other hand, is tangible and fixed even if it changes. It is this fixed dimension that makes space very attractive as a foundation for processes of identity formation because it is important to identify as firm a border as possible in order to construct the firmest identity. Most probably, this unambiguousness has enabled metaphors that refer to space to become increasingly popular as the cultural scientific tendency from the 1980s has come to dominate more and more over the social scientific tendency. Unambiguous metaphors make it so much easier to establish clear distinctions. The reality of space also explains why people, who long to experience coherence in their lives – a Gemeinschaft, which in their time seems to have been lost – participate in the reconstruction of *landscapes* as arenas for scenes and experiences. This occurs for instance when people use their vacations to enact stone or iron age tableaus in the Danish archaeological experimental center in Zealand or when persons congregate to participate in the reconstruction of historical scenarios at battlefields (Handler and Saxton 1988). The joining of movement in the tangible space with the transitory character of the now – "the simultaneousness of all events, the synchronic knowledge of All," as he writes – is what for instance awarded Claudio Magris' epic on the Danube and Central Europe its success (1989:61).[20] Similar understandings are also found in the work of cultural historian, Dane Hugo Matthiessen, who found it necessary to wander old roads or crisscross by boat the Limfjord in order to be able to write about them (Linde-Laursen 1989a:41-45). And "being there" is what generally affords ethnographic accounts their authenticity as it is difficult to negate experience as a valid source of information. Both Bjarne Kildegaard and Barbro Klein have asked that ethnologists take nostalgia seriously (Kildegaard 1990). *Nostalgia*, which today refers to a longing for another time, stems from the Greek word *nostos*, which means home travel or homecoming. Nostalgia, thus, literally means the longing for a place – the home or maybe the homeland: homesickness. The word thereby describes exactly the feeling for the tangibility of space – that border crossings are reversible. Kildegaard and Klein's focus on history and the dimension of time thus, have to be supplemented so that the dimension of space, of geography, can be analyzed – and its foundation in and legitimization through

---

20 On time see especially Magris (1989:40, 87) and on place (1989:252f).

time can be clarified. It is through working with these issues that the spatial and the nostalgic gain analytical significance.[21]

It is through the crossing of borders that it is possible to draw them accurately. The crossing enables an experience of personal bordering, and of the association of the national with the personal. This is exactly what is so fascinating with the geographical and state border. Even when it is not evident from thoughts and practices, it is always indisputable whether one is on one or the other side of the border. This, the border's obviousness, limits and consequently shapes our memory of space. As Orvar Löfgren has pointed out, it is when we have crossed a border that we start looking for differences (Ehn, Frykman, and Löfgren 1993:81ff). Even insignificant differences are awarded much attention as they are transformed into symbols for the difference between nations. Such signifiers are found in the surroundings (for instance the colors of street signs and markings) as well as on the body (as hair), in the body (as the bodily performance of dishwashing), into the body (as food, Schwartz 1991), and out of the body (as language). Signifiers for the border are bestowed mythological importance (Barthes 1972) and have consequences for where and how we move. By memorizing space, differences are accentuated and a foundation for bordering, as the formative process of national identities, becomes possible.

The journey across the border in Øresund becomes in this way a *rite de passage* that signifies the start or the end of an experience abroad. "During the boat ride across Øresund an adventure starts" (student essay from Malmö). To maintain the sensation of the tangible border crossing and differences, bordering rites during the stay in Copenhagen have been developed which signifies that one is really on the other side. "It is the same with hot dogs and beer. Without eating a hot dog and drinking a beer, one has not really been in Denmark. One has to accustom oneself with the habits wherever one goes!" (Malmö). To eat hot dogs at a stand on the Town Hall Square in Copenhagen or to describe the Stockholm Exhibition are in this sense both ritualized practices that point to differences and strengthen identities. However, the hot dog can again and again form a recollected bordering experienced in the memorized space, while the exhibition can only be remembered.

---

21   That nostalgia today refers to a longing in the dimension of time is evident from material in the collections of the *Svenska Akademiens Ordboksredaktion* (the editorial collections of the Swedish Academy's dictionary). The older examples are so few that it is impossible from them to judge if the word earlier was used *only* to refer to space or if it also could be used as a reference to the dimension of time. It is here interesting to note that the meaning of a word like *experience*, which indeed is used a lot in current ethnological and anthropological texts, originally did not refer to age and learning but to travel and movement. Experience, thus, was something gained from movement rather than time (Leed 1991:5ff). It is of additional interest to note that while in sciences we have recently experienced a growing interest in geographical aspects and metaphors, history by some researches has been declared "ended" (Fukuyama 1992, Mørch 1996).

Personal Bordering

As suggested here, personal and collective bordering in particular situations merge and this merger encompasses a strong morally and emotionally charged *we*. It is likely that such situational applications of the unified I/we in most situations are used on private arenas and that they therefore often are inconsequential to larger socio-political processes. Furthermore, it is likely that these applications are most often designed to diminish potential conflicts between individuals who know each other face-to-face and who have a mutual interest in maintaining social bonds. This is why many individuals who express distain for a certain group of Others readily excuse individuals from this group because they know them personally: "I don't like 'X', but 'x' is hardworking/nice/a good family member/etc." or "when Swedes drink, they drink a lot but I have never seen Stefan drunk." That the application of bordering resituates power within private arenas, at least temporarily, is obvious. Being corrected for doing dishes in Danish, even if completely excused by the mutual recognition of natural difference, is disempowering. However, it is likely that in other situations, the correction will be at the other end of the bordering relation and that the Swede(s) involved will have to amend practices. A kind of correctional balance, thus, is likely to exist among people with extensive and symmetric social bonds.

In other situations, such mergers are situated in public arenas and have the potential of becoming important with wider societal consequences. These consequences are often of a benign character and happen for instance, when nations perform with success in sports arenas (Ehn 1989). Through being represented in the competition, all the "I"-s simultaneously experience that "we" are qualified as participants because of what we have been all the time – one in the national crowd. Recollections of this are swiftly appropriated into the national memory as something collective and shared. Through mass media, the strengths and weaknesses of our imagined community, as represented through the performance of our athletes, are discussed (Schoug 1992). In situations like these, where heroes and heroines are celebrated, the national collective is easily expandable. Everyone who decides to associate can be included through acting in synchrony with the *we*. Thus, it is not a contradiction that a Kenyan born athlete represents the Danish community at the Olympic games. This resembles the inclusive character of demonstrations of ethnic affiliation that Barbro Klein has discussed from her studies of celebrations among Swedes in the US (1988).

The unification of individuals with their nation can, however, also lead to exclusion. If a coincidence between bordering narrative and experience exists, or can be made likely to exist, the imagined community can be mobilized to exclude those who in narratives are defined as culturally different. This can potentially lead to legitimized violence (Drakulic 1994, Frykman and Povrzanovic Frykman

2000) and result in what Hulsey and Frost term "moral cruelty" (2004).[22] It is, thus, through the coincidence, or the postulated coincidence, between memory and recollection, that a national idea of space is filled with – sometimes incalculable – societal energy (Mead 1970:273ff). The invocation of the imagined community, what Benedict Anderson terms "official nationalism" (1991: 83ff), is thus a powerful political tool. That individuals in situations filled with contrasting images can hope for or fear being unified with their imagined community explains the attraction of nations, not least in populist politics. The readiness to die and kill for the "we-the-nation" is ultimately what confirms the coincidence of narrative and experience, the collective and individual, the memorized and recollected (A. Knudsen 1989:31).

Bordering is consequently employed in a multitude of arenas where decisions are made concerning who is or ought to be part of the community, how it is practiced, and with regard to inclusion and exclusion. This is why potentially, bordering is democratic and discriminatory, generous and malicious. Discussions in the public arena continuously shift between these integrated sides of the imagined communities and their borders. People disassociate themselves from the bad qualities and hope for the good (A. Knudsen 1989, Kristeva 1993). Bordering, thus, always has consequences, good and bad:

> Language, habits, culture are real exciting circumstances. Peoples' daily life, horizon of experiences, and norms are actually different. Micro-nationalism writes – as nationalism in general – political meaning into these realities. (A. Knudsen 1991:37)

Because of its societal energy-potentials, bordering is pursued in politics. Nothing indicates that we are ready to relinquish this tool – or that it will abandon us. As a complex process, bordering will still in the foreseeable future be available as a process that can be made topical under particular historical conjunctures. As argued above, the imagined *we* is not a deeper or more important identity; yet, through bordering it is a process available in specific sets of texts and contexts to question and heat up borders, as happened after the Fall of the Wall where it resulted in the breakup of the European states, the Soviet Union, Czechoslovakia, and Yugoslavia. As will be argued in the following chapter, similar heating attempts have been tried along the border in Øresund.

---

22 It is interesting that Hulsey and Frost identify the banality of everyday life as important to their argument (2004:21). However, an ethnographic perspective on everyday life is absent from their analysis.

Chapter 5
Bordering Paradoxes

Now the iron curtain is gone. Shall we then merely conclude that they now have become like us? That is logically impossible, unless we find out who we are.

We can probably all agree that one cannot say: We are not like those, who now have become similar to us.

(Vibeke Sperling 1990)

Opposing Danish and Swedish Metamorphosis

The conditions for the Danish-Swedish relation have changed so dramatically that I personally cannot and will not relate to any general Danish comprehension of Sweden and Swedes, which can be nothing but a relic from times prior to the age of information society. (Student essay 1992, Virum)

In 1990, Mogens Berendt's book *Tilfældet Sverige* (The Swedish Case) was reprinted. Consequently, the author appeared on a Danish national television program where he debated Swedish author and public intellectual Per Olov Enquist over Swedish and Danish. Neither the book nor the program caught much public attention. What happened between 1983 and 1990 to make the previously so well established and topical picture of modern, "total-democratic," Swedish society so relatively inconsequential to Danes?

Around 1930, the narrative on all that constituted Swedish ruptured and was recreated. The Stockholm Exhibition became the glowing symbol for this transformation and the new gaze on the future. The narrative on Danish, which was anchored in history and served as a framework for comprehending everyday life and identity, was not ruptured. The history of PH's film, however, illuminated how the narrative became rooted in a continuously regenerated and more modern *Gemeinschaft*. As has been discussed in previous chapters, change and notions of single events that symbolize rupture is common to narratives about nations. Around 1930, the bordering narratives between the two cultural communities evolved and illustrated that they were seriously out of step. Since the late 1980s, there have been many indications that a new reorganization of the bordering narratives about Danish and Swedish is in progress. In the 1990s, the relation between them was again transformed because of opposing developments in economy and politics that tended to result in narrations, which increasingly depicted the two countries similarly.

West of Øresund, a change in political dialogue took place. Through the 1970s (from the Danish referendum on the affiliation with the EEC and the oil-crises in 1972, through the start of Poul Schlüter's first term as Prime Minister starting 1982), a broad spectrum of political actors agreed on a narrative about Danish economy as something that teetered on the brink of the abyss; in no way did this mean any truce in politics, which continued to be turbulent. Among other things, more new parties were elected to parliament. However during the 1980s, Danish politics increasingly became more tranquil (Christiansen 1984). On the one hand, governments started to complete their terms without repeatedly calling new elections. On the other hand, governments had to take it lying down when a majority in parliament was against it or demanded particular measures contrary to the government's politics. For ten years (1982-1993), Denmark ended up having a right-wing government under the same Prime Minister, Poul Schlüter, and Secretary of State, Uffe Elleman Jensen. After a period with a government led by the Social Democrat Poul Nyrup Rasmussen, which had to find its majority across the political center, the right again came into power in 2001.

Parallel with this calmer political life, the narrative on Denmark's economic structures was fundamentally altered. For many years small and medium-size enterprises had been pointed out as the backbone of Danish economic life by Danish as well as by international reports (for instance, by the OECD and EEC). This was in accordance with the narrative's focus on family enterprises as the core of the economy: small was superb! This situation was changed from the mid-1980s. Against the background of the establishment of the Single European Market from 1992, new recommendations came in reports from the same organizations: enterprises were to merge, so they would hold enough economic potential to expand in the international market. Potentiality was now laid in big corporations with highly developed products – and their aim set firmly on the Single European Market. Suddenly, biggest was best! Protests against the concentration of money and power with a minority, which inevitably would have surfaced in the 1970s, were largely absent. The narrative about Danish economy was thus recreated, and businesses began to realize it. Enterprises in food industry, banking, and insurance merged, and other sectors followed in the footsteps of the Danish cooperative movement, which since the 1960s led the wave of rationalizing fusions. However with no recognition of this altered reality, the enterprises of the cooperative movements continued for some time to be recognized as a democratic element in Danish economy where the participating peasants presumably had equal influence without regard to the size of their enterprise.[1]

---

1   Danish landholdings also were concentrated as individuals bought more land which meant both that the single enterprise was often no longer run by a family and that the number of farms declined. From 1982 to 1995, in thirteen years, one out of every three farms disappeared, the number dropping from 98,000 to 66,000 (numbers from Ritzaus Bureau January 15, 1997).

Similarly through the 1980s, Danish farmers lost their privileged position in the national narrative. In 1972, attention to farming, peasants, and the agricultural export to England could still be mobilized as a significant factor in the Danish referendum on joining the EEC. At about the same time the budding debate on the environment focused on emissions from factories and in more cases, stimulated significant public sentiment; the case of Cheminova, a chemical plant in northern Jutland that leaked toxic waste into the environment, particularly caught the public's interest. However, from the 1980s, both environmental debates and politics targeted farming instead and their use of fertilizers and pesticides; increasingly, the focus became the development of oxygen deficiencies in the fjords and sea around the country, which were caused every summer by algae growth and fueled by runoffs from agricultural areas. Farmers were no longer a sacrosanct element in the cultural community. In the transformed narrative, Danes recognized that industry, for a majority of people and in economic terms, had for a long time been the country's most important commercial activity. Denmark no longer recognized itself as an agricultural community. Interestingly, Germany was again the most important market for Danish agricultural products as was the case before the Danish military defeat in 1864. A century long intermezzo, during which Denmark internationally situated itself mostly in relation to Britain, closed.

From the 1980s, the cultural narrative on Danishness ruptured. Politicians and professionals used mass media to continuously highlight improvements in the balance of trade with foreign countries and decreases in the budget deficit of the state; gradually the picture of a proud, urbanized, industrialized nation emerged. The earlier narrative on Danish privileged the historical process, *Gemeinschaft*, and a critical perspective on the modern project. This new narrative instead emphasized that the country had already recovered from, or at least had a good grasp on, any problems associated with modernity. Thus, a Danish cultural community that was urbanized, highly educated, environmentally conscious, and ready to meet the challenges of information society materialized. That this narrative rupture also was associated with a division of the people into those who have work and resources and those who do not, was rarely mentioned in public debates. Such warnings (Elmquist 1992:51) by others were seen as old-fashioned. In other words, they were in accordance with a discarded hegemonic national narrative. In 1992, the new cultural narrative was performed in cooperation by the museums of Copenhagen in an exhibition: *Dansk, danskere...* ("Danish, more Danish...").[2] The exhibition put on display international trade, industry, environment and global responsibility, new media, and city life as noticeably Danish. In educational materials for schools, there were discussions of food, shipping, tradition and religion, dwellings, nature, and professions – but nothing was mentioned about peasants and agriculture.

---

2   In connection with this exhibition in Nicolaj, an exhibition-hall in Copenhagen, the debate-book *Danskere* (1992), an exhibition-newsletter, educational materials for schools, etc. were produced.

A crucial dimension of the chaotic conditions of politics in Denmark in the 1970s was the collapse in the earlier equilibrium between countryside and town and between popular movements and the state. This provided space outside the established political parties for an anti-state movement as seen in *Fremskridtspartiet*, The Progressive Party (a populist, right wing party), which in periods received considerable backing in elections.[3] During the subsequent quarter century, a new balance within the sketched new national narrative has slowly been emerging, enabling Danes to continue regarding politics as an arena in which "given the choice between two contrary points of view perpetually [we] are able to prefer both" (Fonsmark 1990:317). This new balance is dependent on a fundamental new understanding of what constitutes political opposites. One of the signs of this is that the old liberal party, Venstre, has exchanged its earlier rural profile for an urbanized modern one. The old right, centered on peasants has largely vanished. As early as 1987, Venstre supported a measure in the Danish parliament to recreate the earlier, natural flow of a river, Skjern Å, in Jutland, going against local agricultural interests (Tolnov Clausen 2002). Paradoxically, this new understanding of balanced democratic Danishness is negotiated between two old poles: the Social Democrats and Venstre. However, the Venstre in question is a renewed, urbanized, and professionalized party.

Even though the cultural narrative was transformed through a rupture, the development was depicted as a continuation of something long present in Danish society. A new comprehension of the period 1930-1980 was disseminated. Opposed to the situation within the earlier cultural narrative, the Social Democrats were ascribed now a decisive role for the development of Denmark. It was – for better and worse – their responsibility.[4] As the renewed narrative was presented as a succession, a symbolic representation for the rupture was not necessary or desirable. The new cultural interpretation presented itself as an extension of the part of the narrative on Denmark in which transformations are depicted as continuations. The introduction of absolutism was depicted as a continuation of Christian IV's power and the weakness of the nobility, the thorough agricultural reforms around 1800 were depicted as a recovery for the free peasant of pre-modern times (from the stone-age to the Vikings), the first democratic constitution of 1849 was depicted as picking up where absolutism and the assembly of the ranks let go, and so forth. The Danish project was preserved as in the stage of the Faustian "lover." Still it was a renewed – and now urbanized – *Gemeinschaft* that was missed. Therefore it was no coincidence that from the 1980s, the people who in

---

3   The party has a tumultuous history. Parts of the party are still represented in the Danish parliament as *Dansk Folkeparti* (Danish People's Party).

4   As early as 1982, the Danish historian Søren Mørch published an alternative account of Denmark's history emphasizing modernization 1880-1960 (1982). He evaluated the role of the Social Democrats positively. Other debaters, characterizing the development after 1930, have however been critical of the Social Democrats and the Welfare State (Fonsmark 1990 and Silberbrandt 1993).

the 1970s moved to the countryside to live closer to "nature" started moving back, if they could afford it, to the old city-centers, which increasingly were regarded as very desirable habitats. However, the sketched discontinuity was not total. Some elements of the previous hegemonic narrative were maintained still, and much in the same way as Danishness was continuous through its bordering from a much changed comprehension of Swedishness after 1930. Democracy, understood as proximity between the voters and the elected, and Danishness, imagined as something homogeneous and different from the outside world, persisted and became for instance employed as arguments for Danish exemptions from parts of EU agreements and programs (Jenkins 2000). When new governments under Venstre leadership came into power from 2001, the focus on the Social Democrats continued as that party was often condemned in public discourse for presenting old fashion Welfare State solutions to current problems.

In Sweden, developments were different and to some extent opposite. As Goethe's Faust was standing on his balcony, contemplating the completion of developments and therefore ready to die, four spectral females in gray approached him: Need, Want, Guilt, and Care. What Faust had succeeded in banishing from the real world returned to haunt his mind. He managed to drive three of them away but the vaguest and deepest, Care, continued to haunt him (Berman 1988:70f). In the same way the perception of Sweden was troubled. Around 1970, when the admiration for the Swedish project culminated – also abroad – it signalled that this country had attained modernity and the Welfare State. But at the same time, fissures in the project and serious internal concerns started to appear.

In 1976, right wing parties came into power but without leaving any significant traces lost power in 1982. The 1980s were a time of economic and political anxieties in Sweden (Ruth 1984). A sixteen percent devaluation of the Swedish crown treated economic problems, and in 1982, the Social Democrats came back into power. In 1989, a large tax reform followed. At that time, the political crisis was already apparent (Bjurulf and Fryklund 1994, Edgerton, Fryklund, and Peterson 1994). New parties in the parliament – the Greens at the general election in 1988, the Christian Democrats and the right-wing, popular New Democracy at the elections in 1991 – had discontinued the Swedish tradition of functioning majorities behind governments even when the they held only a nominal minority. The election in 1991 marked "the end of the People's Home as a hegemonic imaginary in Swedish politics" (Fryklund 1994:147). As had been common in Denmark, the Swedish government had to work its way forward with changing majorities across the political center if an alliance with the new fringe parties was not maintained. The Swedish parliament was no longer the "Transport Company" (Huntford 1972:112) that automatically approved the proposals that had been tabled by the government on behalf of "the corporate state." The parliament was transformed into the space where political conflict and debate were performed. This became evident in association with negotiations of the tax reform in 1989. Far into the last night, the Social Democratic minority government and the Liberals (*Folkpartiet*) continued negotiating, discussing different percentages that each

Selvom den svenske nation er i krise,
byder Skånes natur stadig på en
rigdom af oplevelser.

Smut til vidunderlige Lund, kun en god
times rejse med vores dejlige overfart
fra det idylliske Dragør. Kursen er helt
lav, og tilbudene i Lund er fantastiske
- kulturelt, handelsmæssigt og kulinarisk -
og den omliggende natur er pragtfuld.

Få mere information om Lund!
Ring 32 53 15 85, eller få det
udleveret ved Drive-In, når du
kører om bord.

Notes: Towards the end of 1992, the Swedish crown fell drastically against other currencies. Previously, and for quite some time, the flow of consumers across the Øresund had nearly exclusively consisted of Swedes going to Denmark to shop. With the new exchange rate, it would be very beneficial for Danes to go buying a range of goods in Malmö, Landskrona or Helsingborg. However, it was not until spring 1995 that an advertising campaign in Danish media pointed to these possibilities. It might be that Sweden joining the EU in January 1995 psychologically built a bridge for consumers to go from Denmark to Sweden to take advantage of the weak Swedish currency. The ad campaign in May 1995, was well rooted in mutual understandings of the relation between Danish and Swedish – Skåne's history and Swedish nature. The utilization of these stable elements of national narratives, combined with an ironic distance to Sweden's earlier strong economy, illustrates that the relation between Danish and Swedish was uncertain at the time. The economic transformations in both countries had yet to be manifest as new, established narratives on the relation across the Øresund. Ads from the advertising agency Brandt and Falk highlighted the ferry connection between Dragør and Limhamn, printed in Danish dailies *Berlingske Tidende*, *Politiken* og *Søndagsaviserne* in May 1995.

After the opening of the Øresund Bridge in July 1, 2000, shopping on the other side of the border has become more commonplace, also for Danes. In spring 2009, when the exchange rates again heavily benefitted Danish shoppers in Sweden, the Øresund Bridge company ran an advertising campaign in Danish media (among other places on the web-edition of Danish daily *Politiken*), which pointed to how transgressing the border for some parts of the population had become routinized. The ad simply read, "Everything is cheaper in Sweden!"

Figure 5.1   Advertisement for Dragør-Limhamn, 1992

Source: Reprinted with permission from Full Circle Reklamebureau in Copenhagen. Photo courtesy of Björn Andersson, Historiska Media.

party expected to be recognized by their voters as what best served the totality. The economic means and goals employed in the Swedish Model turned out to not work any longer in a still more internationalized economy as large Swedish corporations already regarded themselves parts of the Single European Market. When it became clear in the summer of 1990 that a Swedish application for joining the EU was on its way, it appeared on many levels as a confirmation of a development already well under way.

For more than half a century, the modern was identified as the Swedish aim and the Social Democrats were the carriers of the state, in part because right wing parties could not create a plausible, alternative program to modernization and rationalization. The political and economic crises led to a savage showdown with "the developer." Prominent persons such as the author Jan Myrdal and the film-director Ingmar Bergman wrote memoirs that viciously penetrated the lives, aspirations, and the intellectual framework of their parents: the modern, masculine, imagined community (Behrendt 1987). Similarly, Jan Troell created films (*Sagolandet* and *Il Capitano*) that completely dismantled the imagination of a People's Home (Qvist 1995:140f). The tragedy of development, as an execution of power onto the individual in the modern project, was strongly accentuated in these accounts. The Swedish utopian idea was turned upside down and depicted as a dystopia. This turn seemingly was confirmed when in the fall of 1992, the Swedish crown fell out of the (West) European Monetary System. The result was "that Sweden is becoming a perfectly ordinary European country" (*Sverige i utländsk press* 1994:6).[5]

However, simultaneously, this showdown launched a new imaginary of a Faustian tragedy of development. A new metamorphosis started from "the dreamer," who seeks contact with people in everyday life and evidence of community, to "the lover," who craves the already destroyed. Attempts to create a new national cultural narrative were first and foremost historical. The former editor of the political section of the daily *Expressen* stated that there was "nothing bad today that cannot be cured by turning your attention to something in the good Swedish tradition of yesterday" (Lindroth 1994:114). What was seen as lacking in the Swedish project was reinstalled. Swedish loss of history – apparently after 1930 – was compensated through a renaissance for national history. One manifestation was an initiative that made 1993 "The Year of Swedish History." Everywhere in the country museum exhibitions were open at noon on March 21. Sten Rentzhog, the leader of the project, expressed in his speech, which was broadcast on television nation-wide during the opening ceremony in Stockholm:

---

5   The picture of the political and economic development was clear in the yearly summaries of foreign news media's coverage of Sweden (*Sverige i utländsk press* 1990, 1991, 1992, 1993, 1994). The country's ordinariness was stressed, "It is recognized both near and far how Sweden now has become an ordinary country, a country with problems to deal with" (1991:2).

In this magnificent weather we are celebrating a feast; we are to celebrate the
return of history, the return of history in our museums, in our society, in our
daily life. If it happens, as I hope, we are celebrating the return of history to our
hearts.

The organizers aimed at restoring the cultural community in a period when
borders in Europe were disintegrating and Sweden had become a country with
many immigrants (*Kulturrådet* 1992:10).[6] The showdown with the modern
and Social Democratic project was turned into a declaration of affection for
history. It was though, a celebration of a new interpretation of history. In the
public sphere, accounts appeared which rejected 1930 as the year of a crucial
rupture and that diminished the role of the Social Democrats. Instead debaters
depicted the development as an unbroken progression towards an even greater
project: Europe (see for instance P.T. Ohlsson 1993, Lindroth 1994).[7] Even if a
continuous evolution was accentuated, an era – the Swedish Model – had ended.
A symbol for this transformation, or more correctly, a symbol for the resumption
of something unfinished, was also established. On Friday evening, February 28,
1986, the Swedish Social Democratic Prime Minister Olof Palme was shot in the
streets of Stockholm on his way home from the cinema. This murder, as well as
the fact that no guilty person or organization was ever found and convicted, was
soon regarded as an indication that the development was continuing and as a
strong reminder that Sweden was not outside or above the rest of the world. The
modern (EUropean) welfare state was not secured yet but its Swedish modern
incorruptibility, as Europe's imagined virginity, was indeed lost forever (see for
instance Belt 1993, Berendt 1990:I-X, Enquist 1992:114-179). This loss was
repeated when popular Social Democratic Secretary of State, Anna Lindh, died
on September 11, 2003, after being stabbed the day before during a visit to a
Stockholm department store. The normality of Sweden was confirmed when an
alliance of right wing parties in the election in September 2006 narrowly defeated
the Social Democrats and their supporting parties.

---

6   The exhibitions were vehemently debated in media and among professionals both
before and after the opening. For an outline of the idea, see *Den Svenska Historien* (no year);
for a presentation of the result, see Lindqvist (1993); and on the debate, see for example
Kulturrådet (1992). Even if it was not the merit of the exhibitions, the interest for history
grew, as expressed through the number of Swedes visiting museums. An investigation
confirmed that from 1988-1995, the number of Swedes older than 15 years who had been
in a museum throughout the previous year grew from one out of every three to one of
two (the investigation: *Kulturbarometern i detalj*, mentioned in *Sydsvenska Dagbladet* July
21, 1995). Another similar sign of this turn to history and self is the elevation first of the
Swedish Flag Day to a "national day" in 1983 and then, the declaration of the national day
as a national holiday from 2005.
7   Also abroad, observers appeared who adopted the understanding that the Swedish
project had no ruptures (see for instance the Dane T. Knudsen 1994).

From the late-1980s, the cultural narratives about that which is considered Danish and Swedish are converging as they each, as imaginaries, incorporated prospects of globalization (Bauman 1998, Castells 1996).[8] Both national projects have the character of the Faustian "lover." In the Danish version the role of the Social Democrats from 1930, which was earlier only depicted as participation, was first emphasized then declared obsolete. In the Swedish case, the role of Social Democrats was diminished from having been regarded as the sole agent for development. Contrary to the 1970s, Danish politics were narrated as relatively settled and the national economy was improving as the result of growing industrial exports. In the Swedish narrative the stability of politics was ruptured and the economy was in grave trouble, not least due to growing public deficits and an open unemployment of unprecedented proportions in the modern Swedish context (though still below ten percent).[9]

The reprint of Mogens Berendt's book on the Swedish "total democrats" and their dystopic, modern project was therefore nearly neglected in 1990. It was out of step with the new era. When the book was written in 1983, it relied on a conception of bordering between Danish and Swedish which had existed for half a century but that was already in transition. The transformations and converging developments had infused uncertainty into Danish-Swedish cultural bordering processes. The earlier central political and economic aspects were seriously shattered and processes of identity formation could not be energized from relations between partners who increasingly imagined themselves alike. Other elements of the relational narratives still prevailed; Sweden continued to be described as natural and introverted people, Denmark remained a southern relaxed vacation land.

It is understandable from this development that Denmark, which had been through economic and political crises, from the early 1990s again enjoyed growing attention in Sweden as was the case around 1900. Danish solutions and experiences within the EEC/EU were of interest to Sweden, which increasingly turned its gaze towards Western Europe. Correspondingly, Danish interests in Sweden were stagnating. Danish news media continued far into the 1980s to print optimistic articles on Sweden as the country of good economy and low unemployment but then alarming reports appeared. Later they were turned into commentaries about how a modern model fell apart. Attention in Denmark was directed more towards what happened in the south and southwest than towards what was occurring in the northeast; this was reflected when Danish newspapers and Danish national radio discontinued having their own correspondents in Stockholm.[10]

---

8   See Bernd Henningsen (2005) for an interesting overview of the processes discussed here.

9   In 2003, statistics from OECD indicated that Denmark was again wealthier that Sweden, as was the case a century earlier (sydsvenskan.se, May 15, 2003).

10   Since 2000, Swedish newspapers and radio have more permanent correspondents in Copenhagen; and journalists from southern Sweden intensively cover events in Denmark.

From around 1990, converging narratives eroded the central issues that had fuelled cultural bordering processes between Denmark and Sweden. The similarities between political and economic imaginings cancelled the use of the "Øresund Other" in politics; simply put, no reflective potential could come from sameness. However, this does not imply that the two cultural communities began to appear as identical. Rather, Øresund bordering soon came under renewed pressure and of course, future ruptures with new directions in the developments are likely to make themselves available as signifiers of regenerated national difference and Identity. Such a developing rupture focused immigration policies where in public discourse, Danish and Swedish increasingly came to measure themselves against the "Øresund Other" (Hedetoft, Petersson and Sturfelt 2006). This is in particularly true after 2000 as immigration policies in Denmark became more and more restrictive. Thus, as old political potentials related with different feelings towards Modernity waned, new political possibilities were invented and imagineered across the border.

European Confusions

In Europe, confusion held a prominent place in the political developments following 1989 (Tunander 1995, Tägil 1992, 2001).[11] As pointed out above, general depictions of the integration of states in the West (EU) and the disintegration of states in the former East (Soviet Union and Warsaw Pact countries) paints far too simple of picture. The fall of the Wall was turned into a symbol for discrete eras as earlier ostensibly stable supranational political communities in the East and the West disappeared with this geographical event. Afterwards, the Western EUropean community sought to rematerialize itself in a new form through extensions of the EU, the development of an EU constitution, and so forth (see Shore 2000). Of course, Denmark and Sweden were not unaffected by these occurrences. For the Swedish, and to an even higher degree the Finnish, admission to the EU is both a consequence of the transformations and at the same time a cause for European confusion and additional transformation.

Hobsbawm (1990) explained the growing interest in nations as an indication of the end of the era of imagined communities. Partly because national economies in the future cannot be self-sufficient (1990:173ff), he forecast an era in which nations as political, economic, cultural, or linguistic communities could no longer affect the course of history (1990:181ff). His book was written

---

Danish media cannot in the same way cover what happens in Stockholm from home. Without permanent representation, Danish media often rely on articles bought from international news agencies and free-lance reporters.

11   For an overview of how to interpret economic and political developments in Denmark, Sweden and the other Nordic countries in the first half of the 1990s in relation to European issues, see Ingebritsen (1998).

in a confused time and it is likely that Hobsbawm today would have expressed himself more cautiously on the possible future of nations. This is not least true for his ironic comments on the instability of nationalist sentiments in European states (1990:176f). Hobsbawm, who perceives nations as a phenomenon of conjunctures, thus diminished their future significance. Smith, who believed in nations as something endless and decisive for the identification of individuals, on the contrary accentuated the influence they will have on the course of history in times to come (1991:143-177). Gellner pragmatically expected that nations in the future would play a role albeit possibly a changed and less spiteful one (1983:110-122).[12] Obviously, no agreement exists on the reality for or the possible future effects of nations. The converging narratives about the Danish and the Swedish can be read as expressions of transformations of our complete conceptions of coherence, community, and belonging. A decisive question is if there are in the present confusions any signs that the imagined communities will disappear as foundations for recognition and acknowledgment of differences. Will completely new frameworks for experiences and thoughts replace the modern maze? Or, are the confusions expressions of the reorganization of spaces and relations within the maze? Within the developments around and after 2000, are there any indications of an epistemological reconfiguration in which a new episteme – maybe post-modern, but at the very least non-modern – will format our lives in new ways?

Since the 1990s, cultural tendency (the concept of identity) dominates social tendency (the concept of culture) in discussions within the cultural and social sciences as well as debates in the public sphere. This transformation denotes that certain imagined natural criteria for bordering are accentuated and marked geographically. Other criteria for the delimitation of groups of peoples from the repertoire of the social sciences – for instance, class or age group – are regarded as less significant than they were previously. Scholars are writing less about class today than they did in the 1970s; this is an expression of a decline in the idea of the spatiality of classes, which only conceals that the existence of classes might be a central element in our present and future. Future classes however, might very well not be defined through their relation to the means of production, but could for instance be associated with different patterns of mobility (Bauman 1998). Ulf Hannerz suggests distinguishing between groups of locals and globals though both are employees (1993), and he recognizes the different contexts of the travels of cosmopolitans and locals (1996:102-111). Distinguishing classes from one another in the future might therefore be a problem of recognizing the differences among the culturally competent travellers (cosmopolitans), the tourists, the exiled, and the immovable.

---

12   In a series of essays, written between 1987 and 1994 on studies of nations and nationalism, Gellner maintained his understanding of the historical development and the open-ended future of nations. He also comments on the changed political condition of the phenomenon in the post-Soviet era (1994).

There has been the tendency since 2000, with regard to the concept of identity, to stress that nation states are more comprehendible as cultural communities than as political ones. Or more in agreement with reality: the cultural community, which since the Second World War has been concealed within a political conception of nations, has through these developments appeared before the public and legitimized itself across political landscapes. The cultural community becomes a point around which conflicts evolve over where boundaries ought to be drawn. In the Nordic countries it is at the center of a cultural/national-conservative critique of the development within the EU, also from the left side of the political spectrum. And, it is at the center for the perception, within a broad political specter, that previously culturally homogeneous communities are now going multicultural because of the presence of immigrant groups. The part seen as the "multicultural Other" is most often impersonated by Muslim immigrants; this was also apparent in the newest historical novels about the wars between Denmark and Sweden.[13] One reason some people disassociate themselves from immigrants is an opposition to the condition of the centralized Welfare State (Bjurulf and Fryklund 1994, Edgerton, Fryklund, and Peterson 1994). As Steve Sampson has argued, this point of view appears in a broad political spectrum outside the far right from the early 1990s onwards and illustrates reflective confusions over the ongoing reformulation of hegemonic cultural narratives about the nation (1993).

Bordering has continuously been a part of conjunctures in which it has been used more or less in itself or in combination with other ideas of spatiality in processes of identity formation. The peace ending the First World War can be described as the triumph of the imagined community as it divided the multi-cultural empires (Austria-Hungary, Russia, and the Ottoman Empire) into nation-states. As Uffe Østergård has argued, both cultural and political bordering were employed during the drawing of these new borders in order to provide the victorious powers with the biggest possible territorial gains (1995:176ff). It is this map of Europe that is mutating after the Fall of the Wall. In these confusions, changes will occur in the close relationship between the nation and the territorial principle that advanced and used it as a vehicle – the state. Political structures in the future can possibly be tied less to certain borders as demarcations of states (Tunander 1995). But even if (parts of) Europe in the future can be analyzed as networks of urban centers, Mafias, and enterprises, this does not mean necessarily that bordering towards others no longer takes place. Being inside or outside the demarcations of the networks potentially will play decisive roles for identity formation processes. The establishment of new border-transcending and border-creating networks, thus emphasizes the complexity of today's very modern world.

In Scandinavia, both larger and smaller alternatives offered themselves to the present states as possible spaces for consolidation through bordering. These

---

13   There are plenty of examples: reactions to the murder of Dutch film-maker Theo van Gogh in 2004, riots in France in the summer of 2005, and international reactions to cartoons depicting the Prophet Mohammed first published in Denmark in 2005.

alternative ideas of spatiality followed in the footsteps of the nation through the modern maze. Like states since the nineteenth century, they supported themselves by employing both political and cultural bordering.

On March 31, 1993, a group of experts presented its report on how the EEC (the organization and the European Commission, but especially its European message) could be made more visible and popular in the member states. The European flag ought to be used at military operations, at sport events and championships, and during the handing out of European meritorious medals. European political parties, a European library and museum ought to be created, and stamps (with values in the European artificial currency ECU) should be printed. All citizens ought to have a European birth certificate, and school books ought to emphasize the European dimension especially in history, geography, and social studies. While maintaining respect for cultural diversity, the European Union had to signify progress and prosperity, security and peace:

> The construction of Europe started in 1957 [...] We are striving for Europe. We strive, as the founders of Europe did, to ensure peace between all Europeans; to strengthen the prosperity and social progress of the nations of Europe; to maintain and promote the cultural dimension of Europe's heritage, and its historical contribution to human civilization and welfare; to stimulate solidarity, cohesion and greater understanding between Europeans without damaging our cultural wealth and diversity; to open new perspectives for young Europeans, and develop new hopes and confidence in people's creative abilities to build a better future together. (De Clercq et al. 1993:Foreword)

Developed by a group of experts who already provided the community with a legitimizing history, EUrope was promoted as a larger edition of the political community in this Faustian dream. However, this sketch for a program was not meet with enthusiasm. Politicians and officials distanced themselves from the proposed initiatives that were described as April Fool's jokes, one day premature. In Denmark, where a referendum on the Treaty of Maastricht and Agreement of Edinburgh was scheduled a few weeks later, EU-favorable parties and individuals immediately disassociated themselves from the airy plans that would penetrate the borders of the imagined community. Obviously there was no space in the debate leading up to the referendum for a bigger alternative to the existing national project.[14]

---

14   A Danish referendum in 1992 refused the Treaty of Maastricht. Subsequently, a so-called National Compromise was negotiated between Danish political parties and later with the EU's other member states. This led to the so-called Agreement of Edinburgh that limits Denmark's participation in some areas of EU cooperation: the Euro, EU citizenship, defense and some legal matters. This agreement was approved by a Danish 1993 referendum. That the proposal appeared at an inconvenient time in Denmark is clear from the letter mailed with the report by the Office of the European Commission in Copenhagen.

The sketch however, can be seen as an expression of a frustration, accumulated within the institutions of EU, over the nation (member) states incessant rejections of surrendering authority to the alternative political community that recognized itself as the sanctuary for progress.[15] In Faustian terms, the EU as "the developer" was restrained by the nation-states as "lovers" of their own authority and their "national cultures." The frustration of the EU has taken many forms over the years and was for instance also highlighted when French and Dutch voters in the June 2005 rejected the proposed so-called EU constitution. The fact that the rejection was very much discussed in public as a consequence of national politics is both an ironic comment to and illustration of this point. Crossing and crushing existing bordering processes has turned out to be difficult.

As the larger alternative pursued legitimized authority across Europe, it entered into a paradoxical alliance with smaller alternatives to the existing states that similarly tried to relocate decision-making and power away from nation-state institutions and networks. While the larger alternative, in general, tried to trot the path of a political community, the smaller alternatives, often called "regions," in their pursuit of influence across Europe, frequently dressed themselves (also) as cultural communities. The EU directly called for these "regions" to become the lowest level on which EU-policies were to be executed – the so-called "principle of subsidarity." However, an operational definition of a region was never developed (Linde-Laursen 2001). Everyone therefore could suggest a definition that suited her or his aims the best. This confusion is illustrated during the period when Sweden was considering joining the EU; a review was conducted which suggested that Sweden consisted of between eight and twelve regions (*Regionala roller. En perspektivstudie* 1992). At the same time a Swedish historian argued (in an oft quoted article) that the whole Baltic area constituted only one European region (Gerner 1990). However all over EUrope, regions appeared as they were promoted as subnational entities. By calling attention to themselves, regions posed questions to borders. As a geographical term, regions demanded that lines around them be acknowledged as divides and thusly, they called for the recognition of new borders. There also soon developed a hierarchy of regions; while the subnational was good, regions that crossed recognized national borders were best as they directly questioned the validity of existing borders and thus, the political legitimacy of the existing nation-states. In general, regions institutionally have not become well

---

The Office directly disassociates itself form the suggestions of the report: "But one thing is evident. It is information and not advertising or propaganda we need. Therefore, the *suggestions* of the De Clercq-group cannot be the ground for any future suggestions from the Commission!" (letter by assistant secretary Jens Nymand Christensen).

15   This frustration is apparent in the introduction to the report where the authors on behalf of the organization complain that national governments take the credit for successes while blaming "Brussels" for all failures (De Clercq et al. 1993:1). For a discussion of the development of EU institutional culture, see Shore (2000), Wilson and Smith (1993) and Zabusky (2000).

developed (Le Galès and Lequesne 1998). Such weaknesses however should not lead to the conclusion that regions and regionalization are bound to fail (Amin and Thrift 1994).

At Øresund, regionalization has taken different forms. As mentioned above, the Swedish referendum about joining the EU in 1994 materialized as an erosion of the imagined community. The vote promoted an understanding of separateness and difference from the rest of the nation in Skåne and in parts of the north. In the north, parts of the inland tried to establish narratives about "The Republic of Jämtland" as a political tool for promoting special awareness of its situation in the double (both national and European) periphery (K. Hansen 1998, 1999). In Skåne, an understanding of the landscape's situation in the national periphery, far from the national center in Stockholm, combined with its closeness to the rest of EUrope or the center (in fact "the Continent") became a vehicle for regionalist sentiments. Had such dissent surfaced earlier, it is likely that the Swedish central authorities would have opposed them strongly. Thus, modern *Gesellschaft* managed in the 1980s, with relative ease, to marginalize the Skåne Party, a local regionalist populist party in Malmö (Peterson et al. 1988). However, with the center of power in Stockholm weakened by the above- discussed economic and political turbulence, "dissidents" had the opportunity to argue openly and often uncontested for their cause. While the Swedish central authorities after 2000 again are more secure in their role, opposition to regionalist tendencies became clear. Sven-Erik Österberg, secretary for local governments in the Swedish national government, for instance openly declared that he regarded regions as a potential threat to a strong state (sydsvenskan.se, December 4, 2004).

Two different movements that marketed themselves as proponents of cultural bordering and Skånianess should be mentioned here. From the early 1990s, the organization that most distinctly fought for the recognition of Skåne as a region with some sort of exceptional position within Sweden was the Institute for the Future of Skåne (SSF – *Stiftelsen Skånsk Framtid*; http://www.scania.org/).[16]

---

16 Architect Peter Broberg was a person who for long tried to promote a Skånian, regionalist perspective. Even if his points of view seem unaltered, the tone of his argumentations became harsher over the years (Broberg and Spangenberg Schmidt 1970, Broberg and Kadefors 1981, Broberg 1994). Rather than fundamentalism, this must be seen as a process of fundamentalization. It seems identical with the process of rejection that led to a referendum in a small town in southern Sweden where people, in 1988, voted not to accept refugees in their community (Alsmark 1992). The Institution for the Future of Skåne was established in 1989. In 1990 it was complemented by an association open to members: *Föreningen Skånsk Framtid*. From 1992, the organization was represented in Denmark, and from 1995, it also was represented in Göteborg through local groups of supporters. Later a split between the Institution and the Association happened. In 1993, the Institution for the Future of Skåne applied and was accepted as a representative for Skåne in UNPO (Unrepresented Nations and Peoples Organization), which has observer status with the UN (L.G. Gustafsson 1993, LUF M:21247, and *Sydsvenska Dagbladet* January 17, January 22, January 29, and February 10, 1993).

Defining Skåne as a cultural community, the Institution for the Future of Skåne produced publications and materials for media in an attempt to correct what it regarded as erroneous information about Skåne disseminated by the Swedish state. The organization promoted Skåne as a cultural community with "its own history, its own historical language and an existing specific tone, a special folklore and building tradition as well as a well developed popular mythology" (Broberg 1991:201). According to the Institution for the Future of Skåne, "it is not possible for eternity to live with roots cut off" (SSF 1991:13); the organization argued from the position of the Faustian lover that the Skånians were deprived of their real, historical identity. Through its arguments for a lesser alternative to Sweden as the natural cultural community, SSF made the same projection of a territory onto history as they accused others/Sweden of doing. Thus, they understood suffering in conjunction with the moving of the border in the 1600s as a repression of the region and they ignored – as the writers of the history of the nation – that at that time society was organized due to other state patriotic principles than the states of the nineteenth and twentieth centuries (Broberg 1994, Röndahl 1993). The Institution for the Future of Skåne argued that when people were informed about the truths of the past, they too would become regionalists:

> It is, though, not unthinkable that informed people have a sense of belonging towards their surrounding country, their forefathers, and their own history. In fact, *it is their cultural obligation* to take care of and improve this heritage. (Broberg 1994:13; my italics)

This emphasis on culture and duty is similar to the opinion expressed by Claus Eskildsen in 1936 of: "*The blessing of harmony between nature and disposition.*"

The Institute as a small, closed network of individuals promoted itself as a cultural institution with obvious political-cultural undertones. However, from the late 1990s, administrative changes swept regional cultural bordering arguments onto the central stage of politics in Skåne (Linde-Laursen 2001). As one answer to the integration of Sweden into the EU and in an attempt to address some of the above discussed uncertainties of the political and economic developments, the Swedish central state initiated experiments with local and regional government structures. January 1, 1997, the two counties in Skåne merged (G. Larsson 1997). For individuals in the Institute as well as for many others, this was publicly interpreted as a resurrection of the "historical region Skåne." Politicians from across the political specter strongly supported and promoted the development of Skåne as a "region" with more – and ever more – independence from the national center in Stockholm; the choice by the Swedish state to call the created unit a "region," instead of "län," the ordinarily used term for this political-administrative level, facilitated this development. In 1998, a newsletter, informing the population in Skåne about the conditions of the region, was distributed to all households. The image of a young, blonde person with the regional flag painted on his or her face decorated the cover of the first issue.[17] In addition, on page three of the same issue, the regional flag was used as the backdrop for a series of statements emphasizing the cultural uniqueness and separateness of Skåne from the rest of Sweden. Under the heading: "The Skånish for the people of Skåne" this *official* pamphlet emphasized that:

"Maybe one can say that Skåne at last has become, or been allowed to become, what it has always been: a region" or,

"Historically Skåne is a Danish landscape that became a part of Sweden. But never became truly Swedish. There is a Skånish identity" or,

"Skåne does not look like other Swedish landscapes. Somewhat more flat, somewhat more fertile, somewhat more beautiful if you want."[18]

From the mid-1990s, both Europe and Skåne, as two bordering Faustian alternatives offered themselves as foundations for processes of identity formation in a time when the nation as an idea of spatiality (as always) was under pressure. As earlier there were alternative paths towards the dynastic states and their successors – the nation-states – spatial alternatives existed in the 1990s, prepared to compete. These alternatives employed history, respectively the future (EU) and

---

17   *Region Skåne* (Tidning utgiven av Regionförbundet Skåne, Skånelandstingen och Malmö Stad. Nr. 1/1998). The Skånian flag is red with a yellow cross. It was constructed as a "landscape-flag" perhaps as early as around 1870, but it was not used much prior to the late 1960s. In the 1980s the flag was closely associated with a local political organization in Malmö, the Skåne Party, and the flag became a signifier for xenophobia and prejudice. No other group therefore flew the flag and it more or less disappeared from public spaces as the Skåne Party increasingly vanished from public politics (Peterson et al. 1988). However, with the creation of Region Skåne in 1997, the same flag gained new meanings and appeared everywhere. As a signifier for efforts to emphasize the particularity of Skåne, the flag from the late 1990s flew at public buildings, on town squares, in front of office buildings, and so forth throughout the landscape. From being a signifier for a small group of right-wing populists, the flag became the central symbol for an increasingly regionalist mainstream in Skåne.

18   A regional political organization, *Skånes Väl* (The Wellbeing of Skåne), an umbrella for already existing local populist right wing parties, among them The Skåne Party, was also created. In the first election to the regional representatives, often talked about as the regional "parliament," on September 20, 1998, the Wellbeing of Skåne gained 4.1 percent of the votes on a political program clearly built on cultural bordering and xenophobia. In the party's program for the election (distributed in Lund and probably throughout Skåne) it among other things stated:"Stop the immigrant 'refugees' – only convention and quota refugees are admitted. Others are stopped and sent back. Those already here are offered help to return – and if they stay here, they must assimilate to our culture. The state [as opposed to the regional authorities] pays the expenses."

the past (the region/Skåne), to promote themselves as the appropriate framework for the development and execution of power. History had been deemed finished as a consequence of both an accelerating transience in the cultural processes of societies and the stretching of liberal democracy to spaces where other ideologies had prevailed earlier (Baudrillard 1985, 1994, Fukuyama 1992). Yet history seemed to be alive and well when the millennium approached. In the former Yugoslavia for instance, people demonstrated that the fear of a merger of the remembered with the recollected, and of the collective with the personal, prepared them to get killed or to kill others for the sake of a specific space legitimized by time – and tragically with the destruction of the social and economic structures of the same geographical space as consequences (Østergård 1995:157ff).

Bridging Bordering

When the border between Denmark and Sweden was moved to the Øresund in 1658-1660, the international powers' interest in this particular drawing of the line lay in securing their lands' trade in and out of the Baltic. The powers in Western Europe wanted to guard their access to the provisions coming from the area, and the states in the Baltic sought safe passage to the international market, which at that time had its pivotal center in the Netherlands. Thus, the concern of the era was with the movement *through* Øresund. This external political concern with the flow through Øresund remained prominent for centuries and influenced the actions of the Danish and Swedish states. This was the case, for instance, when they abandoned war as a means to move the border, as discussed above, or when the passages between the Baltic and the North Sea during the First World War were mined.

During the Cold War, from the 1950s to the 1980s, only one border in Europe was really hot – the great divide between East and West. It was as if all thermal energy was gathered along one huge ditch. In this situation, international complications were associated with any discussion about constructing (built) connections across Øresund. While technical abilities to build a bridge had existed for a long time it is doubtful whether the Soviet Union would have accepted a fixed link across the international waterway. The Soviet Union could reason that a bridge would enable NATO to block this connection between the Soviet naval bases in the Baltic and the high seas if the hot border flared. The political logic internal to the nation state and the "map as logo" furthermore prescribed that a bridge across Øresund would have to wait until the different parts of the Danish imagined community were connected. It would be unimaginable that the Danish capital should be connected with the periphery of another country before infrastructural connections effectively tied together all major parts of Denmark. Obviously, both internal and external logics deferred building an Øresund bridge to a post-Soviet era. Elaborate plans for a connection across Øresund had existed for more than a century when in 1991, the Danish and Swedish governments and parliaments decided to build a bridge from Malmö to Copenhagen (Idvall 1997, 2000, Tangkjær and Linde-Laursen 2004, Wieslander 1997). Work on a bridge across the Great Belt (*Storebælt*) separating Zealand (where Copenhagen is situated) from Funen and Jutland had already commenced.

After 1989, political and economic developments in Denmark and Sweden, as well as in the two countries' European context, challenged Øresund as a cold, natural ditch between Danes and Swedes. Compared to the preceding decades, Denmark experienced relative political and economic stability. Coupled with the new narrative about Denmark as an urbanized, industrialized, and capitalist nation with a firm focus on development within the EU, this perception opened a discursive space for a bridge across Øresund as an instrument for expanding the market for Danish products and services eastwards.

Simultaneously with this development, there existed in political, economic, and scientific circles another understanding of the geographical organization of the world that not only tried to cool down but also to freeze away borders. It was forecast that the idea of spatiality of the nation would diminish as a framework for international political and economic processes. Instead of nation-states, it was argued that networks of metropolises would appear to lead developments (see Castells 1996, MacLeod 2001, Tangkjær and Linde-Laursen 2004).[19] Promoting border transgressions, advanced technology, and of course metropolitan lifestyles – such thoughts suggested the freezing of both old and possible new borders. This metropolitanism was immanent in much of the thinking in EUrope since the introduction of the soon famous "Blue Banana," the industrial/urban core that runs southeast from the Benelux countries towards Catalonia. Cross-border networks were thought to undermine the homogenizing effects that the bureaucracies of the nation-states had within each political-geographical unit, and at the same time erode differences, or "barriers"[20] as they were most described, between imagined communities. From such perspectives, an appropriate infrastructure connecting metropolises was important for the circulation of goods, persons, and information. Across Europe and supported by the EU, ideas of linking emerged which challenged and changed perceptions of relations and identities. Paris and London were connected by the Channel Tunnel (Darian-Smith 1999), and with the fading of the East-West division, Øresund and the Danish Great Belt were made accessible to constructions that connected spaces that were formerly perceived as disassociated. A group called the European Round Table identified

---

19 There also existed a discussion of such processes as a type of return to conditions in Medieval Europe (Ambjörnsson and Sörlin 1992).

20 "Barrier," it is interesting to note, was another geographical term that was used in Scandinavia to conceptualize differences in legislation, bureaucratic structures, and "culture." It was used much in public debate and also became the popular name of several reports written on the development around Øresund after the decision to build a bridge. This was, for instance, the case with a report written by scholars at Lund University for the Swedish government (*Integration och utveckling i Öresundsregionen* 1999).

"Missing Links" in many places, invoking the future of a new, connected and continuous continent from Sicily to the North Cape. Volvo's executive director Pehr G. Gyllenhammar led the group in Scandinavia, and suggested in the mid-1980s investments in infrastructure including bridges across Great Belt, Øresund, and Fehmarn Belt (Tangkjær 2000a). This group established an agenda that soon became noticeable in Scandinavian traffic policy. More importantly, this group and its project signified that the time had passed when passage through the Øresund was the most important. Now, it was the speedy transport *across* the water that was crucial. Even if the group's ideas did not materialize at the pace it suggested, its efforts decisively contributed to reorganizing the understanding of the Øresund border. It legitimized thinking of the ditch no longer as a shield against "others" but as an obstacle to "connectedness."[21]

Thus, from the maze of integrative and disintegrative forces in Europe and the Scandinavian nation states, the establishing of both supranational and subnational ideas of spatiality, and the heating and cooling of European borders, a discourse of an Øresund Region emerged. The bridge as a physical structure in itself had limited effect upon social relations and narratives around Øresund. The prediction was that the bridge would shorten travel time between for instance Lund and downtown Copenhagen by around seven minutes. However, the bridge became a central metaphor for developments that emphasized the above-mentioned confusing directions of ongoing changes in perceptions of self and other. The professional, urban, and post-industrial character of Malmö and Copenhagen repeatedly were stressed locally as well as nationally. Around Øresund, this "region" was imagineered as the spearhead of development in both countries. People who regarded this new cross-border region as an economic unit and argued for a globalized future channelled through metropolitan networks were often enmeshed arguing within the idea of spatiality of imagined communities (see for instance *"The Birth of a Region"* 1997 and *Øresund – en region bliver til* 1999). For example, the trade organizations in Denmark and southern Sweden set up a joint body in 1999, the purpose of which was to promote integration and help erase barriers. The chairman for the Øresund Industry and Trade Organization, Svend Holst-Nielsen, who was born in Denmark but lived in Sweden, explained that "naturally the more general idea [is] to create a regional identity" (*Sydsvenska Dagbladet*, November 1, 1999). From the beginning, there was an imbalance between the two sides' situations within their respective national contexts. It was often assumed in Copenhagen that the region would spur a functional integration where the movement of persons, commodities, and information would be facilitated. In Malmö and Skåne, not least politicians believed that the cross-

---

21 It can for instance be argued that efforts by environmental movements to stop the Øresund Bridge were significantly hampered as "green arguments" in public debates became identified with the flow (of salty water from the North Sea into the Baltic) *through* Øresund. Thus, "the Greens" promoted a text and directional movement that was out of step with the contemporary context.

border region would propel forward their bordering attempts against Stockholm and that their cultural particularity would be recognized. Thus, the formation of such a regional identity demanded that borders be created which demarcated the region from both the rest of Denmark and Sweden. While in Skåne such a division was regarded as beneficiary, from Copenhagen's perspective there was no reason to erect barriers that separated the capital from parts of Denmark for which Copenhagen historically and functionally was already the center. For Copenhagen to disassociate itself from Funen and Jutland and to embark on an uncertain endeavor to form a "region" together with Skåne did not make much sense.

However, many different understandings of such a new Øresund Region were circulating generally as many agents established new bordering practices and performed "*Ørespeak*" (Berg, Linde-Laursen and Löfgren 2000; see also Falkheimer 2004) in many different arenas in what Christian Tangkjær has described as an "open house" strategy (2000a, 2000b). From the beginning no criteria for what defined this region, its economy, history, functionality as integrated infrastructure, as well as its expanse were set out; "everyone" was invited to participate in discussions about what the region could/should be. In this process, individual actors often drew upon antagonistic understandings as they set out to promote this new region as one metropolitan center in the emerging European network.[22] One group that deliberately set out to propel this region forward stands out among all the performers of Ørespeak. The group can be regarded as a loose network of mostly middle-aged male, civil servants working in different capacities in public relations. With the support of so-called *Interreg* funding from the EU, the group developed a strategic branding process for Øresund.[23]

---

22 Regional programs and projects targeted a great variety of fields such as writing new history books, providing Skåne with its own literary history (Vinge 1996-1997), linking schools across the border, creating an umbrella organization for the universities in the area, making jobs on the other side more accessible, integrating and possibly homogenizing policies on social services, distributing regional statistics including a "Gross Regional Product" (*Øresund. The Region in Figures* 1999), or encouraging private enterprises to create cross-water and cross-border structures for themselves (see *Our New Region* 1999). In many different contexts and on many arenas, bordering as rhetoric production reached the population around the Øresund and was also performed by local, emerging regional, and national media.

23 I have since 2000 worked closely with Christian Tangkjær, then at Copenhagen Business School, in cooperative efforts to study this network. I thank Christian for many inspiring discussions over the years on region building, branding, and the future of network society. Branding as a strategic tool in making new places is an interesting and in the early 2000s, a much used strategic tool; see Buhl Pedersen, Tangkjær, and Linde-Laursen (2003), Olins (1999, 2002) and Tangkjær and Linde-Laursen (2004). An interesting aspect of this network is that it consisted of civil servants and elected politicians of each state. Thus, a serious attempt to regionalize across borders was initiated from within the states, even if the predictable result would be, at least to some extent, the erosion of the states. As none of the participants held positions where they would benefit economically from such

The informal network started its work in early 1997 under the auspices of the Øresund Committee, a cross-border public agency with regional and national political representatives. In October the same year, visions and strategies of the process called "The Birth of a Region" were publicized in a document with the same name. "The Birth of a Region" (1997) in many ways was an important materialization of the regional idea of spatiality. It conjured a substantial vision for the region that was shaped as a strategic process, and was developed in cooperation with Wolff Olins in London, a recognized international consultancy company specializing in marketing and branding of geographical places. The document identified critical decisions to be made as well as marketing as central to the strategy. Furthermore, it cemented the metaphor "Birth" as central to the understandings and communications of the regional vision. Mass media soon recognized the branding efforts to be of some interest. One contributing factor probably was the estimated costs of bringing the region to global audiences; Wolff Olins calculated the cost at almost 90 million USD. The participants in the network, most of them working in public organizations, of course realized the problem of raising such an amount of funds.

Still, after the strategy was published, a strategic branding process was initiated as "the Birth of a Region" continued and was transformed. The network of organizations supporting the project was expanded and a formal 16-member steering committee, which included observers from the Danish and Swedish national governments, was established.[24] The project committee secured additional financing for "the Birth of a Region" from EU's *Interreg II* program, which provided co-financing for transnational region building. The branding project started with a budget of 500,000 USD. With those funds, proposals for the development of the strategic branding platform were solicited; Wolff Olins eventually won this competition. Through the work of the consultancy company, the concept *the Human Capital* was developed which would be the regional essence that should be

---

developments, there is a very interesting aspect of "altruistic" manipulations of (collective and individual) self and other in this process which could erode the very state that enabled the network participants to partake in the project, as it had afforded them jobs and positions as platforms from where they could take part in the discussions. This is not the place to discuss this aspect of these processes. However, two things should be noted. Contrary to the involvement of Swedish intellectuals in the modern transformations of the state, which has been discussed in previous chapters, the network's involvement was associated directly with the market through public relations and marketing. And second, the network willingly opened itself at least to some researchers from the outside as well as listened to the same researchers; Christian Tangkjær, myself, and other researchers were invited to partake and provide input into the branding process. Thus to some extent, albeit most likely to a very limited extent, we affected what we studied in a very direct way.

24 Interestingly, and in line with what could be predicted from the different understandings of the nation and region as discussed above, the institutional perspective taken in the two countries was different. The Swedish observer represented the State Department while the Dane represented the Department of Industry and Trade.

implemented and communicated to a wider, global audience. Agents and agencies in the region found this concept appropriate and corresponding to how they thought about the region. People representing businesses and the research sector also endorsed this concept, expressing that this concept reflected the impression of people visiting the region.

Wolff Olins arranged a one-day workshop for the Human Capital in London in October 1998. This workshop had a profound impact on the network's/ regionalists' understanding of the branding process; it is constantly referenced in interviews made with people from the steering committee and reference group of the "Birth of a Region" project who attended the workshop. The participants went to London with the purpose of formulating a vision for the Øresund region 2008 – what had to be done and who had to do it. Their discussions provided reflections for *the Human Capital*, which needed to be translated into a new concept that was communicable to a global audience and therefore, had to be translated into a global aesthetic (Tangkjær and Linde-Laursen 2004).

As the carrier of the individuality of the region, finding the right name was essential. The commodity to be branded was a regional vision that was to be institutionalized in parts of two separated countries. It was not a geographic functionality but a geographical imagination, politically produced for consumption by both local and global audiences. Wolff Olins produced three suggestions for a name along with three designed logos to carry *the Human Capital*. The three suggestions – "Copenhagen Scania Region," "Eko Region," and "The Sound Region" – were tested in focus groups in London, New York, and Frankfurt; none did well. However, the focus group in New York came back with the suggestion to call the region Øresund; the group found it to be a good brand that also, due to the Danish "Ø," communicated something very Scandinavian. The Øresund Committee stepped onto the arena and took responsibility. After lengthy negotiations during which the name and logo were redone and redesigned many times over, the Committee accepted Øresund and a version of the suggested logo for the Eko Region as the region's new brand. After the Committee's meeting on December 3, 1999, it was announce that the "Birth of a Region" had used it funds and that the process had identified that what was hitherto know as Øresund should now be known as: Øresund. To people external to the process, and possibly in particular to journalists, it was a capital blunder; how could anyone use $500,000 to establish something so obvious? To people partaking in the process, the process had been very productive as it clarified both the branding process and the potential benefits of the brand.

Through this branding process a vision for Øresund, a transnational region encompassing parts of Denmark and Sweden, was established as economically competitive and as an ideal of how late-modern lives ought to be lived. *Brandbook*, which was published by the Birth of a Region project late in 1999, suggested proper ways of using the regional logo and understanding the region.

A region that offers more of everything. More people, more choice, more interaction, more international attention. Soon, more will become better. Better market opportunities, faster growth, higher investment, a stronger competitive position and a higher standing in Europe. (*Brandbook* 1999: 2)

Furthermore it was stated that Øresund was a *cool* place where business was *hot*, and everything was done with a *human* touch. To international tourists and investors, this obviously was a place not to bypass, while the people already living there should feel content. Visiting the homepage of Øresund (www.oresund. com), one was met by a map that located the region and welcomed the visitor to "the Human Capital of Europe." Øresund then, was not only a discernible territory or place but also a mind-set that transcended place; it was a mind-map for life, emphasizing a human approach to being a person there whether living in the region, being a tourist, managing an enterprise, or managing a public organization. As such a branded place, the region was a matter of opportunities and potentialities formed around an idea of "more-ness." The region promised progress without restraint for an imagined community organized as a strategic tool based on its attractiveness as a place for future lifestyles, investments, atmosphere, and production.

After the Birth of a Region project was finalized, negotiations about the future of the brand were started. The organization *Øresund Network* was created in the spring 2000 to maintain and administer the brand whose ownership rights remained with the Øresund Committee. The idea was that private and public organizations from the region would pick up the Øresund brand and use it together with their own logos and brands. Doing so they would disseminate the content, *the Human Capital*, and promote the region as a good and attractive place for visits, everyday lives, and investments.

Bordering Paradoxes

The Swede proves that there is a God who has a sense of humor; that he generated water between Sweden and Denmark indicates that he is loving and considerate as well. (The winner of Danish IKEA's contest on jokes about Swedes, 1992)

On July 1, 2000, the new Øresund Bridge opened with a carefully choreographed event that included the two royal houses as well as many leaders from organizations, politics, and businesses. Over many hours, both Danish and Swedish media directly transmitted the event so the imagined communities could partake in it from their couches. In the event, many reflections on the historic existence of the national idea of spatiality were provided even if the event itself was choreographed as a launch pad for the Øresund Region. The Danish Prime Minster at the time, Poul Nyrup Rasmussen, told the freezing crowd of VIPs who were seated on a wind swept cape south of Malmö, next to the eastern bridgehead:

Today, we're beginning a new chapter in the joint history of Denmark and Sweden. The best chapter [is] the one where we're building a bridge between us.

For a short moment, world attention was on the region as international news agencies, CNN, the *New York Times*, etc. carried the story (Linde-Laursen 2002, Berg, Linde-Laursen, and Löfgren 2002). However, the next day global awareness turned elsewhere.

Obviously, several or many alternatives existed at this moment in history for people to think and practice bordering. No wonder that the discussions in the public were confusing as past and future, Skåne and Øresund, EU and nation-states, economy and emotion, locals and globals, etc. often were juggled in the same statements. What is interesting to contemplate is if and how all these discourses changed bordering practices. Research at the Øresund border around 2000 was privileged by not knowing what was going to happen. As pointed out earlier, when historians, anthropologists, and so forth have studied nations, the direction and current result of such processes – the modern nation-state – has been known and often taken for granted. However, the future of Øresund was unknown; any of the (often contradictory) suggestions presented were theoretically possible as the outcome of these relatively new and mostly confusing bordering processes. While of course all outcomes were not equally likely, it was impossible to distinguish among them, as no rear view mirror was available for gazing at "realities."

In the years since the opening of the bridge, all the invoked alternatives have undergone troubles of different kinds. The following is not a catalogue but rather an indication of the very existence of bordered identity instabilities. The greater alternative, the EU, has gone through enlargement processes as the number of member states went from 15 to 27. In the summer of 2005 that process resulted in the defeat of the new EU constitution and an immediate crisis over budgeting principles. The smaller alternatives, too, have been in trouble. The Swedish region Skåne had problems soon after it was created not least as a consequence of continuous deficits on its budget, which raised questions to some about the very logic of its construction. In Denmark, regionalization was not discussed as a term. Rather, a perceived need for new local and regional governmental structures led to prolonged discussions that blocked discussions of the cross-Øresund region; this was also something Swedish-Skånish politicians remarked on as they requested that Denmark modernized its governmental structures to meet its partners east of the channel on an equal footing (a structural reform of Danish municipalities and counties was implemented from 2007). In relation to the branding process initiated by the network and Ørespeak this was not of great significance; the interesting thing was that the branded region nearly collapsed after the bridge was opened. As Orvar Löfgren has discussed (2002), Ørespeak worked very well and was satisfying as long as it was exactly that – talk about future opportunities that did not request specific commitments except maybe a relatively minor economic contribution as a co-payment to an otherwise EU/Interreg financed project. However with the bridge's opening to traffic, regional realities rather than dreams and expectations came into focus. In such realities, diverging interests of different organizations and levels of governments soon significantly limited the interest in and ability of the same to commit to joined, cross-border projects and processes.

While Øresund Network continued to exist, albeit on a very slim budget compared to the original estimates, the regional logo, the Human Capital, which was to carry the brand to locals and globals alike soon mostly disappeared. Until 2006, the logo could at least still be found on metal posters in the train-station under Copenhagen's Airport, which was built as part of the bridge project; the posters however, from before the opening of the bridge, were decaying. Most other organizations regarded the logo of Øresund Network as a signifier of that institution and consequently, did not comply with suggestions to include it in their own advertising and information materials. Likewise there was reluctance among already existing and established organizations that promoted tourism and investments to use the logo as they existed before Ørespeak and were thus organized and financed within the national idea of spatiality. Why should they suddenly include the Others, their competitors? With the failure of the logo to materialize the region on many arenas, Ørespeak suffocated. This of course did not prevent particular projects, which from the start were associated with the cross-border region, continued working and became successful. However, their success contributed neither to the sustenance not the reinvigoration of the cross-border regional idea of spatiality. For instance, a 64-page infomercial: "Special report: *Denmark: A vibrant setting*" that accompanied *Gulf News* (English daily in Dubai, United Arab Emirates, June 12, 2005), did not use the Øresund Region logo in its promotion either of Danish businesses in the Middle East or Denmark as a tourist destination. Without mentioning the regional idea at all, it does identify "the Øresund Region" as "one of Europe's most stirring places for R&D [research and development]," although it only discusses facilities on the Danish side of the border (page 22). Similarly, it does mention "Medicon Valley," a spin-off of the regionalizing idea of spatiality, as an "impressive life sciences cluster, situated in Greater Copenhagen on the Danish side, and in Skaane, Sweden" (page 16). However, in this context it does not mention the region by name.

In 2002, the Øresund Bridge company, which built and runs the bridge, and is jointly owned by the Danish and Swedish states, held a lunch to mark the publication of a volume on the bridge opening event in 2000. In the book, researchers from Copenhagen Business School and Lund University discuss many different approaches to the event (Berg, Linde-Laursen, and Löfgren 2002). Interestingly, during this lunch it became apparent that this was the first time since the bridge opening event that all the central individuals in the above discussed network had gathered under one roof. While assessing the development of the region, a key figure in the network suggested during lunch that it was doing well – the name of the region, Øresund, was now in common use, and traffic across the bridge was growing. However, while it is probably true that Øresund is mentioned more often in public than before Ørespeak seriously started in 1997, the region remains undefined; many confusions remain concerning what is actually meant when "Øresund" is mentioned. Likewise, while a significant growth in cross-border traffic can be observed, it is unclear what that means. The above discussed encounters with drunken Swedes in the streets of Copenhagen should warn against regarding one-day tourists, still a significant portion of the bridge traffic, as promoting mutual affection; rather they work very well as providers of bordering narratives and experiences.[25]

In 1999, before the bridge opened, the man who initiated the start of the Birth of a Region process explained in an interview the importance of "thinking big." When a company launches a new car, it will spend 40 percent of the development money on marketing. He wanted to do the same with the region and to employ the same branding tool. When asked what would happen if the region did not materialize as established cultural practices in five years for instance, he answered "well, then we have to let it fall. What can substitute, I don't know; we will have to come up with something else." With a post-bridge, regional history of nearly a decade, it is possible to gauge that the animated visions of Ørespeak have largely failed. One of the principal reasons is that the promoters of the region failed to embrace bordering as a constitutive principle. The logic of the region as an identity providing spatiality needs to have clear distinctions as ditches around it; however, such ditches were never pursued. National governments clearly were not committed to accept distinctions that would separate a cross-border region from the rest of the two imagined national territories.[26] The failure of recognizing bordering as the basis for identity processes in regionalization has led to the demise of the Øresund region as a strategically driven process and response to processes of globalization.

Consequently, archaeological remnants from different levels of geographical belonging now surround Øresund: the EU, the nation-state, the region, historic and current counties, parishes.... The connections and tensions between the different levels have left an arena open for peoples' experimentations with everyday lives that, disconnected from Ørespeak, employ the new infrastructures and its possibilities. This situation feeds a number of paradoxes that highlights fractures between regional practice and rhetoric, Ørespeak.

Most discussions that were part of Ørespeak addressed the need for regional "integration." In a pre-bridge report, the two national governments pointed to the integration of education, work, housing, social services, taxation, language, and culture as significant for the region (*Øresund – en region bliver til* 1999). The discussion of integration was, as discussed above, preconditioned on the

---

25   Already prior to the opening of the bridge, I warned against using travel statistics as a measure of the success of the regional idea of spatiality (Linde-Laursen 1999c). However, as the only "objective" measure, traffic numbers, essential to the economy of the toll bridge, are frequently used as indicators of success by regionalists.

26   This, as explained by Swedish under-secretary of state, Sven-Eric Söder, made sense both for the states and for the citizens who had come to expect national solutions to things like social security (see "Fortsatt krångligt för Sundspendlarna," sydsvenskan.se June 25, 2001).

understanding that "barriers" existed, which as differences across the border hampered and challenged the dynamic development of the region. In the report's concluding part on "Culture and Identity," which discusses among other things language and organizational culture, this was all summarized:

> The cultural differences are, however, by far not as great as the common understanding suggests. The myths thrive on both sides of the Sound, and false conceptions can be as damaging to integration as the real differences.
>
>   Through the integration across Øresund, which the bridge will stimulate and strengthen, the Danish and Swedish cultural identity will be moved closer together. Commuting will increase. People will progressively regard available jobs, homes, goods, services, social services – and culture [here mostly understood as theater, opera and other culture with a capital C] – as parts of the same integrated market. This will not quickly, but over some time, increase mutual understandings and diminish the differences. (*Øresund – en region bliver til* 1999:82)

The whole discourse of integration within the idea of spatiality of the region constituted an arena on which "integration" was understood in a case-specific way. First, it was obviously associated with an understanding of a market as noted above. Second, it constituted the region as a particular kind of a *limited* imagined community. As mentioned above, in Danish and Swedish public debate "integration" usually referred to assimilation processes of people recognized as "immigrants" – identified as non-Western, non-modern, and most often both non-white and Muslim.[27] Thus, the regional idea of spatiality created an arena where "integration" had a special meaning – that of assimilating Danes and Swedes. The imagineering of this discursive arena in itself excluded many from participating in the regionalization of the Øresund area, which is the area in the Nordic countries with the most non-Nordic citizens (Nilsson 1999). Thus, in the branding process the existence of two imagined communities was taken for granted; however, the process itself was organized for a limited imagined community not only in a geographical sense but as an imagined community for only a limited part of the inhabitants who could and would practice a very particular rendition of this

---

27 This focus on Muslims as the non-national Other has been most noticeable in Denmark, as illustrated by the "Mohammed Caricature" case from late 2005 through early 2006 (see Linde-Laursen 2007). Before that however, attacks on immigrants had become a normal part of Danish political discourse – for instance in the context of the national elections in November 2001 (see for instance: "Voldsom kritik af Danmark" politiken.dk, November 15 and "Dansk debat ryster udlandet" November 19; news.bbc.co.uk, "Danes undecided as pools loom" November 19 and "Immigration dominates Danish vote" November 20). Of particular interest in this context are also changes to Danish immigration laws that from 2002 made it very difficult for Danish citizens to bring a spouse to the country if the person was not from an EU country (see the postscript).

geographical space. From this perspective, the branded region subscribed to a limited cultural community and simultaneously negated any understanding of the nation-states as political communities that ideally were available as imagined communities to all their inhabitants.

Within this limited imagined community, but largely outside the discursive framework of Ørespeak, an increasing number of people cross the water and through their practices as commuters, tourists, consumers, and so forth, obtain experiences with people from the other side (Linde-Laursen 2005). According to the idea discussed above, integration should limit cultural differences. However, the opposite resulted. As people in growing numbers have experiences with and recollections of real differences from more and more arenas, the narratives that support national cultural bordering are multiplying. Not surprisingly a growing number of reports from researchers and consultants document such differences in more and more detail (for instance politiken.dk, May 20, 2005); in doing so the reports most often reflect already known stereotypical understandings (Swedes are introverts, Danes are cozy, etc. See for instance Berglund et al. 1999).[28] Thus, the border reacts against these attempts to prescribe integration and new and increased bordering becomes the paradoxical result. Regardless if these narratives reflect real differences or what the government report terms "false conceptions," they stress and legitimize that the border today is a cultural reality. Such narratives thereby provide acceptable explanations to situations and experiences in peoples' everyday lives that can be explained with reference to nationality but might not otherwise need explanation or could in other situations be explained with references to other collective foundations for identity. Furthermore, as the bridge facilitates cross-border social interaction and as people in specific situations erase conflicts they experience by referencing the national idea of spatiality, this does not solve the increasing number of problems experienced. Thus, many cross-border projects in organizations and businesses end in frustrated narratives about "the people from the other side of the border." Even if problems and failures were caused by incompetence, lack of financing, or other factors that involve people themselves, nationality is invoked as an explanation that relieves all involved from personal responsibility – "Oh, that is so Swedish/Danish." Consequently, the combination of integration attempts and infrastructural development has resulted in an increased number and sophistication of narratives about national differences. In more and more situations there exists a potential correspondence between on the one hand, any random cultural criterion by which one can distinguish oneself from another and on the other hand, the increasing diverse narratives of cultural bordering. Increasingly, experiences of bordering at Øresund (in specific situations) lead to

---

28   Such research results can be reproduced endlessly (see for instance Berglund et al. 1999 and "Der *er* forskel på svenskere og danskere" politiken.dk, May 20, 2005). Reports about economic life even replicated the stereotypes that Sundbärg (1911) had pointed out a century earlier (see "Nu är kulturkrockarna kartlagda" sydsvenskan.se, August 22, 2006).

coincidences between individual recollection and collective memory that confirm that the Other, for sure, is of a completely different kind.

One of the markets identified by the governments for integration is the market for homes (an arena that in itself has produced a number of reports, for instance *Danske erfaringer med at bo i Skåne*, 2005). As discussed above, people also prior to the bridge and Ørespeak moved across the border. While some people moved their whole everyday life, others – maybe most – moved to obtain structural advantages from differences in taxation, social services, and so forth while they continued to have part or most of their life in the place where they originated. In the mid-1990s, many Danes moved to southern Sweden, and especially to Landskrona, a smaller town north of Malmö. The Danish state and municipalities regarded the campaign that Landskrona ran to attract Danes as organized tax evasion. In 1997, the bilateral tax agreement was renegotiated and the tax advantage that could be obtained from living in Sweden and working in Denmark disappeared. Soon many Danes moved back to Copenhagen (*Sydsvenska Dagbladet* December 7, 1997, and January 19, 1998). It is thus obvious, that the integration of the market for dwellings historically as well as currently is founded on differences. In the mid-2000s, many young, childless Danes moved to dwellings in Malmö that were cheaper than what they could find in Copenhagen (Linde-Laursen 2005).[29] The paradox of course is that if the governments manage integration well and the differences in the future disappear, then there will be no such structural reasons for people to move across the water. Even if that will eliminate protracted discussions of the direction of the "advantage" obtained from such moves, it will also eliminate the incentives to move altogether. One way of mitigating this could be to deliberately change the direction of the advantage every five years; however, such a suggestion and obvious manipulation of peoples' lives is of course politically unacceptable.

From the late 1990s, Ørespeak and the resulting activities, supported by the economic incentives provided through the EU-sponsored Interreg program, resulted in many different cross-border programs and initiatives in all the different areas suggested by the governments. Universities as well as kindergartens started

---

29   Two differences dominated public discourse about Øresund in the early 2000s. There was a large group of Swedes who were going to Copenhagen to work, as jobs were easier to find in Denmark; in September 2006, the group was said to consist of around 2,350 persons, up 69 percent from the year before. And there was a large group of Danes moving to Southern Sweden and especially to Malmö to find cheaper dwellings; in September 2006 this group was estimated to be around 4,700 for the year, a number 52 percent larger than the year before and, of course, accumulating a rather large Danish minority in Southern Sweden ("Vi flytter til Sverige i stor stil" politiken.dk, September 26, 2006). Both these groups contributed significantly to the traffic on the bridge which reports strong growth year by year; the growth from mid-2005 to mid-2006 was 17 percent for road traffic and 18 percent for train traffic (osb.oeresundsbron.dk, 2006 mid-year report). It should be noted that an estimated one in five of the Danes moving to Sweden did so under pressure from Danish immigration laws preventing them from settling in Denmark with a spouse from outside the EU (*Danske erfaringer med at bo i Skåne* 2005:4).

to visit their counterparts, businesses and social services met and identified areas where cross-border cooperation would be beneficial, and so forth. As the history about the Øresund Region logo illustrates, there were also attempts to provide symbolic representations that could unite Øresunders and thus, separate them from other Danes and Swedes. In their report, the governments directly "assume that private and regional agents will make an effort to develop regional symbols that can stimulate integration" (*Øresund – en region bliver til* 1999:87). Thus, from the symbolic to the functional, from everyday lives to economics, the governments rhetorically supported the integrative ideas and practices of the Øresund regionalists. Paradoxically though, it is obvious that the successful completion of integration would be a very bittersweet victory for the regionalists. The erasure of the border would be a completion that would also end a generator of money and power to which the same people and their organizations subscribe. Therefore and not surprisingly, each concluded project at Øresund ends with the identification of new and necessary steps to complete before "the integration" is finalized and the differences between people and societies on both sides of Øresund have become reduced or totally erased.[30]

Thus, if the neighboring nation-states do not make a concerted effort for integration, it can be expected that not much will change as the border remains unchallenged. However, if the nation-states make an effort for cross-border integration, the border seems, at least in the short-term perspective, to "hit back" as both existing and previous unprovoked potentials for bordering are released with force. The paradox of course is that bordering processes that have maintained the nation-states and have worked well for more than a century are today well embedded in cultural ideas of spatiality and practices. Such history is impossible to neglect; thereby Øresund is a very real border regardless if neglecting it would be beneficial and would be an attractive local adaptation to globalization.

Identity by Prescription, Negotiation, and Neglection

The paradoxical nature of current bordering should not lead to the mistaken conclusion that nothing is changing. Obviously if increasing (or decreasing) numbers of people, goods and information are moving across a border, something is happening and that "something" is more likely than not contributing to changes – changes to bordering and consequently to foundations for identity formation. What remains is to outline how engagement in such changes might be conceptualized as strategies.

---

30 This, of course, is also true for researchers. Many researchers have studied the regionalizing process at Øresund. Together with partners in Lund and Copenhagen in the five years from 1998 through 2003, I received about $800,000 from Swedish municipalities, from Interreg, and from The Bank of Sweden Tercentenary Foundation for studying such issues.

Jonathan Murdoch (1998) has suggested distinguishing between two spatial kinds of actor networks. He suggests that "spaces of prescription" and "spaces of negotiation" differ according to the amount of remote control and autonomy found in an actor network even if they should not be seen as two ends of a classificatory dichotomy but as "two sides of the same coin" (1998:364). Murdoch suggests that spaces of prescription are formalized and standardized spaces that are imposing fairly rigid and predictable forms of behavior. Within the actor network, power is distributed to a center that holds the license to act. Power, while still a composition of many actors, is attributed to one of them (Murdoch and Marsden 1995:372).[31] "Spaces of negotiation" are spaces in variation and flux wherein reality is constantly being re-negotiated, new actors enrolled, and new alliances practiced. Because links are provisional and divergent, these spaces allow modes of resistance and processes of translation. Thus, when standards and norms are established they are primarily established from resistance and compromise.

From such perspectives, the national idea of spatiality in Scandinavia for about a century starting from when the states became stabilized around 1900 through the merger of the political with the cultural communities and continuing until the 1980s materialized as "spaces of prescription." Power was situated and the structures of each state very formalized and standardized. The current larger alternative and the smaller alternatives (not least the Øresund Region) are signified as examples of "spaces of negotiation." All kinds of actors were invited to participate with their rendition of the Øresund Region forming "open houses" where many different kinds of regions coexisted and were possible. There was the idea that the region would form through its practices.[32] Or similarly, the idea, as described in the government report, that "the road materializes, while one walks on it" (*Øresund – en region bliver til* 1999:18), allowed for both immense and largely uncontrolled creativity as well as, after the opening of the bridge, a lack of focus and institutional commitment. This situation has resulted in the establishments of a great many projects and institutions designed to affect the cultural formation of belonging around and across Øresund, many of which are ready to duplicate processes that are also basic to the nation-states even if the geographical scopes and limitations are obviously different. If the EU and the regions developed towards "spaces of prescription," people's loyalty would be called upon in the same way by different bodies. Thus, well-intentioned identity politics could end up heating the border in Øresund rather than the opposite.

----

31   It is this kind of process that as an extreme can take the form of "ethnic cleansing" or a similarly violent process on which very exclusive bordering practices can be built. The invented character of the imagineered basis for such practices has been discussed above; again it is necessary to understand that discursive inventions form very real frameworks for peoples' everyday lives.

32   Institutions created for the regionalization efforts were thus largely undefined and had to define their purposes as they worked. See for instance Rahbek Rosenholm's analysis of the Øresund Committee (1997).

An interesting aspect discussed earlier is that both the modern nation-state governments in 1999 and the state of Carl XI in 1678 were interested in creative experimentation to sustain and propel desired bordering processes and that both expected a certain amount of time and contact would provide them with what was desired. However, there is of course (at least) one crucial difference. Carl XI could do with targeting and to some extent substituting three socially and politically limited groups in his society – the nobility, bureaucrats, and clergy that formed the connectors in state patriotic society between the center of power and the areas in which bordering should be achieved. In the modern cultural community, where "we all hold the truth both about ourselves and the others" (Harbsmeier 1986:53), the whole society needs to be involved or at least, a very significant part of it in terms of numbers and/or power potentialities must be active and strategizing players. This makes the process more difficult to guide and facilitates paradoxical processes that counteract the desired bordering outcome. Even if these paradoxes could have been predicted (Linde-Laursen 1999c), they would have been very hard to avoid.

So far, Murdoch's concepts are satisfying and point to interesting similarities and differences across time and space. However, there is the precondition that people not only are acting and being active agents for bordering, but also being conscious and deliberate in their practices. What is obvious at Øresund is that many people in their everyday lives are not interested in obtaining identity from bordering along national or regional ideas of spatiality. Practicing what can be termed "spaces of neglection" suggests that bordering processes can be the result of pragmatic practices disassociated from ideas of spatialities and where leadership and control is difficult if not impossible to both establish and define. Tom O'Dell (2002) discusses how groups of people (such as the "real" immigrants within the Øresund Region discussed above) very well can practice the region through their movements in relation to the border but do so with no interest in or knowledge of ongoing bordering. The "real" immigrants might go from Copenhagen to Malmö to attend the mosque there or visit friends or family from their home country yet, they do not use and do not reflect on the region when doing so. While they navigate the region and live it, O'Dell terms them "regionauts," their practices neglect ideas of spatiality and bordering. In relation to regionalization, they constitute an identifiable group that declines or avoids obtaining regional identity through their spatial regional practices. Their behavior exactly neglects – and not resists – bordering, and in this way the "regionauts" act and practice in ways not very different from how a lot of people probably associated with first the political community and then its fusion with the cultural community. Obviously, the women who practiced and developed modern homes in the first half of the 20th century while vigorously pursuing modern and clean everyday lives, did not intend such practices to develop later into arenas for bordering experiences associated with national idea of spatiality.

Bordering practiced by Danish fishermen trawling in restricted waters in Øresund, using the line against Swedish authorities, has persisted.[33] However, to regard bordering as limited to such obvious manipulations of the line obscures how unpredictable and uncertain processes in the past and future have been in encompassing cultural processes at this European border. Here bordering for a long time seemed settled as Danes and Swedes lived peacefully on each side of Øresund, recognizing and acknowledging the mutual differences of their counterpart. Bordering though, is not an ended process. Rather, promoted by European confusions and developments within the two nation-states, bordering unleashes new suggestions and paradoxical developments. Through this, spaces of prescription, spaces of negotiation, and spaces of neglection are layered over each other, forming foundations for ongoing and multifaceted processes of bordering. As nation-states, the EU, regions, and unintended side effects of previous cultural processes interact within the modern maze, the results of such cultural processes at a European border, in the age of globalization, are complex and unpredictable. When assessing which strategy eventually will contribute most to ongoing processes of change, it might be reasonable to remind us of the 4.1 million visitors at the Stockholm Exhibition and the 5,767 catalogs sold. Neglection, to act without seeking to create or obtain identity, might turn out to be a strategy that generally contributes very significantly to bordering.

33　For instance, see new cases against fishermen from Gilleleje in November 2003 and March 2005 (sydsvenskan.se, "Danska tjuvfiskare avslöjades i Öresund" November 10, 2003, and "Åklagarna höjer tonen mot danskt tjuvfiske" March 29, 2005). In the latter case, the prosecutor demanded jail terms for the perpetrators.

Postscript

As illustrated throughout these chapters, bordering processes are inherently both good and bad, liberating and confining, agreeable and malicious. As cultural scientists, we need to come to terms with these confusions. Stressing only one aspect – politics or culture, future or past, recollection or memory – will encourage understandings of borders and the spaces they encompass as one-dimensional and un-reflexive. Because of such limitations, it would be inappropriate for this text to have a singular conclusion.

As it happened, history formed the border in the Øresund during a window-of-opportunity, which enabled the boundary's development and existence under what can be perceived as ideal circumstances in relation to specific understandings of historical processes, conjunctures, and modern ideas and practices. This happened when the state patriotic structures in Denmark and Sweden followed general trends as they developed towards the European central state and absolutism, when those states in the 1800s recognized themselves as political communities, and later when these political communities were supplemented through cultural bordering. In contrast with some other places in the world, this made it possible, at Øresund, to experience ways of thinking about and practicing bordering that dressed the border as uncomplicated and "natural." However, pressure from new ideas of spatiality, Europe, and regions, question if this simplicity can persist. The following are two different concluding reflections on bordering at a European border.

I

The first is written by renowned Danish poet Benny Andersen (1929-) and is called *Closet Swedes* (1995:47ff). I thank Benny Andersen for his permission to include his wit and wisdom on this topic as the first ending to this book. Cynthia La Touche Andersen, Benny Andersen's wife, translated the poem; the explanatory notes are mine. While the intention is to have them connect this postscript with specific aspects present in other parts of this book, my notes are much less poetic than the notes Benny Andersen wrote for the translation, which can be found in "Cosmopolitan in Denmark – and Other Poems about the Danes."

Closet Swedes

Is there anything as Danish as a potato?
The potato stems from South America.

Is there anything as Danish as the Dannebrog itself?
It fell from heaven in Estonia a long time ago
and resembles the Swiss flag.[1]

Does anything sound more genuinely Danish
than the music of the ballad-opera "Elverhøj"?
Composed by a German with frequent use
of Swedish folk tunes.[2]

Be careful
now it becomes difficult:
Is there anyone more Danish than the Danes?
Descendants of the Danes
a tribe in Sweden
who invaded our country sometime in the third century AD
while the original Danes
the Herules
the noble and brave
but outnumbered Herules
were driven to flight by the terrible Swedish Danes
had to roam displaced around ancient Europe
for several centuries until finally
a few thousand of these original Danes succeeded
in reaching Sweden where they settled
under the dubious name *Swedes*

Here is the question once again
and think carefully before you answer:
Is there anyone more Danish than the Danes?

---

1   "Dannebrog" is the name of the Danish national flag. Legend tells that it fell from heaven and granted the Danish king, Valdemar II Sejr, victory over Estonian heathens during his crusade in Estonia 1219. From this story, Dannebrog is claimed to be the oldest national flag in the world.

2   Elverhøj from 1828 is considered a Danish national play. It was authored by Johan Ludvig Heiberg (1791-1860) to music by Friedrich Kuhlau (1786–1832), who was born in today's Germany. Included in Elverhøj is the "Romance" by Johannes Ewald discussed above (p. 75ff.). This piece, later known as "Kong Kristjan," is regarded as one of Denmark's two (sic) national anthems.

The correct answer is:
Yes!
The Swedes!
They are the authentic primeval Danes
Like the Jews in the wilderness
they are constantly drawn to the promised land
which flows with beer and bacon
but has been occupied for seventeen hundred years
By whom?
By the Swedes!
By us!

No wonder that Skåne
demands to have Denmark back
no wonder that many of us Crypto-Swedes
have difficulty speaking proper Danish
cut of suffixes
swallow consonants
choke on syntax
so everything sounds like "Rødgrød med fløde"[3]
no wonder that we hardly understand each other
no wonder that the most frequently used word is "what?"
It is not at all our language
We are not at all us
We are a bunch of bloody Swedish foreign workers
who have wrecked this lovely country[4]
why don't we go back to where we come from
home to Sweden
"Thy sun, thy skies, thy verdant meadows smiling"[5]
where at last we could acknowledge our true identity
show our colours
"We are yellow
We are blue"
Where we could decisively beat ourselves at football

---

3   "Rødgrød med fløde," or red fruit pudding with cream, is what foreigners are most often asked to say in Danish to the amusement of Danes who know that the words are very difficult to pronounce for non-native speakers.

4   This is a reference to the other Danish national anthem, "There is a Lovely Country," written by Adam Oehlenschläger, probably in 1819.

5   A line from the Swedish national anthem: "Thou Ancient, Thou Freeborn," written by Richard Dybeck around the middle of the nineteenth century. In the original Danish version of Benny Andersen's poem, this line is in Swedish as is the first line and title of the anthem appearing below.

and defeat ourselves in the European Song Contest
Oh, how we have needed
and longed for this
at last to be able to sing Bellman's ballads
in their original language
or *språk* as it is actually called[6]
at last to have the exclusive right
of being the only ones in the world
who can faultlessly pronounce
Sjutusensjuhundrasjutisju[7]
without loosing our dentures
finally to be freed from our
frustrating national inferiority complex
and be allowed to spread our wings
and soar to the skies
as the freest Nordic swans[8]
At last we can be rid of continuous blame
for that stupid Jante's Law
which some crazy Norwegian writer
has saddled upon us[9]

At last ourselves
at last free
"Thou ancient, thou freeborn"
at last home where we belong
at last to have a chance
to make a Half a Whole
and generously drink a toast to ourselves
as soon as we have introduced humane conditions
concerning alcohol
The drinking songs are already available

---

6   Carl Michael Bellman (1740-1795) was a famous Swedish writer of popular songs. *Språk* is the Swedish word for language.

7   "Sjutusensjuhundrasjutisju" translates to seven-thousand-seven-hundred-seventy-seven, a phrase non-native speakers of Swedish are asked to say as "sju" is supposed to be very difficult to pronounce.

8   In the 1960s and 1970s, the Nordic countries were often depicted as a group of five swans (Denmark, Finland, Iceland, Norway, and Sweden). Today the number is debatable (what about the autonomous areas: the Faeroe Islands, Greenland, and Åland, and what about the Sami peoples), but the symbol endures as one for the whole area. The symbol of the Nordic Council and the Nordic Council of Ministers, for example, is a swan.

9   Jante's Law is a "law" of ten paragraphs, written by the Norwegian-Danish author, Aksel Sandemose, about the boring average-ness of life in Denmark (or Scandinavia in general).

now is the time to fulfill them
make them trustworthy

Great times ahead
and
If we can make it here
we'll make it everywhere

And after all
there is far more space in Sweden.

II

The second ending describes a foundation for recollections that inform me about the continued existence and importance of the Danish-Swedish border in my autobiography. In April 2003, I was sitting in our apartment in Lund watching the evening news on one of the Danish national television channels. Interviewed that night was a Danish man, who for many years – I believe it was seventeen – had been married to a woman from the US where the couple had also lived. He explained that as his parents were getting older and as he and his wife were concerned with directions of US politics, they decided to move to Denmark. Upon arriving in the country, he was welcomed but his wife was granted only a three months visiting visa. A new law from 2002 on immigration and family reunion meant that in order for his wife to receive residency in Denmark, the couple would have to prove that they together had a more developed connection to Denmark than to any other place in the world.

The new law explicitly aimed to limit family reunifications for "real" immigrants. It was perceived that legal family reunification facilitated the continuous flow of new immigrants into Denmark; cases of Muslim immigrants with Danish residency who married people from their "home" country were obviously considered a cause for concern. The legitimizing argument for the law was to prevent "arranged marriages," which the majority of members in the Danish parliament regarded as unwelcome traditional practices.[10]

The next day I called Danish immigration authorities. On the phone, they confirmed that the number of days a couple had spent together in Denmark would be the implemented measure for their joint connection with the country. And, I realized that even if my family, as far back as is known, has always lived in Denmark

---

10 Later numbers show that people with Danish and Nordic citizenship account for most of the rejections of family reunion with a spouse ("Flest danskere får nej til at hente ægtefælle" politiken.dk, October 4, 2004). The Danish immigration laws since 2002 have been changed several times (see for instance comments to that on http://www.flygtning. dk/). Thus, the fear of "real" immigrants might be unfounded.

since the early 1500s, I could not settle there with my wife who is a US citizen. There was, however, a possible escape route. Since 1988, I had worked at Lund University, Sweden, where I was, in 2003, an Associate Professor of Ethnology. According to Swedish laws, Nordic citizens can notify Swedish authorities of a change of citizenship; if Nordic citizens have lived in Sweden for some time and have fulfilled other requirements, they do not need not to go through an application procedure to become Swedish. And, with Swedish citizenship, Danish authorities could not prevent my wife and me from settling in Denmark, as that would be against EU rules that enforce the free mobility of individuals, capital, and goods.

A few days later I mailed my notification to the Swedish authority, and ten days later (and about $50 poorer) I was a Swedish citizen. I soon received a somewhat antagonistic letter from the Danish consul general in Malmö, requesting that I come to his office and surrender my Danish passport. Thus, pushed by cultural bordering in the Danish imagined community, I had made an individual, political decision to switch my citizenship and join the, as it seemed, more welcoming imagined community on the other side of Øresund.[11] The Danish-Swedish border had with force imposed itself on my biography. This imposition I recollect each time I fill in customs and other forms during international travel – the blank space *nationality* always evokes in me a not disagreeable sensation of hanging in the air. And from up there, I contemplate the evolving realities and imaginings of bordering between Danish and Swedish.

---

11   It should be recognized that many individuals find the Swedish laws beneficial in this area and thus, are actively employing bordering. As mentioned earlier, a whole new category of people has joined the flow of individuals and families moving from Copenhagen to Malmö – Danes who due to Danish legislation are prevented from marrying or living with the person of their choice.

References

Åberg, Alf. 1958. *När Skåne blev svenskt*. Stockholm: LTs förlag.

Åberg, Alf. 1960. *Nils Dacke och Landsfadern*. Stockholm: LTs förlag.

Åberg, Alf. 1975. *I snapphanebygd*. Stockholm: Rabén and Sjögren.

Åberg, Alf. 1981. *Snapphanarna*. Second edition. Lund: Signum.

Åberg, Alf. 1995. *Kampen om Skåne under försvenskningstiden*. Stockholm: Natur och Kultur.

Abu-Lughod, Lila. 1999. The Interpretation of Culture(s) after Television. In: Sherry B. Ortner (ed.): *The Fate of "Culture". Geertz and Beyond*, pp. 110-135. Berkeley: University of California Press.

Adriansen, Inge. 1987. Mor Danmark. Valkyrie, skjoldmø og fædrelandssymbol. *Folk og Kultur*: 105-163.

Adriansen, Inge. 1990. *Fædrelandet, folkeminderne og modersmålet. Brug af folkeminder og folkesprog i nationale identitetsprocesser – især belyst ud fra striden mellem dansk og tysk i Sønderjylland*. Sønderborg: Museumsrådet for Sønderjyllands Amt.

Agersnap, Torben (ed.). 1973. *Kort fra Danmark og andre Danmarksfilm*. Nyt fra samfundsvidenskaberne.

Agger, Gunhild. 1992. *En eventyrromans mission. Gøngehøvdingen i populærfiktionens historie*. Aalborg University: Department of Communication.

Agger, Gunhild; Ib Bondebjerg; Anker Gemzøe; Inger-Lise Hjordt-Vetlesen; Hans Jørn Nielsen; and Anne Birgitte Richard. 1984. *Dansk litteraturhistorie 7. Demokrati og kulturkamp 1901-1945*. Copenhagen: Gyldendal.

Åkerman, Brita. 1983. Idealism och praktiskt handlag. Sveriges Husmodersföreningars Riksförbund. In: Brita Åkerman et al.: *Vi kan, vi behövs! – kvinnorna går samman i egna föreningar*, pp. 134-150. Stockholm: Akademilitteratur.

Åkerman, Brita et al. 1983a. *Vi kan, vi behövs! – kvinnorna går samman i egna föreningar*. Stockholm: Akademilitteratur.

Åkerman, Brita et al. 1983b. *Den okända vardagen – om arbetet i hemmen*. Stockholm: Akademilitteratur.

Åkerman, Brita et al. 1984. *Kunskap för vår vardag – forskning och utbildning för hemmen*. Stockholm: Akademilitteratur.

Åkesson, Folke. 1934. *Danmark kors och tvärs*. Stockholm: Natur och Kultur.

Alonso, Ana María. 1994. The Politics of Space, Time and Substance: State Formation, Nationalism, and Ethnicity. *Annual Review of Anthropology* 23: 379-405.

Alsmark, Gunnar. 1992. The Lessons of Sjöbo. In: Göran Rystad (ed.): *Encounter with Strangers. Refugees and Cultural Confrontation in Sweden*, pp. 37-53. Lund: Lund University Press.

Alver, Brynjulf. 1989. Folklore and National Identity. In: Reimund Kvideland and Henning K. Sehmsdorf (eds.): *Nordic Folklore. Recent Studies*, pp. 12-20. Bloomington: Indiana University Press.

Ambjörnsson, Ronny and Sverker Sörlin. 1992. Medborgarens återkomst. *Moderna tider*, October 1992: 32-35.

Amdrup, Erik. 1993. *Tro, håb og nederlag*. Copenhagen: Gyldendal.

Amin, Ash and Nigel Thrift (eds.). 1994. *Globalization, Institutions, and Regional Development in Europe*. Oxford: Oxford University Press.

Andersen, Benny. 1995. *Cosmopolitan in Denmark – and Other Poems about the Danes*. Translated by Cynthia La Touche Andersen. Copenhagen: Borgen.

Andersen, C.C. 1909. *Geografi for Folkeskolen*. Haslev.

Andersen, P. and M. Vahl. 1910. *Geografi for Mellemskolen* II, third edition. Copenhagen: Gyldendal.

Anderson, Benedict. 1983. *Imagined Communities. Reflections on the Origin and Spread of Nationalism*. London: Verso.

Anderson, Benedict. 1991. *Imagined Communities. Reflections on the Origin and Spread of Nationalism*. Revised and extended edition. London: Verso.

Anderson, James. 1988. Nationalist Ideology and Territory. In: R.J. Johnston, David B. Knight, and Eleonore Kofman (eds.): *Nationalism, Selfdetermination, and Political Geography*, pp. 18-39. Beckenham: Croom Helm.

Andersson, J.O. 1881. *Om nationela svagheters och fördomars inflytande på våra industriela förhållanden*. Stockholm: Lecture in Svenska Slöjdföreningen.

Andreas, Peter. 2000. *Border Games. Policing the U.S.-Mexico Divide*. Ithaca: Cornell University Press.

Arcadius, C.O. 1886. Om Bohusläns införlivande med Sverige. *Bidrag till kännedom om Göteborgs och Bohusläns fornminnen och historia* III: 1-119.

Arctander, Philip. 1959. *Plan i køkkenet*. Copenhagen: Statens Byggeforsknings-institut, SBI-anvisning 46.

Arnstberg, Karl-Olov. 1989. *Svenskhet. Den kulturförnekande kulturen*. Stockholm: Carlssons.

Arrhenius, Sara and Cecilia Sjöholm. 1995. *Ensam och pervers*. Stockholm: Bonnier Alba.

Askgaard, Finn. 1974. *Kampen om Østersøen. Et bidrag til nordisk søkrigshistorie på Carl X Gustafs tid, 1654-60*. Copenhagen: Nyt Nordisk Forlag Arnold Busck.

Askgaard, Finn and Arne Stade (eds.). 1983. *Kampen om Skåne*. Copenhagen: ZAC and Co.

Asplund, Gunnar; Wolter Gahn; Sven Markelius; Gregor Paulsson; Eskil Sundahl; and Uno Åhrén. 1980 (1931). *acceptera*. Stockholm: Tidens förlag.

Asplund, Johan. 1991. *Essä om Gemeinschaft och Gesellschaft*. Göteborg: Korpen.

Bagge, Povl. 1992 (1963). Nationalisme, antinationalisme og nationalfølelse i Danmark omkring 1900. In: Ole Feldbæk (ed.): *Dansk identitetshistorie* III, pp. 443-467. Copenhagen: C.A. Reitzels Forlag.

Balle-Petersen, Poul. 1978. Efternavnet som kulturelt kendetegn. *Folk og Kultur*: 109-131.

Barnouw, Erik. 1983. *Documentary. A History of the Non-Fiction Film*. Revised Edition. Oxford: Oxford University Press.

Balibar, Etienne and Immanuel Wallerstein. 1991. *Race, Nation, Class. Ambiguous Identities*. London: Verso.

Barth, Fredrik. 1969. Introduction. In: Fredrik Barth: Ethnic Groups and Boundaries. The Social Organization of Cultural Difference, pp. 9-38. Boston: Little Brown.

Barthes, Roland. 1972. *Mythologies*. New York: Hill and Wang.

Barthes, Roland. 1981 (1980). *Camera Lucida. Reflections on Photography*. New York: Hill and Wang.

Baudrillard, Jean. 1985. År 2000 kommer inte äga rum. Efter historiens försvinnande: simulationens makt? *Res Publica* 1. Historiens slut?: 23-37.

Baudrillard, Jean. 1994. *The Illusion of the End*. Stanford: Stanford University Press.

Bauman, Zygmunt. 1998. *Globalization. The Human Consequences*. New York: Columbia University Press.

Begtrup, Bodil M. and Kamma Svensson. 1949. *Ægteskab og huslighed*. Copenhagen: Chr. Erichsens Forlag.

Behrendt, Poul. 1987. Hvad er det med Sverige? I anledning af Ingmar Bergmans erindringer. *Kritik* 82: 6-22.

Bell, Colin and Howard Newby. 1971. *Community Studies. An Introduction to the Sociology of the Local Community*. London: George Allen and Unwin.

Belt, Don. 1993. Sweden. In Search of a New Model. *National Geographic* 184 (2, August): 2-35.

Ben-Amin, Shlomo. 1991. Basque Nationalism between Archaism and Modernity. *Journal of Contemporary History* 26(3-4): 493-521.

Berendt, Mogens. 1982. Luk Sverige! *Berlingske Søndag*, 14/3. Copenhagen.

Berendt, Mogens. 1983. *Tilfældet Sverige*. Copenhagen: Chr. Erichsen.

Berendt, Mogens. 1990. *Tilfældet Sverige*. Second edition. Copenhagen: Holkenfeldts Forlag.

Berg, Per Olof; Anders Linde-Laursen; and Orvar Löfgren (eds.). 2000. *Invoking a Transnational Metropolis. The Making of the Øresund Region*. Lund: Studentlitteratur.

Berg, Per Olof; Anders Linde-Laursen; and Orvar Löfgren (eds.). 2002. *Öresundsbron på uppmärksamhetens marknad. Regionbyggare i evenemangsbranschen*. Lund: Studentlitteratur.

Berggreen, Brit. 1991. "Kvinner selv..." Kvinners nasjonale erfaring. In: Anders Linde-Laursen and Jan Olof Nilsson (eds.): *Nationella identiteter i Norden – ett fullbordat projekt?*, pp. 149-174. Stockholm: The Nordic Council, Nord 1991: 26.

Berglund, Sara; Malena Gustavson; Mette Bjerregaard Kirk; Lotte Mortensen Vandel, and Anders Linde-Laursen. 1999. *Svenskt eller danskt? Kulturella variationer i tanke och handling.* Kulturkompetens 1998-99. Karlskrona: Svenska BoProjekt AB.

Bergman, Karl. 2002. *Makt, möten, gränser. Skånska kommissionen i Blekinge 1669-70.* Lund: Nordic Academic Press.

Bergsøe, Flemming. 1946. *Mellem halvbrødre. Skitser fra en rejse i Sverige.* Copenhagen: Thaning and Appels Forlag.

Bergsson, Gudbergur. 1988. Om identitet. *Undr.* Nyt Nordisk Forum Nr. 52. Ultima Thule: 63-69.

Berman, Marshall. 1988 (1982). *All That Is Solid Melts Into Air. The Experience of Modernity.* New York: Penguin Books.

Berntsen, Arent. 1656. *Danmarckis oc Norgis fructbar Herlighed.* Copenhagen.

Billig, Michael. 1995. *Banal Nationalism.* London: Sage.

Björlin, Gustav. 1890. Krigsrörelserna i Bohuslän 1788. Ett hundraårsminne. *Bidrag till kännedom om Göteborgs och Bohusläns fornminnen och historia* IV: 209-295.

Björnsson, Sven. 1946. *Blekinge. En studie av det blekingska kulturlandskapet.* Lund: Gleerup.

Bjurling, Oscar. 1940. Om oxstallningen och exporten av stalloxar från Skåne under årtiondena före och efter år 1700. *Scandia* XIII: 257-276.

Bjurling, Oscar. 1945. *Skånes utrikessjöfart 1660-1720. En studie i Skånes handelssjöfart.* Lund: Gleerup.

Bjurulf, Bo and Björn Fryklund (eds.). 1994. *Det politiska missnöjets Sverige. Statsvetare och sociologer ser på valet 1991.* Lund: Lund University Press.

Bloch, Elisabeth; Marianne Germann; Lise Skjøtt-Pedersen; Inge Kjær Jansen; and Henning Bro (eds.). 2000. *Over Øresund før Broen. Svenskere på Københavnsegnen i 300 år.* Kastrup: Lokalhistoriske Arkiver i Storkøbenhavn.

Blom, K. Arne and Jan Moen. 1987. *Snapphaneboken.* Trelleborg: Skogs.

Boalt, Carin. 1983. Tid för hemarbete. Hur lång tid då? In: Brita Åkerman et al.: *Den okända vardagen – om arbetet i hemmen*, pp. 39-69. Stockholm: Akademilitteratur.

Boalt, Carin and Sten Lindegren. 1987. Bostad och bostadsforskning. En återblick på 1940- och 50-talen. In: Christina Engfors (ed.): *Folkhemmets bostäder 1940-1960*, pp. 59-73. Stockholm: ArkitekturMuseet.

Boende förr, nu och i framtiden. Vällingby: Konsumentverket 1979.

Bøgh, Anders. 1985. Om bondeoprør: Analyseskemaer, hypoteser og teorier. In: Jørgen Würtz Sørensen and Lars Tvede-Jensen (eds.): *Til Kamp for Friheden. Sociale oprør i nordisk middelalder*, pp. 4-25. Aalborg: Aalborg University Press.

Boll-Johansen, Hans and Michael Harbsmeier (eds.). 1988. *Europas opdagelse. Historien om en idé.* Copenhagen: Christian Ejler's Forlag.

Bonde, Christer. 1658. Berättelse om Skånes, Hallands och Blekinges tillstånd och förbättring, strax efter desse landskapers afträdande till Sverige. In: *Handlingar rörande Skandinaviens Historia* VI: 85-174. Stockholm 1818.

Bringéus, Nils-Arvid. 1976. *Människan som kulturvarelse. En introduktion till etnologin.* Lund: LiberLäromedel.

Bringéus, Nils-Arvid. 1979. Byggnadsskick. In: Bringéus, Nils-Arvid (ed.): *Arbete och redskap. Materiell folkkultur på svensk landsbygd före industrialismen*, pp. 290-321. Fourth edition. Lund: LiberFörlag.

Broberg, Peter. 1991. Fyra artiklar. Sett med skånska ögon. In: SSF (Stiftelsen Skånsk Framtid; ed.): *333 års-boken om Skånelandsregionen – historielös – försvarslös – framtidslös*, pp. 179-204. Örkelljunga: Settern.

Broberg, Peter. 1994. Världens lyckligaste omnationalisering – ett påstående om Skåneland. PM, Stiftelsen Skånsk Framtid.

Broberg, Peter and Ulf Kadefors. 1981. *Två systrar* (submitted to the essay-contest "Øresundsregionen – slagord eller realitet"). Copenhagen and Malmö: Øresunds kontakt.

Broberg, Peter and Kaj Spangenberg Schmidt. 1970. *Drömmen om Scantopia. Tanker om människor och framtid i Sydskandinavien.* Täby: Bokförlaget Robert Larson.

Brockdorff, Victor and Sven Tägil. 1985. *Gränsland. En historisk resa i skånsk-danska gränstrakter.* Höganäs: Wiken.

Brubaker, Rogers. 1996. *Nationalism Reframed. Nationhood and the National Question in the New Europe.* Cambridge: Cambridge University Press.

Brück, Ulla. 1984. Identitet, lokalsamhälle och lokal identitet. *Rig* 67: 65-78.

Buhl Pedersen, Søren; Christian Tangkjær; and Anders Linde-Laursen. 2003. *Mellem postkort og politisk strategi – branding af nationer, regioner og byer.* Copenhagen Business School, Department of Management, Politics and Philosophy: Working Paper No. 4.

Bundsgaard, Inge and Sidsel Eriksen. 1986. Hvem disciplinerede hvem? *Fortid og Nutid* XXXIII(1): 55-69.

Burke, Peter. 1994 (1978). *Popular Culture in Early Modern Europe*, Revised reprint. Hants: Scolar Press.

Busk-Jensen, Lise; Per Dahl; Anker Gemzøe; Torben Kragh Grodal; Jørgen Holmgaard; and Martin Zerlang. 1985. *Dansk litteraturhistorie 6. Dannelse, folkelighed, individualisme 1848-1901.* Copenhagen: Gyldendal.

Calhoun, Craig. 1993. Nationalism and Ethnicity. *Annual Review of Sociology* 19: 211-239.

Campbell, Åke. 1928. *Skånska bygder under förre hälften av 1700-talet. Etnografisk studie över den skånska allmogens äldre odlingar, hägnader och byggnader.* Uppsala: Lundequistska Bokhandeln.

Campbell, Åke. 1936. *Kulturlandskapet. En etnologisk beskrivning med särskild hänsyn till äldre svenska landskapstyper.* Stockholm: Albert Bonniers Förlag, Studentföreningen Verdandis småskrifter N:r 387.

Caplan, Pat (ed.). 2003. *The Ethics of Anthropology. Debates and Dilemmas*. London: Routledge.

Carlsson, Ernst. 1904. *Skolgeografi*, eighth edition. Stockholm: Norstedt.

Castells, Manuel. 1996. *The Rise of the Network Society*. Oxford: Blackwell.

Cederborg, C. Aug. 1945 (1900). *Den siste snapphanen. Historiskt-romantiska skildringar från snapphanefejden* I-II. Göteborg: Västra Sverige.

Cederborg, C. Aug. 1948a (1899-1900). *Göingehövdingen. Historiskt-romantiska skildringar från Carl XI:s krig i Skåne* I-II. Göteborg: Västra Sverige.

Cederborg, C. Aug. 1948b (1913). *Snapphaneblod. Historisk roman* I-II. Göteborg: Västra Sverige.

Cederborg, C. Aug. 1948c (1921). *Önnestads präst. Bilder från Skånes sista krig*. Göteborg: Västra Sverige.

Chatterjee, Partha. 1999. *The Nation and its Fragments. Colonial and Postcolonial Histories*. Princeton: Princeton University Press.

Childs, Marquis W. 1936. *Sweden. The Middle Way*. New Haven: Yale University Press.

Christensen, C.C. 1899. *Geografi for Real- og Latinskoler* I. Europa. Copenhagen: Nordisk Forlag.

Christensen, C.C. and A.M. Krogsgaard. 1909. *Geografi for Folkeskolen*, third edition. Copenhagen: Gyldendal.

Christensen, Georg. 1923. Danskes Rejser i Sverige. *Samlaren*. Ny följd 3: 175-207.

Christensen, Georg (ed.). 1924. *Jacob Bircherods Rejse til Stockholm 1720*. Copenhagen: Aschehoug.

Christensson, Jakob. 1991. Clifford Geertz – en antropolog för historiker? *Häften för Kritiska Studier* 24(3): 3-12.

Christiansen, Niels Finn. 1984. Denmark: End of the Idyll. *New Left Review* 114: 5-32.

Christiansen, Niels Finn. 1992. Socialismen og fædrelandet. Arbejderbevægelsen mellem internationalisme og national stolthed 1871-1940. In: Ole Feldbæk (ed.): *Dansk identitetshistorie* III, pp. 512-586. Copenhagen: C.A. Reitzels Forlag.

Clifford, James. 1986. Introduction: Partial Truths. In: James Clifford and George E. Marcus (eds.): *Writing Culture. The Poetics and Politics of Ethnography*, pp. 1-26. Berkley: University of California Press.

Clifford, James and George E. Marcus (eds.). 1986. *Writing Culture. The Poetics and Politics of Ethnography*. Berkley: University of California Press.

Cohen, Stanley. 1980 (1972). *Folk Devils and Moral Panics. The Creation of the Mods and Rockers*. New York: St. Martin's Press.

Colley, Linda. 1992. *Britons. Forging the Nation 1707-1837*. New Haven: Yale University Press.

Connor, Walker. 1973. The Politics of Ethnonationalism. *Journal of International Affairs* 27(1): 1-21.

Conrad, Flemming. 1991. Konkurrencen 1818 om en nationalsang. In: Ole Feldbæk (ed.): *Dansk identitetshistorie* II, pp. 150-252. Copenhagen: C.A. Reitzels Forlag.

Corall, Gerd and Britta Calminder. 1959. *Skolans hemkunskap, årskurs 7*, second edition. Stockholm: AVCarlsons.

Cordua, Jeanne. 1993. *Oprørets Børn*. Allinge: Gornitzkas Forlag.

Crapanzano, Vincent. 1986. Herme's Dilemma. The Masking of Subversion in Ethnographic Description. In: James Clifford and George E. Marcus (eds.): *Writing Culture. The Poetics and Politics of Ethnography*, pp. 1-26. Berkley: University of California Press.

Cronholm, Christopher. 1976. *Blekings beskrivning*. Författad av Christopher Cronholm år c:a 1750-1757. Malmö: Blekingia.

Dahl, Sven. 1942. *Torna och Bara. Studier i Skånes bebyggelse- och näringsgeografi före 1860*. Lund: Meddelanden från Lunds universitets geografiska institution.

Dam, Folmer. (1960). *Statens Husholdningsråd gennem 25 år. 1935-1960*. Copenhagen: Statens Husholdningsråd.

Damgaard, Ellen. 1989. The Far West of Denmark. Peasant Initiative and World Orientation in Western Jutland. *Ethnologia Scandinavica*: 107-127.

Damsholt, Kirstine. 1993. Nationalkarakterens genkomst. *Folk og Kultur*: 68-84.

Damsholt, Kirstine (Tine). 1995. On the Concept of the "Folk". *Ethnologia Scandinavica*: 5-24.

Danske erfaringer med at bo i Skåne. Copenhagen: Øresundsbron and Öresundskomiteen 2005.

Danskere. 17 tanker om danskere og danskheden. Udgivet i anledning af udstillingen "Dansk, danskere…". Copenhagen: Museumsrådet for København og Frederiksberg 1992.

Darian-Smith, Eve. 1999. *Bridging Divides. The Cannel Tunnel and English Legal identity in the New Europe*. Berkeley: University of California Press.

Daun, Åke. 1983. Svensk mentalitet. Oplæg: Nye veje, nye horisonter, nye arbejdsmarker. Nordisk etnolog- og folkloristkongres, Danmark, 30 maj – 3 juni 1983.

Daun, Åke. 1989a. *Svensk mentalitet. Ett jämförande perspektiv*. Stockholm: Rabén and Sjögren.

Daun, Åke. 1989b. Studying National Culture by Means of Quantitative Methods. *Ethnologia Europaea* XIX(1): 25-32.

Daun, Åke. 1992. *Den europeiska identiteten. Bidrag till samtal om Sveriges framtid*. Stockholm: Rabén and Sjögren.

Daun, Åke. 1996. *Swedish Mentality*. University Park: Pennsylvania State University Press.

Daun, Åke; Carl-Erik Mattlar; and Erkki Alanen. 1988. Finsk och svensk personlighet. In: Åke Daun and Billy Ehn (eds.): *Blandsverige. Om kulturskillnader och kulturmöten*, pp. 266-294. Stockholm: Carlssons.

De Clercq, Willy et al. 1993. Reflexion on information and communication Policy. Report by the group of experts chaired by Mr Willy De Clercq. Kommissionen for de Europæiske Fællesskaber.

Den Svenska Historien. Ett utställningsprojekt. Stockholm: Historiska Museet, Nordiska Museet, no year.

Dhillon. 1976. Ni är alldeles för lagom! *Ordets makt* (3): 44-48.

Diskning. HFI meddelanden Nr 1. Stockholm: Hemmens forskningsinstitut 1946.

Donnan, Hastings and Thomas M. Wilson. 1999. *Borders. Frontiers of Identity, Nation and State.* Oxford: Berg.

Douglas, Mary. 1966. *Purity and Danger. An Analysis of the Concepts of Pollution and Taboo.* London: Routledge and Kegan Paul.

Douglass, William A. 1998. A Western Perspective on an Eastern Interpretation of Where North Meets South: Pyrenean Borderland Cultures. In: Thomas M. Wilson and Hastings Donnan (eds.): *Borders Identities. Nation and State at International Borders*, pp. 62-95. Cambridge: Cambridge University Press.

Drakulic, Slavenka. 1994 (1993). *The Balkan Express. Fragments From the Other Side of the War.* New York: HarperPerennial.

Dübeck, Inger. 1987. *Fra gammel dansk ret til ny svensk ret. Den retlige forsvenskning i de tabte territorier 1645-1683.* Studies and sources vol. 3. Copenhagen: Rigsarkivet, G.E.C. Gad.

Dyrbye, Martin. 1990. M. W. Bruns dramatisering af "Gjøngehøvdingen" og opsætningen på Folketeatret 1865. *Historiske Meddelelser om København*: 78-99.

Edgerton, David; Björn Fryklund; and Tomas Peterson. 1994. *"Until the Lamb of God Appears..." The 1991 Parliamentary Election: Sweden Chooses a New Political System.* Lund: Lund University Press.

Edvardsson, Vigo. 1960. *Loshultskuppen 1676. En märklig händelse från snapphanefejden.* Broby: Författarens förlag.

Edvardsson, Vigo. 1974. *Snapphanekriget 1675-1679. Dokument från en orolig tid II.* Broby: Författarens förlag.

Edvardsson, Vigo. 1975. *Snapphanekriget 1675-1679. Dokument från en orolig tid I* (second edition). Broby: Författarens förlag.

Edvardsson, Vigo. 1977. *Snapphanekriget 1675-1679. Dokument från en orolig tid III.* Tyringe: Göinge-bladet.

Edwards, John. 1985. *Language, Society, and Identity.* Oxford: Basil Blackwell.

Egardt, Brita. 1962. *Hästslakt och rackarskam. En etnologisk undersökning av folkliga fördomar.* Stockholm: Nordiska museets handlingar 57.

Ehn, Billy. 1989. National Feeling in Sport. The Case of Sweden. *Ethnologia Europaea* XIX(1): 56-66.

Ehn, Billy. 1992. Kulturalisering på gott och ont. *Kulturella Perspektiv* (1): 3-7.

Ehn, Billy; Jonas Frykman; and Orvar Löfgren. 1993. *Försvenskningen av Sverige. Det nationellas förvandlingar.* Stockholm: Natur och Kultur.

Ehn, Billy and Orvar Löfgren. 1982. *Kulturanalys.* Lund: LiberFörlag.

Ekström, Anders. 1994. *Den utställda världen. Stockholmsutställningen 1897 och 1800-talets världsutställningar*. Stockholm: Nordiska museets handlingar 119.

Elmqvist, Bjørn. 1992. Det flade Danmark...En god model. In: *Danskere. 17 tanker om danskere og danskheden*. Udgivet i anledning af udstillingen "Dansk, danskere...", pp. 49-53. Copenhagen: Museumsrådet for København og Frederiksberg.

Emigrationsutredningen. Bilag XX. Svenskarna i utlandet. Uppgifter rörande svenskarnas ställning i vissa främmande länder. Äfvensom uttalanden angående åtgärder för återinvandringen. Stockholm 1911.

Enquist, Per Olov. 1992. *Kartritarna*. Stockholm: Norstedts.

Enzensberger, Hans Magnus. 1989. *Europe, Europe: Forays into a Continent*. New York: Pantheon Books.

Erfelt, Fredrik. 1991. Kulturism – kulturer och turism. Specialstudie av dansk vintersportturism i Sverige. In: Anders Linde-Laursen and Jan Olof Nilsson (eds.): *Nationella identiteter i Norden – ett fullbordat projekt?*, pp. 279-291. Stockholm: The Nordic Council, Nord 1991: 26.

Eriksen, Anne. 1993. Den nasjonale kulturarven – en del av det moderne. *Kulturella Perspektiv* (1): 16-25.

Erixon, Sigurd (ed.). 1957. *Materiell och social kultur. Atlas över svensk folkkultur, I*. Uddevalla: Niloé.

Ernst, John. 1973. Hvad er en Danmarksfilm? In: Torben Agersnap (ed.): *Kort fra Danmark og andre Danmarksfilm*, pp. 5-16. Nyt fra samfundsvidenskaberne.

Eskildsen, Claus. 1945 (1936). *Dansk Grænselære*. Sixth edition. Copenhagen: C.A. Reitzels Forlag, Grænseforeningen.

Eskilsson, Lena. 1993. Kvinnlig vänsterhållning och radikalism i 30-talets kultur- och samhällsdebatt. In: Bertil Nolin (ed.): *Kulturradikalismen. Det moderna genombrottets andra fas*, pp. 157-169. Stehag: Brutus Östlings Bokförlag Symposion.

Etlar, Carit. 1991a (1853). *Gøngehøvdingen*. Fourth edition. Copenhagen: Gyldendals paperbacks.

Etlar, Carit. 1991b (1854). *Dronningens vagtmester*. Fourth edition. Copenhagen: Gyldendals paperbacks.

Ewald, H.F. 1867. *Svenskerne paa Kronborg. Historisk Roman* I-II. Copenhagen: Gyldendal.

Faber, Peder Ditlev. 1965 (1820). Svend Poulsen, Bonde i Sjælland. En Fortælling. *Historisk Samfund for Præstø Amt. Årbog* 1965, ny række 6(4): 371-422.

Fabricius, Knud. 1972 (1906a). *Skaanes Overgang fra Danmark til Sverige. Studier over Nationalitetsskiftet i de skaanske Landskaber i de nærmeste Slægtled efter Brømsebro- og Roskildefredene. Første Del (1645-1660)*. Copenhagen: Selskabet for udgivelse af kilder til dansk historie.

Fabricius, Knud. 1972 (1906b). *Skaanes Overgang fra Danmark til Sverige. Studier over Nationalitetsskiftet i de skaanske Landskaber i de nærmeste Slægtled efter Brømsebro- og Roskildefredene. Anden del (1660-1676)*. Copenhagen: Selskabet for udgivelse af kilder til dansk historie.

Fabricius, Knud. 1972 (1952). *Skaanes Overgang fra Danmark til Sverige. Studier over Nationalitetsskiftet i de skaanske Landskaber i de nærmeste Slægtled efter Brømsebro- og Roskildefredene. Tredie del*. Copenhagen: Selskabet for udgivelse af kilder til dansk historie.

Fabricius, Knud. 1972 (1958). *Skaanes Overgang fra Danmark til Sverige. Studier over Nationalitetsskiftet i de skaanske Landskaber i de nærmeste Slægtled efter Brømsebro- og Roskildefredene. Fjerde del*. Copenhagen: Selskabet for udgivelse af kilder til dansk historie.

Falkheimer, Jesper. 2004. *Att gestalta en region. Källornas strategier och mediernas föreställningar om Öresund*. Göteborg: Makadam.

Febvre, Lucien. 1973 (1928). Frontière: the word and the concept. In: *A new Kind of History*. From the writings of Febvre. Peter Burke (ed.), pp. 208-218. London: Routledge and Kegan Paul.

Feldbæk, Ole. 1991a. Fædreland og Indfødsret. 1700-tallets danske identitet. In: Ole Feldbæk (ed.): *Dansk identitetshistorie* I, pp. 111-230. Copenhagen: C.A. Reitzels Forlag.

Feldbæk, Ole. 1991b. Skole og identitet 1789-1848. Lovgivning og lærebøger. In: Ole Feldbæk (ed.): *Dansk identitetshistorie* II, pp. 253-324. Copenhagen: C.A. Reitzels Forlag.

Feldbæk, Ole (ed.). 1991-1992. *Dansk identitetshistorie*, vol. I-IV. Copenhagen: C.A. Reitzels Forlag.

Feldbæk, Ole and Vibeke Winge. 1991. Tyskerfejden 1789-1790. Den første nationale konfrontation. In: Ole Feldbæk (ed.): *Dansk identitetshistorie* II, pp. 9-109. Copenhagen: C.A. Reitzels Forlag.

Fernebring, Lars and Mac Robertson. 1993. *C. Aug. Cederborg, snapphanetidens berättare*. Södra Sandby: Vekerum.

Fink, Hans. 1989. Arenabegreber – mellem ørkensand og ørkensand. In: Hans Fink (ed.): *Arenaer – om politik og iscenesættelse*, pp. 7-21. Kulturstudier 5. Århus: Aarhus University Press.

Fink, Hans. 1991. Om identiteters identitet. In: Hans Fink and Hans Hauge (eds.): *Identiteter i forandring*, pp. 204-226. Kulturstudier 12. Århus: Aarhus University Press.

Fink, Hans. 1992. Om grænsers måder at være grænser på. In: Frederik Stjernfelt and Anders Troelsen (eds.): *Grænser*, pp. 9-30. Kulturstudier 15. Århus: Aarhus University Press.

Fischer, Michael M.J. 1986. Ethnicity and the Post-Modern Arts of Memory. In: James Clifford and George E. Marcus (eds.): *Writing Culture. The Poetics and Politics of Ethnography*, pp. 194-233. Berkley: University of California Press.

Floto, Inga. 1985. *Historie. Nyere og nyeste tid*. Videnskabernes historie i det 20. århundrede. Copenhagen: Gyldendal.

Folkmängd 31 dec 1988. Del 3. Fördelning efter kön, ålder, civilstånd och medborgarskap i kommuner m m. Sveriges officiella statistik. Stockholm: Statistiska centralbyrån 1989.

Fonsmark, Henning. 1990. *Historien om den danske utopi. Et idépolitisk essay om danskernes velfærdsdemokrati.* Copenhagen: Gyldendal.

Foster, Robert J. 1991. Making National Cultures in the Global Ecumene. *Annual Review of Anthropology* 20: 235-260.

Foucault, Michel. 1972 (1971). The Discourse on Language. Appendix in: Michel Foucault: *The Archaeology of Knowledge*, pp. 215-237. New York: Pantheon Books.

Foucault, Michel. 1979 (1974). *Discipline and Punish. The Birth of the Prison.* New York: Vintage Books.

Frandsen, Karl-Erik. 1994. *Okser på vandring. Produktion og eksport af stude fra Danmark i midten af 1600-tallet.* Herning: Skippershoved.

Frandsen, Steen Bo. 1996. *Opdagelsen af Jylland. Den regionale dimension i Danmarkshistorien 1814-1864.* Aarhus: Aarhus University Press.

Fredriksson, Berndt. 1983. Krigets krav och konsekvenser. Ekonomiska och sociala aspekter på Skånska kriget. In: Finn Askgaard and Arne Stade (eds.): *Kampen om Skåne*, pp. 335-378. Copenhagen: ZAC and Co.

Fryklund, Björn. 1994. Avslutande Sammanfattning. In: Bo Bjurulf and Björn Fryklund (eds.): *Det politiska missnöjets Sverige. Statsvetare och sociologer ser på valet 1991*, pp. 145-154. Lund: Lund University Press.

Frykman, Jonas. 1988a. *Dansbaneeländet. Ungdomen, populärkulturen och opinionen.* Stockholm: Natur och Kultur.

Frykman, Jonas. 1988b. Folklivsarkivet, frågelistorna och forskningen. In: Nils-Arvid Bringéus (ed.): *Folklivsarkivet i Lund 1913-1988. En festskrift till 75-årsjubileet*, pp. 91-107. Lund: Skrifter från Folklivsarkivet i Lund 25.

Frykman, Jonas. 1989. Social Mobility and National Character. *Ethnologia Europaea* XIX(1): 33-46.

Frykman, Jonas. 1990. What People Do But Seldom Say. *Ethnologia Scandinavica*: 50-62.

Frykman, Jonas and Orvar Löfgren. 1987 (1979). *Culture Builders. A Historical Anthropology of Middle-Class Life.* New Brunswick: Rutgers University Press.

Frykman, Jonas and Orvar Löfgren (eds.). 1985. *Modärna tider. Vision och vardag i folkhemmet.* Malmö: Liber.

Frykman, Jonas and Orvar Löfgren (eds.). 1991. *Svenska vanor och ovanor.* Stockholm: Natur och Kultur.

Frykman, Jonas and Orvar Löfgren (eds.). 1996. *Force of Habit. Exploring Everyday Culture.* Lund: Lund University Press.

Frykman, Jonas and Maja Povrzanovic Frykman. 2000. Europa, Balkan och krigets alltför många offer. *Kulturella Perspektiv* (4): 24-30.

Fukuyama, Francis. 1992. *The End of History and the Last Man.* New York: The Free Press.

Gallén, Jarl. 1968. *Nöteborgsfreden och Finlands medeltida östgräns.* Helsingfors: Svenska Litteratursällskapet i Finland.

Gallén, Jarl and John Lind. 1991. *Nöteborgsfreden och Finlands medeltida östgräns*, andra delen. Helsingfors: Svenska litteratursällskapet i Finland.

Geertz, Clifford. 1975. *The Interpretation of Cultures. Selected Essays*. London: Hutchinson.

Gellner, Ernest. 1983. *Nations and Nationalism*. Oxford: Basil Blackwell.

Gellner, Ernest. 1994. *Encounters with Nationalism*. Oxford: Blackwell.

Gerholm, Lena. 1993. Fältarbete, intervju och text. *Nord-Nytt* 52: 15-19.

Gerholm, Lena and Tomas Gerholm. 1989. Om inte kultur fanns, så skulle man behöva uppfinna den. *Nord-Nytt* 37: 8-16.

Gerner, Kristian. 1990. Två hundra års europeisk felutveckling – karaktäriserat av nationalismen, rasismen och territorialstaten – går mot sitt slut. *Nord Revy* (3): 11-16.

Gertten, Magnus and Arne König. 1985. *Unga om unge. 17 intervjuer med unga vid Öresund*. Copenhagen and Malmö: Øresunds Kontakt.

God bostad. Stockholm: Kungl. Bostadsstyrelsens Skrifter 1954.

Gregory, Derek. 1994. *Geographical Imaginations*. Cambridge: Blackwell.

Gudmundsson, Gestur. 1991. Rockmusik som producent af moderne islandsk identitet. In: Anders Linde-Laursen and Jan Olof Nilsson (eds.): *Nationella identiteter i Norden – ett fullbordat projekt?*, pp. 195-225. Stockholm: The Nordic Council, Nord 1991: 26.

Guerrina, Roberta. 2002. *Europe. History, Idea, and Ideologies*. London: Arnold.

Gullestad, Marianne. 1989. Small Facts and Large Issues: The Anthropology of Contemporary Scandinavian Society. *Annual Review of Anthropology* 18: 71-93.

Gundelach, Peter. 2002. *Det er dansk*. Copenhagen: Hans Reitzels Forlag.

Gustafsson, Harald. 1994a. Conglomerats or Unitary States? Integration Processes in Early Modern Denmark-Norway and Sweden. *Wiener Beiträge zur Geschichte der Neuzeit* 21: 45-62.

Gustafsson, Harald. 1994b. Vad var staten? Den tidigmoderna svenska staten: sex synpunkter och en modell. *Historisk tidskrift* (2): 203-227.

Gustafsson, Harald. 2000a. *Gamla riken, nya stater. Statsbildning, politisk kultur och identiteter under Kalmarunionens upplösningsskede*. Stockholm: Atlantis.

Gustafsson, Harald. 2000b. Præsten som velfærdsforvalter i tidligt moderne tid. In: Tim Knudsen (ed.): *Den nordiske protestantisme og velfærdsstaten*, pp. 87-97. Aarhus: Aarhus Universitetsforlag, Center for Europæisk Kirkeret og Kirkekundskab.

Gustafsson, Harald and Hanne Sanders (eds.). 2006. *Vid gränsen. Integration och identitet i det förnationella Norden*. Göteborg: Makadam.

Gustafsson, Lars G. 1993. *Öresundsregion eller Skåneland? Två regionala framtidsvisioner*. Lund: Seminar-paper, ETN 003, Department of European Ethnology.

Hagberg, Jan-Erik. 1986. *Tekniken i kvinnornas händer. Hushållsarbetet och hushållsteknik under tjugo- och trettiotalen.* Linköping Studies in Arts and Science 7. Malmö: Liber.

Hallenberg, Aug. 1940. Från 1600-talets Bleking under danskt välde. *Blekingeboken:* 171-186.

Haller, Dieter and Hastings Donnan (eds.). 2000. Borders and Borderlands. An Anthropological Perspective. Special Issue, *Ethnologia Europaea* 30(2).

Hamilton, Hugo. 1830. *Teckningar ur Skandinaviens Äldre Historia.* Sveriges och Norriges Arffurstar Carl, Gustav och Oscar i underdånighet tillägnade. Stockholm.

Hammerich, Paul. 1986. *Lysmageren. En krønike om Poul Henningsen.* Copenhagen: Gyldendal.

Handler, Richard. 1983. The Dainty and the Hungry Man. Literature and Anthropology in the Work of Edward Sapir. In: George W. Stocking (ed.): *Observers Observed. Essays on Ethnographic Fieldwork,* pp. 208-231. Madison: University of Wisconsin Press.

Handler, Richard and William Saxton. 1988. Dyssimulation: Reflexivity, Narrative, and the Quest for Authenticity in "Living History". *Cultural Anthropology* 3(3): 242-260.

Hannerz, Ulf. 1992. *Cultural Complexity. Studies in the Social Organization of Meaning.* New York: Columbia University Press.

Hannerz, Ulf. 1993. The Withering away of the Nation? An Afterword. *Ethnos* 58(3-4): 377-391.

Hannerz, Ulf. 1996. *Transnational Connections. Culture, People, Places.* London: Routledge.

Hansen, Georg. 1944. *Degnen. Studie i det 18. Aarhundredes Kulturhistorie.* Copenhagen: J.H. Schultz Forlag.

Hansen, Georg. 1947. *Præsten paa Landet i Danmark i det 18. Aarhundrede. En kulturhistorisk Undersøgelse.* Copenhagen: Det danske Forlag.

Hansen, Kjell. 1998. *Välfärdens motsträviga utkant. Lokal praktik och statlig styrning i efterkrigstidens nordsvenska inland.* Lund: Historiska media.

Hansen, Kjell. 1999. Det föreställdas politiska makt. In: Tine Damsholt and Fredrik Nilsson (eds.): *Ta fan i båten. Etnologins politiska utmaningar,* pp. 26-43. Lund: Studentlitteratur.

Hansgaard, H.C. 1956. *Af gøngernes (snaphanernes) og gøngehøvdingernes saga. Sabotage- og partisankrig for Danmark gennem 150 år.* Århus: Søren Lunds Forlag.

Hanssen, Börje. 1973a. Kulturens permanens och förändring. *Nord-Nytt* 3-4: 34-44.

Hanssen, Börje. 1973b. Common Folk and Gentlefolk. *Ethnologia Scandinavica:* 67-100.

Hanssen, Börje. 1977 (1952). *Österlen. Allmoge, köpstafolk och kultursammanhang vid slutet av 1700-talet i sydöstra Skåne.* Stockholm: Gidlunds.

Hansson, Göran. 1994. *EU och regionerna. Sverige – ur takt med tiden.* Marieholm: Hansson.

Harbsmeier, Michael. 1986. Danmark: Nation, kultur og køn. *Stofskifte. Tidsskrift for Antropologi* 13. Danmark: 47-73.

Hastrup, Kirsten. 1988. Kultur som analytisk begreb. In: Hans Hauge and Henrik Horstbøll (eds.): *Kulturbegrebets kulturhistorie*, pp. 120-139. Kulturstudier 1. Århus: Aarhus University Press.

Hauge, Hans. 1991. Identitetens trussel. In: Hans Fink and Hans Hauge (eds.): *Identiteter i forandring*, pp. 185-203. Kulturstudier 12. Århus: Aarhus University Press.

Hechter, Michael. 1975. *Internal Colonialism. The Celtic Fringe in British National Development, 1536-1966*. Berkeley: University of California Press.

Hechter, Michael and William Brustein. 1980. Regional Modes of Production and Patterns of State Formation in Western Europe. *American Journal of Sociology* 85(5): 1061-1094.

Hedetoft, Ulf; Bo Petersson; and Lina Sturfelt (eds.). 2006. *Bortom stereotyperna? Invandrare och integration i Danmark och Sverige*. Göteborg: Makadam.

Heiberg, Edvard; Eske Kristensen; and Bent Salicath. 1950. *Planlægning af køkkener i etagehuse*. Copenhagen: Fællesorganisationen af almennyttige danske boligselskaber.

Heiberg, Steffen. 1993. *Enhjørningen Corfitz Ulfeldt*. Copenhagen: Gyldendal.

von Heidenstam, Verner. 1920 (1896). *Svenskarnas lynne*. Stockholm: Skolan för bokhantverk.

Henningsen, Bernd. 2005. Crumbling solidarity and the need for reforms. Transformations in the "old" and "new" Europes. In: Lars-Folke Landgrén and Pirkko Hautamäki (eds.): *People, Citizen, Nation*, pp. 172-189. Helsinki: Renvall Institute.

Henningsen, Poul. 1930. Stockholms Udstillingen. *Nyt Tidsskrift for Kunstindustri* 3(6): 81-92.

Henningsen, Poul. 1933. *Hvad med Kulturen?* Copenhagen: Monde.

Henningsen, Poul. 1994. *På Hundredåret. Tekster 1918-1967*. Second revised edition, Olav Harsløf (ed.). Copenhagen: Hans Reitzels Forlag.

Hervik, Peter (ed.). 1999. *Den generende forskellighed. Danske svar på den stigende multikulturalisme*. Copenhagen: Hans Reitzels Forlag.

Hirdman, Yvonne. 1989. *Att lägga livet tillrätta – studier i svensk folkhemspolitik*. Stockholm: Carlssons.

Hirsch, E.D., Jr. 1987. *Cultural Literacy. What Every American Needs to Know*. Boston: Houghton Mifflin Company.

Hobsbawm, Eric. 1969. *Bandits*. Delacorte Press.

Hobsbawm, Eric. 1990. *Nations and Nationalism since 1780. Programme, myth, reality*. Cambridge: Cambridge University Press.

Hobsbawm, Eric and Terence Ranger (eds.). 1992 (1983). *The Invention of Tradition*. Cambridge: Canto.

Høiris, Ole. 1988. Kulturbegrebet i antropologien. In: Hans Hauge and Henrik Horstbøll (eds.): *Kulturbegrebets kulturhistorie*, pp. 95-119. Kulturstudier 1. Århus: Aarhus University Press.

Højrup, Ole. 1963. *Levnedsløb i Sørbymagle og Kirkerup kirkebøger 1646-1731* I. Biografierne. Copenhagen: Udvalget for udgivelse af kilder til landbefolkningens historie.

Højrup, Thomas. 1983a. *Det glemte folk. Livsformer og centraldirigering.* Hørsholm: Statens Byggeforsknings-institut.

Højrup, Thomas. 1983b. The Concept of Life-Mode. A Form-Specifying Mode of Analysis Applied to Contemporary Western Europe. *Ethnologia Scandinavica*: 15-50.

Holboe, Manita. 1962. *Vore nordiske naboer. Over sø og land* I, second edition. Copenhagen: Gjellerup.

Holm Joensen, Poul. 1967. *Norden og Nordsølandene.* Geografi 2 i Jorden-rundt serien. Copenhagen: Gad.

Holst, Olaf. 1941. Folket og dets Egenart. In: Hakon Stangerup (ed.): *Det moderne Sverige. Evne, Indsats, Vilje*, pp. 30-51. Copenhagen: Forlaget af 1939.

Honko, Lauri (ed.). 1980. *Folklore och nationsbyggande i Norden.* Åbo: NIF Publications No. 9.

Hoy, Suellen. 1995. *Chasing Dirt. The American Pursuit of Cleanliness.* New York: Oxford University Press.

Hulsey, Timothy L. and Christopher J. Frost. 2004. *Moral Cruelty. Ameaning and the Justification of Harm.* Lanham: University Press of America.

Huntford, Roland. 1972. *The New Totalitarians.* New York: Stein and Day.

Hylland Eriksen, Thomas. 1993. *Ethnicity and Nationalism. Anthropological Perspectives.* London: Pluto Press.

Idvall, Markus. 1997. Nationen, regionen och den fasta förbindelsen. Ett hundraårigt statligt projekts betydelse i ett territoriellt perspektiv. In: Sven Tägil, Fredrik Lindström, and Solveig Ståhl (eds.): *Öresundsregionen – visioner och verklighet. Från ett symposium på Lillö*, pp. 126-151. Lund: Lund University Press.

Idvall, Markus. 2000. *Kartors kraft. Regionen som samhällsvision i Öresundsbrons tid.* Lund: Nordic Academic Press.

Ingebritsen, Christine. 1998. *The Nordic States and European Unity.* Ithaca: Cornell University Press.

Ingesman, Per. 2000. Kirke, stat og samfund i historisk perspektiv. In: Tim Knudsen (ed.): *Den nordiske protestantisme og velfærdsstaten*, pp. 65-86. Aarhus: Aarhus Universitetsforlag, Center for Europæisk Kirkeret og Kirkekundskab.

Integration och utveckling i Öresundsregionen. Möjligheter och utmaningar. En utredning gjord av Lunds universitet på uppdrag av Utrikesdepartementet. Lund: Lund University, 1999 (see also: http://www.lu.se/lu/oresund/).

Jeansson, Alice. 1936. *Det genomtänkta hushållsarbetet. En bilderbok för den tröttkörda husmodern.* Stockholm: Skattebetalarnas förenings budgetbyrå.

Jenkins, Richard. 2000. Not Simple at All: Danish Identity and the European Union. In: Irène Bellier and Thomas M. Wilson (eds.): *An Anthropology of the European Union*, pp. 159-178. Oxford: Berg.

Jeppsson, Gert. 1983. En omstridd befolkning. In: Finn Askgaard and Arne Stade (eds.): *Kampen om Skåne*, pp. 83-109. Copenhagen: ZAC and Co.

Johannesson, Gösta. 1981. *Danmarks historie. Skåne, Halland og Blekinge.* Copenhagen: Politikens Forlag.

Johnsson, Pehr. 1908. *Paul Enertsen. Den sidste danske prästen i Skåne.* Landskrona.

Johnsson, Pehr. 1909. *Bidrag till snaphanerörelsen i Halland 1676-79.* Falkenberg.

Johnsson, Pehr. 1949. *Östra Göinge domsaga. Broby som tingsplats.* Broby: Östra Göinge Domarekansli.

Johnston, R.J.; David B. Knight; and Eleonore Kofman. 1988. Nationalism, Self-determination and the World Political Map: An Introduction. In: R.J. Johnston, David B. Knight, and Eleonore Kofman (eds.): *Nationalism, Selfdetermination, and Political Geography*, pp. 1-17. Beckenham: Croom Helm.

Jørgensen, Gerd Anja. 1994. Ren i skind er ren i sind. *Folk og Kultur*: 99-111.

Kaae, Lars. 1986. Ikkun som voxne Menneske-Børn: Grundtvig og frihed. In: Tønnes Bekker-Nielsen (ed.): *Stykkevis og delt. 5 essays om Grundtvig og grundtvigianismen*, pp. 75-122. Århus: Antikva.

Karlsson, Svenolof (ed.). 1994. *Norden är död. Länge leve Norden! En debattbok om de nordiska länderna som en "megaregion" i Europa.* Stockholm: The Nordic Council, Nord 1994: 11.

Kayser Nielsen, Niels. 1993. *Krop og oplysning. Om kropskultur i Danmark 1780-1900.* Odense: Odense University Press.

Kayser Nielsen, Niels. 1994. Dem fra Farre med det røde V. Den belyste krop og kroppens udstråling. In: Anders Linde-Laursen and Jan Olof Nilsson (eds.): *Möjligheternas landskap. Nordiska kulturanalyser*, pp. 180-194. Copenhagen: Nordic Council of Ministers, Nord 1994: 21.

Kellas, James G. 1991. *The Politics of Nationalism and Ethnicity.* London: Macmillan.

Kierkegaard, Søren. 1988 (1845). *Stages on Life's Way.* Studies by various persons. Edited and Translated with Introduction and Notes by Howard V. Hong and Edna H. Hong. Princeton: Princeton University Press.

Kildegaard, Bjarne. 1990. Mourning the Past, Hoping for the Future (with commentary by Barbro Klein). *Ethnologia Scandinavica*: 34-39.

Kirk, Hans. 1999. *The Fishermen.* Iowa City: Fanpihua Press.

Kjær, Gitte. 1992. *Svend Poulsen Gønge – i virkeligheden.* Ebeltoft: Skippershoved.

Kjær Hansen, Max. 1941. Økonomi og Erhvervsliv. In: Hakon Stangerup (ed.): *Det moderne Sverige. Evne, Indsats, Vilje*, pp. 74-104. Copenhagen: Forlaget af 1939.

Kjærgaard, Thorkild. 1979. Gårdmandslinien i dansk historieskrivning. *Fortid og Nutid* XXXVIII(2): 178-191.

Kjellman, Gunilla. 1993. *Varats oändliga tinglighet. En studie om föremål som kulturbärare.* Stockholm: Carlssons.

Klein, Barbro. 1988. Den gamla hembygden eller vad har hänt med de svenska folktraditionerna i USA? In: Åke Daun and Billy Ehn (eds.): *Blandsverige. Om kulturskillnader och kulturmöten*, pp. 43-67. Stockholm: Carlssons.

Klein, Barbro. 1989. Ett eftermiddagssamtal hos Elsa. In: Billy Ehn and Barbro Klein (eds.): *Etnologiska beskrivningar*, pp. 207-230. Stockholm: Carlssons.

Knudsen, Anne. 1989. *Identiteter i Europa*. Copenhagen: Christian Ejlers' Forlag.

Knudsen, Anne. 1991. Mikronationalismens dannelseshistorie. In: Anders Linde-Laursen and Jan Olof Nilsson (eds.): *Nationella identiteter i Norden – ett fullbordat projekt?*, pp. 19-38. Stockholm: The Nordic Council, Nord 1991: 26.

Knudsen, Tim. 1994. Den svenske model fra ideal til krise. *Den jyske historiker* 68: 89-109.

Knudsen, Tim and Bo Rothstein. 1994. State Building in Scandinavia. *Comparative Politics* 26(2): 203-220.

Kök. Planering. Inredning. Stockholm: Hemmens forskningsinstitut 1954.

Kök – planering, inredning. En bok för den som planerar, bygger eller bygger om kök. Stockholm: Konsumentverket 1981.

Kondo, Dorinne K. 1990. *Crafting Selves. Power, Gender, and Discourses of Identity in a Japanese Workplace*. Chicago: University of Chicago Press.

Koskinen, Lennart. 1980. *Tid och evighet hos Sören Kierkegaard*. Lund: Doxa.

Krantz, Claes. 1951. *Att resa i Danmark*. Stockholm: Wahlström and Widstrand.

Krause-Jensen, Esbern. 1978. *Viden og magt. Studier i Michel Foucaults institutionskritik*. Copenhagen: Rhodos.

Kristensen, Frode. 1938. *Sverige. Erhvervs- og samfundsliv*. Copenhagen: Martins Forlag.

Kristeva, Julia. 1993. *Nations Without Nationalism*. New York: Columbia University Press.

Kristiansen, Kristof K. and Jens R. Rasmussen (eds.). 1988. *Fjendebilleder og fremmedhad*. Copenhagen: FN-forbundet.

Kryger, Karin. 1986. *Frihedsstøtten*. Copenhagen: Landbohistorisk Selskab.

Kulturrådet 1-2. Den svenska historien m.m. Statens kulturråd 1992.

Kürti, László. 2001. *The Remote Borderland. Transylvania in Hungarian Imagination*. Albany: State University of New York Press.

Lagerkvist, Pär. 1913. *Ordkonst och bildkonst. Om modärn skönlitteraturs dekadans. Om den modärna konstens vitalitet*. Med ett förord av August Brunius. Stockholm: Bröderna Lagerströms Förlag.

Lakoff, George and Mark Johnson. 1980. *Metaphors We Live By*. Chicago: University of Chicago Press.

Larsson, Göran. 1997. Skånska län 1526-1996. *Ale* (1): 1-8.

Larsson, Lars-Olof. 1964. *Det medeltida Värend. Studier i det småländska gränslandets historia fram till 1500-talets mitt*. Lund: Gleerup.

Larsson, Lars-Olof. 1974. *Historia om Småland*. Växjö: Diploma.

Larsson, Lars-Olof. 1992 (1979). *Dackeland*. Second edition. Växjö: Diploma.

Laurin, Carl G. 1911. Svensk og dansk. Nogle Ord i Anledning af Gustav Sundbärgs "Det svenska folklynnet". *Tilskueren*. Maanedsskrift (2): 361-368.

Lauring, Palle. 1961 (1952). *Danmark i Skåne. En usentimental rejse.* Copenhagen: Steen Hasselbalchs Forlag.

Le Galès, Patrick and Christian Lequesne (eds.). 1998. *Regions in Europe.* London: Routledge.

Leed, Eric J. 1991. *The Mind of the Traveller. From Gilgamesh to Global Tourism.* New York: Basic Books.

Lefebvre, Henri. 1991 (1974). *The Production of Space.* Cambridge: Blackwell.

Leizaola, Aitzpea. 2000. Mugarik ez! Subverting the Border in the Basque Country. *Ethnologia Europaea* 30(2): 35-46.

Levin, Paul. 1911. Sundbärg – Laurin. *Tilskueren.* Maanedsskrift (2): 433-437.

Liljenberg, Carl-Gustav. 1963. Krig och ofredsår vid riksgränsen. *Osby Hembygdsförening, Årsbok*: 131-197.

Lindberg, Folke. 1928. Bondefreder under svensk medeltid. *Historisk tidskrift* 48(1): 1-32.

Lindeberg, Lars. 1985. *Arvefjenden. Dansk-svenske landskampe på slagmarker og fodboldbaner fra sagntid til nutid.* Copenhagen: Forlaget Danmark.

Linde-Laursen, Anders. 1988. Fra u-land til storebror. Danskeres syn på Sverige i det 20. århundrede. Lund: Seminar-paper, Department of European Ethnology.

Linde-Laursen, Anders. 1989a. *Hugo Matthiessens kulturhistorie. Belysninger og baggrunde.* Højbjerg: Hikuin.

Linde-Laursen, Anders. 1989b. Danske skvatmøller – "Fup" eller "Fakta"? *Fortid og Nutid* XXXVI(1): 1-28.

Linde-Laursen, Anders. 1993. The Nationalization of Trivialities: How Cleaning becomes an Identity Marker in the Encounter of Swedes and Danes. In: Ulf Hannerz and Orvar Löfgren (eds.): Defining the National. *Ethnos* (3-4): 275-293.

Linde-Laursen, Anders. 1994. En bro gör ingen Copmalubo. In: *Sydsvenska Dagbladet* 22/6. Malmö.

Linde-Laursen, Anders. 1995a. *Det nationales natur. Studier i dansk-svenske relationer.* Copenhagen: Nordic Council of Ministers.

Linde-Laursen, Anders. 1995b. Danish fields forever. In: Anders Linde-Laursen and Jan Olof Nilsson (eds.): *Nordic Landscopes. Cultural Studies of Place*, pp. 68-97. Copenhagen: Nordic Council of Ministers, Nord 1995: 15.

Linde-Laursen, Anders. 1997. Främmande böjningsformer av det danska. Marknadsföring och nationell identitet i Solvang, Kalifornien. In: Gunnar Alsmark (ed.): *Skjorta eller själ? Kulturella identiteter i tid och rum*, pp. 174-198. Lund: Studentlitteratur.

Linde-Laursen, Anders. 1999a. Taking the National Family to the Movies. Changing Frameworks for the Formation of Danish Identity. *Anthropological Quarterly* 72(1): 18-33

Linde-Laursen, Anders. 1999b. Danes in the Middle of the World. In: Åke Daun and Sören Jansson (eds.): *Europeans. Essays on Culture and Identity*, pp. 181-192. Lund: Nordic Academic Press

Linde-Laursen, Anders. 1999c. Fremtidens Øresund – en kulturel region? In: *Integration och utveckling i Öresundsregionen. Möjligheter och utmaningar.* En utredning gjord av Lunds universitet på uppdrag av Utrikesdepartementet, pp. 145-154. Lund: Lund University (see also: http://www.lu.se/lu/oresund/).

Linde-Laursen, Anders. 2001. The Borders between Denmark and Sweden and the Question of Skåne. In: Michael P. Barnes (ed.): *Borders and Communities*, pp. 95-111. London: Centre for Nordic Research, University College of London.

Linde-Laursen, Anders. 2002. Broer til omverdenen. In: Per Olof Berg, Anders Linde-Laursen and Orvar Löfgren (eds.): *Öresundsbron på uppmärksamhetens marknad. Regionbyggare i evenemangsbranschen*, pp. 187-202. Lund: Studentlitteratur.

Linde-Laursen, Anders. 2005. Kalkyler ersätter visioner. *Invandrare and Minoriteter* 2005/3: 22-24.

Linde-Laursen, Anders. 2007. Is Something Rotten in the State of Denmark? The Muhammad Cartoons and Danish Political Culture. *Contemporary Islam* (1)3: 265-274.

Linde-Laursen, Anders and Jan Olof Nilsson (eds.). 1995. *Nordic Landscopes. Cultural Studies of Place.* Copenhagen: Nordic Council of Ministers, Nord 1995: 15.

Lindemann, Kelvin. 1943. *Den kan vel Frihed bære.* Copenhagen: Steen Hasselbalchs Forlag.

Lindroth, Bengt. 1994. *Sverige och Odjuret. En essä om den goda svenska traditionen.* Stockholm: En moderna tider bok.

Lindqvist, Herman. 1993. *En vandring genom Den Svenska Historien.* Höganäs: Wiken.

Linge, Lars. 1969. *Gränshandeln i svensk politik under äldre vasatid.* Lund: Gleerups.

von Linné, Carl. 1959 (1751). *Skånska resa, förrättad år 1749.* Stockholm: Wahlström and Widstrand.

Löfgren, Orvar. 1986. Om behovet for danskhed. *HUG!* 47. Danmarksbilleder: 49-56.

Löfgren, Orvar. 1989. The Nationalization of Culture. *Ethnologia Europaea* XIX(1): 5-24.

Löfgren, Orvar. 1990. Medierna i nationsbygget. Hur press, radio och TV gjort Sverige svenskt. In: Ulf Hannerz (ed.): *Medier och kulturer*, pp. 85-120. Stockholm: Carlssons.

Löfgren, Orvar. 1992. Mitt liv som konsument. Livshistoria som forskningsstrategi och analysmaterial. In: Christoffer Tigerstedt, J.P. Roos, and Anni Vilkko (eds.): *Självbiografi, kultur, liv. Levnadshistoriska studier inom human- och samhällsvetenskap*, pp. 269-288. Järfälla: Brutus Östlings Bokförlag Symposion.

Löfgren, Orvar. 1993. Swedish modern: Konsten att nationalisera konsumtion och estetik. In: Christa Lykke Christensen and Carsten Thau (eds.): *Omgang med tingene*, pp. 159-180. Kulturstudier 17. Århus: Aarhus University Press.

Löfgren, Orvar. 1999a. *On Holiday. A History of Vacationing*. Berkeley: University of California Press.

Löfgren, Orvar. 1999b. Crossing Borders. The Nationalization of Anxiety. *Ethnologia Scandinavica* 1999: 5-27.

Löfgren, Orvar. 2002. Begravningen av Pylonia. In: Per Olof Berg, Anders Linde-Laursen and Orvar Löfgren (eds.): *Öresundsbron på uppmärksamhetens marknad. Regionbyggare i evenemangsbranschen*, pp. 229-240. Lund: Studentlitteratur.

Lönnroth, Erik (ed.). 1963. *Bohusläns historia. Utarbetad på uppdrag av Göteborgs och Bohusläns landsting*. Stockholm: Almqvist and Wiksell.

Lövgren, Britta. 1993. *Hemarbete som politik. Diskussioner om hemarbete, Sverige 1930-40-talen, och tillkomsten av Hemmens Forskningsinstitut*. Stockholm: Almqvist and Wiksell.

Lukács, Georg. 1969 (1936-37). *The Historical Novel*. Harmondsworth: Penguin Books.

Lundgreen-Nielsen, Flemming 1992. Grundtvig og danskhed. In: Ole Feldbæk (ed.): *Dansk identitetshistorie* III, pp. 9-187. Copenhagen: C.A. Reitzels Forlag.

Lundkvist, Artur. 1968. *Snapphanens liv och död. En prosaballad*. Stockholm: Bonniers.

MacLeod, Gordon. 2001. New Regionalism Reconsidered: Globalization and the Remaking of Political Economic Space. *International Journal of Urban and Regional Research* 25(4): 804-829.

Magris, Claudio. 1989 (1986). *Danube*. New York: Farrar Straus Giroux.

Malkki, Lisa H. 1995. *Purity and Exile. Violence, Memory, and National Cosmology among Hutu Refugees in Tanzania*. Chicago: The University of Chicago Press.

Malling, Ove. 1992 (1777). *Store og gode Handlinger af Danske, Norske og Holstenere*. Edited by Erik Hansen. Det danske Sprog- og Litteraturselskab. Copenhagen: Gyldendal.

Månsson, Anna. 2002. *Becoming Muslim. Meanings of Conversion to Islam*. Lund: Lund University.

Marchiavelli, Niccolò. 2004 (1532). *The Prince and The Art of War*. New York: Barnes and Noble Books.

Marcus, George E. 1998. *Ethnography through Thick and Thin*. Princeton: Princeton University Press.

Matthiessen, Hugo. 1910. *Bøddel og Galgefugl. Et kulturhistorisk Forsøg*. Copenhagen: Gyldendal.

Matthiessen, Hugo. 1919. *De Kagstrøgne. Et Blad af Prostitutionens Historie i Danmark*. Copenhagen: Gyldendal.

Mbembe, Achille. 1992. The Banality of Power and the Aesthetics of Vulgarity in the Postcolonial. *Public Culture* 4(2): 1-30.

Mead, George H. 1970 (1934). *Mind, Self, and Society. From the Standpoint of a Social Behaviorist*. Chicago: University of Chicago Press.

Medborgarboken. Stockholm: Svenska landskommunernas förbund, 1950.

Mielche, Hakon. 1941. *Paa Krydstogt gennem Sverige*. Copenhagen: Steen Hasselbalchs Forlag.

Miller, Daniel. 1998. *A Theory of Shopping*. Ithaca: Cornell University Press.

Møller, Theodor. 1903-1905. Henrik Gerners Afskedstale. *Kirkehistoriske Samlinger* V(II): 144-160.

Møller, Viggo Sten et al. 1949. *Køkkenundersøgelse. Metode, Materiale og Resultater*. Publikation Nr. 1. Copenhagen: Fællesudvalget for Boligundersøgelser.

Møllgaard, Johannes. 1988. Det "mørke" Jylland og "verdensmarkedet". *Folk og Kultur*: 61-99.

Montgomery, Henry. 1989. Myter om svensken. *Tvärsnitt* (3): 38-44.

Mørch, Søren. 1982. *Den ny Danmarkshistorie 1880-1960*. Copenhagen: Gyldendal.

Mørch, Søren. 1996. *Den sidste Danmarkshistorie. 57 fortællinger af fædrelandets historie*. Copenhagen: Gyldendal.

Mosse, George L. 1985. *Nationalism and Sexuality. Middle-Class Morality and Sexual Norms in Modern Europe*. Madison: University of Wisconsin Press.

Mrázek, Rudolf. 2002. *Engineers of Happy Land. Technology and Nationalism in a Colony*. Princeton: Princeton University Press.

Munch-Petersen, Erland. 1990. Den levende Gøngehøvding. *Bogens Verden. Tidsskrift for dansk biblioteksvæsen* (4): 318-322.

Murdoch, Jonathan. 1998. The Spaces of Actor-Network Theory. *Geoforum*, 29(4): 357-374.

Murdoch, Jonathan and Terry Marsden. 1995. The spatialization of politics: local and national actor-spaces in environmental conflict. *Transactions of the Institute of British Geographers, New Series* 20(3): 368-380.

Myrdal, Janken. 1988. *1500-talets oxdrifter*. Stockholm University: Department of Economical History.

Neander-Nilsson, S. 1946. *Är svensken människa?* Stockholm: Fahlcrantz and Gumælius bokförlag.

Nederveen Pieterse, Jan. 2004. *Globalization and Culture. Global Mélange*. Lanham: Rowman and Littlefield.

Nielsen, Harald. 1912. *Svensk og dansk*. Copenhagen: Gyldendal.

Nilsson, Fredrik 1997. "The Floating Republic". On Performance and Technology in Early Nineteenth-Century Scandinavian Politics. *Journal of Folklore Research* 34(2): 85-103.

Nilsson, Fredrik. 1999. *När en timme blir tio minuter. En studie av förväntan inför Öresundsbron*. Lund: Historiska Media.

Nilsson, Jan Olof. 1991. Modernt, allt för modernt. Speglingar. In: Anders Linde-Laursen and Jan Olof Nilsson (eds.): *Nationella identiteter i Norden – ett fullbordat projekt?*, pp. 59-99. Stockholm: The Nordic Council, Nord 1991: 26.

Nilsson, Jan Olof. 1994. *Alva Myrdal – en virvel i den moderna strömmen.* Stehag: Brutus Östlings Bokförlag Symposion.

Nilsson, Jan Olof. 2000. *Berättelser om Den Nya Världen.* Lund: Lund University, Research Report in Sociology 2000:3.

Nilsson, Olof. 1874-1905. *Danmarks uppträdande i den Svenska Tronföljarefrågan, åren 1739-1743.* Malmö.

Nilsson, Sven A. 1977 (1958). Reduktion eller kontribution. Alternativ inom 1600-talets svenska finanspolitik. In: Göran Rystad (ed.): *Svenskt 1600-tal.* Problem i äldre historia, pp. 144-183. Lund: Studentlitteratur.

Nissen, Henrik S. 1992. Folkelighed og frihed 1933. Grundtvigianernes reaktion på modernisering, krise og nazisme. In: Ole Feldbæk (ed.): *Dansk identitetshistorie* III, pp. 587-673. Copenhagen: C.A. Reitzels Forlag.

Nolin, Bertil (ed.). 1993. *Kulturradikalismen. Det moderna genombrottets andra fas.* Stehag: Brutus Östlings Bokförlag Symposion.

Nordens läroböcker i historia. Föreningarna Nordens Historiska Publikationer I. Ömsesidig granskning verkställd av Föreningarna Nordens facknämnder. Helsingfors, 1937.

Nordens samhällen i Nordens skolor. Historik. Nutid. Framtid. Utgiven av föreningarna Nordens historiska facknämnder under redaktion av Nils Andrén, Helge Lundin, and Inga Löwdin. Stockholm: Föreningarna Nordens Förbund 1983.

Nordling, Erik; Rune Wittenstam; Marianne and Rune Fröroth. 1982. *Vida världen 2. Geografi. Naturkunskap. Samhällskunskap. Grundboken.* Stockholm: Almqvist and Wiksell Läromedel.

Nordström, Ludvig. 1930. *"Svea Rike".* Stockholms-utställningen 1930. Stockholm: Stockholms-utställningen.

Nordström, Ludvig. 1934. *Pyramiden Sverige. Om den nationella enheten.* Stockholm: Albert Bonniers förlag.

Nordström, Ludvig. 1938. *Lort-Sverige.* Stockholm: Kooperativa förbundet.

Nothin, Torsten. 1956. *Svenskar under ämbetsmän och fogdevälde.* Stockholm: Wahlström and Widstrand.

Numelin, Ragnar. 1935. *Danska dagar. Kulturgeografiska skisser.* Stockholm: Natur och kultur.

Nyblom, Helena. 1900. Den svenska och den danska nationalkarakteren. *I Vår Tids Livsfrågor*, skriftserie utgifven af Sydney Alrutz, N:o XV. Uppsala.

Nyrop, Fredrik P. 1929. *Sverige og Svenskerne set med danske Øjne.* Copenhagen: Radiolytternes Forlag.

Nyström, Per. 1983. *I folkets tjänst. Historikern, journalisten och ämbetsmannen Per Nyström.* Artiklar 1927-83 i urval av Anders Björnsson i samarbete med författaren. Utgivna till Per Nyströms 80-årsdag den 21 november 1983. Stockholm: Ordfronts förlag.

Nyt køkken. Praktisk and sikkert. Pjece 1. Copenhagen: Statens Husholdningsråd, 1986.

O'Dell, Tom. 1992. Myten om "amerikanaren" och Sverige. *Kulturella perspektiv* (3-4): 32-39.

O'Dell, Tom. 1997. *Culture Unbound. Americanization and Everyday Life in Sweden.* Lund: Nordic Academic Press.

O'Dell, Tom. 2002. Regionauterna. In: Per Olof Berg, Anders Linde-Laursen and Orvar Löfgren (eds.): *Öresundsbron på uppmärksamhetens marknad. Regionbyggare i evenemangsbranschen.* pp. 97-113. Lund: Studentlitteratur.

Ohlsson, Per T. 1993. *Gudarnas ö. Om det extremt svenska.* Stockholm: Brombergs, En moderna tider bok.

Ohlsson, Stig Örjan. 1978-1979. *Skånes språkliga försvenskning,* 1-2. Lund: Walter Ekstrand Bokförlag.

Ohlsson, Stig Örjan. 1993. Den språkliga förändringen. In: Erik Osvalds (ed.): *Spelet om Skåne,* pp. 87-97. Malmö: Malmö Museer.

Olins, Wally. 1999. *Trading Identities. Why Companies and Countries are Taking on Each Others' Roles.* London: Foreign Policy Centre.

Olins, Wally. 2002. Branding the Nation. The Historical Context. *Journal of Brand Management,* Vol. 9, No. 4/5.

Olrik, Hans. 1912. *Svenskt och danskt folklynne.* Stockholm: Aktiebolaget Ljus.

Olsson, Lena. 1986. *Kulturkunskap i förändring. Kultursynen i svenska geografiläroböcker 1870-1985.* Malmö: Liber.

Olwig, Kenneth R. 1995. Landscape, *landskap,* and the body. A British Italy in the North. In: Anders Linde-Laursen and Jan Olof Nilsson (eds.): *Nordic Landscopes. Cultural Studies of Place,* pp. 154-169. Copenhagen: Nordic Council of Ministers, Nord 1995:15.

Omstridda spörsmål i Nordens historia I. Föreningarna Nordens Historiska Publikationer II. Wilhelm Carlgren, K. Kretzschmer, A. Mickwitz, and Haakon Vigander (eds.). Helsingfors, 1940.

Omstridda spörsmål i Nordens historia II. Föreningarna Nordens Historiska Publikationer III. Wilhelm Carlgren, K. Kretzschmer, A. Mickwitz, and Haakon Vigander (eds.). Stockholm, 1950.

Omstridde spørsmål i Nordens historie III. Foreningene Nordens historiske publikasjoner IV. Norsk-islandske problem. Oslo: Universitetsforlaget, 1965.

Omstridte spørsmål i Nordens historie IV. Foreningene Nordens historiske publikasjoner V. Reviderte utgaver av avhandlingene i bind I om de nordiske unioner 1380-1523 og den dansk-norske forbindelsen 1536-1814. Oslo: Foreningene Nordens Forbund 1973.

Øresund – en region bliver til. Rapport udarbejdet af den danske og svenske regering, May 1999.

Øresund. The Region in Figures. Statistics Denmark, Statistics Sweden, The Oresund consortium, June 1999.

Ortner, Sherry B. 1999a. Introduction. In: Ortner, Sherry B. (ed.): *The Fate of "Culture". Geertz and Beyond*, pp. 1-13. Berkeley: University of California Press.

Ortner, Sherry B. (ed.). 1999. *The Fate of "Culture". Geertz and Beyond*. Berkeley: University of California Press.

Osbak, Edith. 1993. *Anderledes – og dog. En svensk kontordame bliver dansk.* Copenhagen: Foreningen Danmarks Folkeminder.

Österberg, Eva. 1971. *Gränsbygd under krig. Ekonomiska, demografiska och administrativa förhållanden i sydvästra Sverige under och efter nordiska sjuårskriget.* Lund: Gleerup.

Österberg, Eva. 1979. Agrar-ekonomisk utveckling, ägostrukturer och sociala oroligheter: de nordiska länderna c:a 1350-1600. *Scandia* 45(2): 171-204.

Österberg, Eva. 1989. Bönder och centralmakt i det tidigmoderna Sverige. Konflikt – kompromiss – politisk kultur. *Scandia* 55(1): 73-95.

Österberg, Eva. 1992. Folklig mentalitet och statlig makt. Perspektiv på 1500- och 1600-talens Sverige. *Scandia* 58(1): 82-102.

Österberg, Eva. 1995. *Folk förr. Historiska essäer.* Stockholm: Atlantis.

Østergård, Uffe. 1984. Hvad er det "danske" ved Danmark? Tanker om den "danske vej" til kapitalismen, grundtvigianismen og "dansk" mentalitet. *Den jyske historiker* 29-30. Kultur, mentalitet, ideologi: 85-137.

Østergård, Uffe. 1990. Begrundelser for nationalitet. To definitioner af nationen i det 19. århundredes politiske tænkning. *Scandia* 56(1): 79-88.

Østergård, Uffe. 1991a. National og etnisk identitet. In: Hans Fink and Hans Hauge (eds.): *Identiteter i forandring*, pp. 144-184. Kulturstudier 12. Århus: Aarhus University Press.

Østergård, Uffe. 1991b. Hvorfor hader vi svenskerne? Danmarkshistorierne mellem svensk og tysk. In: Anders Linde-Laursen and Jan Olof Nilsson (eds.): *Nationella identiteter i Norden – ett fullbordat projekt?*, pp. 117-147. Stockholm: The Nordic Council, Nord 1991: 26.

Østergård, Uffe. 1992a. *Europas ansigter. Nationale stater og politiske kulturer i en ny, gammel verden.* Copenhagen: Rosinante.

Østergård, Uffe. 1992b. Peasants and Danes. The Danish National Identity and Political Culture. *Comparative Studies in Society and History* 34(1): 3-27.

Østergård, Uffe. 1992c. What is National and Ethnic Identity? In: Per Bilde, Troels Engberg-Pedersen, Lise Hannestad, and Jan Zahle (eds.): *Ethnicity in Hellenistic Egypt.* Aarhus: Aarhus University Press.

Østergård, Uffe. 1994. Norden – europæisk eller nordisk? *Den jyske historiker* 69-70. De Nordiske Fællesskaber. Myter og realitet i det nordiske samarbejde: 7-37.

Østergård, Uffe. 1995. Det habsburgske og det osmanniske imperium. Mure, krige og fælles historie. In: Ola Tunander (ed.): *Europa och Muren. Om "den andre", gränslandet och historiens återkomst i 90-talets Europa*, pp. 157-194. Aalborg: Nordic Summer University.

Østergård Andersen, Jørgen. 1992. Kulturens grænser. In: Frederik Stjernfelt and Anders Troelsen (eds.). 1992. *Grænser*, pp. 81-103. Kulturstudier 15. Århus: Aarhus University Press.

Osvalds, Erik (ed.). 1993. *Spelet om Skåne*. Malmö: Malmö Museer.

Our New Region. Vår nya region. Vores nye region. Copenhagen and Malmö: City of Copenhagen and City of Malmoe, 1999.

Paulsson, Gregor. 1930. Stockholms Udstillingen. *Nyt Tidsskrift for Kunstindustri* 3(2): 17-18.

Peterson, Tomas; Mikael Stigendal; and Björn Fryklund. 1988. *Skånepartiet. Om folkligt missnöje i Malmö.* Lund: Arkiv.

Petersson, Emil. 1940-1943. *Allmogens i Blekinge besvär inför Skånska kommissionen 1669-1670* I-IV. Karlskrona: Svenssons Eftr:s bokindustri.

PH – see: Henningsen, Poul.

Ploug, Niels. 1990. *Befolkningsvandringer mellem Norden og EF* – en statistisk belysning af udenlandske statsborgere i og vandringerne mellem de nordiske lande og EF-landene. Arbejdsnotat 6. Copenhagen: Socialforskningsinstituttet.

Pontoppidan Thyssen, A. 1980. The Rise of Nationalism in the Danish Monarchy 1800-1864, with Special Reference to its Socio-economic and Cultural Aspects. In: Rosalind Mitchison (ed.): *The Roots of Nationalism: Studies in Northern Europe*, pp. 31-45. Edinburgh: John Donald Publishers.

Porskær Poulsen, Poul E. 1985. Afholdsbevægelsen som disciplineringsagent. En skitse til belysning af afholdsbevægelsens ideologi på lokalt plan i Silkeborg og i bevægelsens propaganda. *Fortid og Nutid* XXXII(3): 163-182.

Porskær Poulsen, Poul E. 1986. Et svar til Inge Bundsgaard og Sidsel Eriksen. *Fortid og Nutid* XXXIII(2): 128-130.

Poulantzas, Nicos. 1980 (1978). *State, Power, Socialism.* London: Verso.

Pred, Allan. 1992. Pure and simple lines, future lines of vision: The Stockholm Exhibition of 1930. *Nordisk Samhällsgeografisk Tidskrift* 15: 3-61.

Pred, Allan. 1995. *Recognizing European Modernities. A Montage of the Present.* London: Routledge.

Pred, Allan. 2000. *Even in Sweden. Racisms, Radicalized Spaces, and the Popular Geographical Imagination.* Berkeley: University of California Press.

von Pufendorf, Samuel. 1915 (1696). *Sju böcker om Konung Carl X Gustavs bragder.* Stockholm: Wahlström and Widstrand.

Qvist, Per Olov. 1986. *Jorden är vår arvedel. Landsbygden i svensk spelfilm 1940-1959.* Uppsala: Filmhäftet.

Qvist, Per Olov. 1995. Paradise Lost and Regained. The Countryside as Dream and Reality in Swedish Film. In: Anders Linde-Laursen and Jan Olof Nilsson (eds.): *Nordic Landscopes. Cultural Studies of Place*, pp. 126-143. Copenhagen: Nordic Council of Ministers, Nord 1995: 15.

Rabinow, Paul. 1986. Representations Are Social Facts. Modernity and Post-Modernity in Anthropology. In: James Clifford and George E. Marcus (eds.): *Writing Culture. The Poetics and Politics of Ethnography*, pp. 234-261. Berkley: University of California Press.

Rahbek Christensen, Lone. 1986. Funktionalismens sociale program – et etnocentrisk selvbedrag. *Folk og Kultur*: 45-64.

Rahbek Rosenholm, Christian. 1997. *Øresundskomiteen. En analyse af en konstitutionelt ukendt transnational regionalpolitisk organisation.* Copenhagen: SAMS Research Reports 97:15.

Rasmussen, Ebbe Gert. 1967. Begivenhederne på Bornholm under Sveriges besiddelse af øen 1658. *Bornholmske Samlinger* 2. række (3).

Rasmussen, Ebbe Gert. 1972. *Kilder til Bornholm 1658.* Rønne: Bornholms Historiske Samfund.

Rasmussen, Ebbe Gert. 1982. Dette gavebrev. Det politiske spil omkring den bornholmske opstand og Peder Olsens indsats i løsrivelsesværket 1658-59. *Bornholmske Samlinger* 2. række (15-16).

Rasmussen, Ebbe Gert. 1985. *Bornholm og arveriget. En oversigt.* Rønne: Rønne Statsskole.

Rasmussen, Ebbe Gert. 1990. Modstanden i Skåneland 1658-59. Konfrontation og oprør – perspektiver – hjælpemidler. *Bornholmske Samlinger* 3. række (4): 45-84.

Rasmussen, Holger. 1979. *Dansk museumshistorie. De kulturhistoriske museer.* Odense: Dansk Kulturhistorisk Museumsforening.

Rasmussen, Holger. 1992. *En husmand i Treårskrigen 1848-50. Underjæger Hans Pedersens optegnelser fra 1849 og hans visesamling.* Copenhagen: Foreningen Danmarks Folkeminder.

Reddy, G. Prakash. 1993. *Danes are like that. Perspectives of an Indian Anthropologist on the Danish Society.* Mørke: Grevas Forlag.

Redfield, Robert. 1960. *The Little Community and Peasant Society and Culture.* Chicago: The University of Chicago Press.

Regionala roller. En perspektivstudie. Betänkande av regionutredningen. Stockholm: Civildepartmentet, Statens offentliga utredningar 1992: 63.

Renan, Ernest. 1990 (1882). What is a Nation? In: Homi K. Bhabha (ed.): *Nation and Narration*, pp. 8-22. London and New York: Routledge.

Rerup, Lorenz. 1980. The Development of Nationalism in Denmark, 1864-1914. In: Rosalind Mitchison (ed.): *The Roots of Nationalism: Studies in Northern Europe*, pp. 47-59. Edinburgh: John Donald Publishers.

Rerup, Lorenz. 1991. Fra litterær til politisk nationalisme. Udvikling og udbredelse fra 1808 til 1845. In: Ole Feldbæk (ed.): *Dansk identitetshistorie* II, pp. 325-390. Copenhagen: C.A. Reitzels Forlag.

Rerup, Lorenz. 1992. Folkestyre og danskhed. Massenationalisme og politik 1848-1866. In: Ole Feldbæk (ed.): *Dansk identitetshistorie* III, pp. 337-442. Copenhagen: C.A. Reitzels Forlag.

Rerup, Lorenz. 1994. Nationalisme og skandinavisme indtil Første Verdenskrigs udbrud. *Den jyske historiker* 69-70. De Nordiske Fællesskaber. Myter og realitet i det nordiske samarbejde: 79-87.

Richman, Michèle. 1992. On the Power of the Banal: (Un)common Categories in Recent Social Thought. *Public Culture* 5(1): 113-122.

Riis, Ole and Peter Gundelach. 1992. *Danskernes værdier*. Copenhagen: Forlaget Sociologi.

Rockstroh, K.C. 1907. Kronborg 1658-59. *Fra Frederiksborg Amt*: 47-91.

Rockstroh, K.C. 1908. Hans Rostgaard og Kronborg-Anslaget 1659. *Fra Frederiksborg Amt*: 98-100.

Rogers, Susan Carol; Thomas M. Wilson; and Gary W. McDonogh (eds.). 1996. *European Anthropologies. A Guide to the Profession*. Volume I: Ethnography, Ethnology, and Social/Cultural Anthropology. Arlington: American Anthropological Association/Society for the Anthropology of Europe.

Röndahl, Uno. 1986. *Skåneland II. På jakt efter historien*. Sölvesborg: Lagerblads.

Röndahl, Uno. 1993 (1981). *Skåneland utan förskoning. Om kungahusens och den svenska överklassens folkdråp och kulturskövling i Skåneland. En studie av omnationaliseringens tragik*. Second edition. Sölvesborg: Lagerblads.

Rosander, Göran. 1986. The "nationalisation" of Dalecarlia. How a special province became a national symbol for Sweden. *Arv. Scandinavian Yearbook of Folklore* 42: 93-142.

Rosén, Jerker. 1943. *Hur Skåne blev svenskt*. Det levande förflutna. Stockholm: Hugo Gebers Förlag.

Rosén, Jerker. 1977 (1946). Statsledning och provinspolitik under Sveriges stormaktstid. En författningshistorisk skiss. In: Göran Rystad (ed.): *Svenskt 1600-tal*. Problem i äldre historia, pp. 222-183. Lund: Studentlitteratur.

Ruth, Arne. 1984. The Second New Nation: The Mythology of Modern Sweden. *Dædalus* 113(2): Nordic Voices: 53-96.

Rysén, Elisabeth. 1993. Det osynliga bagaget. Svenska bilder av Danmark och danskarna. Lund: Seminar-paper, ETN 003, Department of European Ethnology.

Rystad, Göran. 1965. Gränskrig och bondefred. *Ale. Historisk tidskrift för Skåneland* (2): 31-42.

Sahlins, Marshall. 1993. Goodbye to *Tristes Tropes:* Ethnography in the Context of Modern World History. *Journal of Modern History* 65: 1-25.

Sahlins, Peter. 1989. *Boundaries. The Making of France and Spain in the Pyrenees*. Berkeley: University of California Press.

Sahlins, Peter. 1998. State Formation and National Identity in the Catalan Borderlands During the Eighteenth and Nineteenth Centuries. In: Thomas M. Wilson and Hastings Donnan (eds.): *Borders Identities. Nation and State at International Borders*, pp. 31-61. Cambridge: Cambridge University Press.

Said, Edward W. 1978. *Orientalism*. New York: Vintage Books.

Said, Edvard W. 2000. *Out of Place. A Memoir*. New York: Vintage Books.

de Saint-Exupéry, Antoine. 1943. *The Little Prince*. New York: Reynal and Hitchcock.

Salje, Sven Edvin. 1958. *Man ur huse*. Stockholm: LTs Förlag.

Salje, Sven Edvin. 1960. *Kustridaren*. Stockholm: LTs Förlag.

Salje, Sven Edvin. 1968. *Natten och brödet*. Stockholm: LTs Förlag.

Sampson, Steve. 1993. "The Threat to Danishness": Danish Culture as Seen by Den danske Forening. Esbjerg: Paper, Nordic Seminar for Migration Research.

Sanders, Hanne and Ole Vind (eds.). 2003. *Grundtvig – nyckeln till det danska?* Göteborg: Makadam.

Schepelern, Peter. n.d.: *Danmarksfilmene.* Copenhagen: Statens Filmcentral.

Schivelbusch, Wolfgang. 1993. *Tastes of Paradise. A Social History of Spices, Stimulants, and Intoxicants.* New York: Vintage.

Schlee, Günther (ed.). 2002. *Imagined Differences. Hatred and the Construction of Identity.* New York: Palgrave.

Schmidt, Erik Ib. 1993. *Fra psykopatklubben. Erindringer og optegnelser.* Copenhagen: Gyldendal.

Schmidt, Lars-Henrik and Jens Erik Kristensen. 1986. *Lys, luft og renlighed. Den moderne socialhygiejnes fødsel.* Copenhagen: Akademisk Forlag.

Schoug, Fredrik. 1992. Olympiska spel och nationell virilitet. Sambandet mellan den norska succén och den svenska katastrofen i Albertville. *Lunda Linjer* 108: 53-58.

Schultz, Sigurd. 1930. Kunstindustriens Stilling paa Stockholmsudstillingen. *Nyt Tidsskrift for Kunstindustri* 3(9): 133-135.

Schwartz, Jonathan Matthew. 1991. Om at dechifrere en dansk frokost. Mødet mellem natur og kultur i en dansk ceremoni. *Højskolebladet* 116(22): 341-345.

Segerstedt, Torgny T:son. 1944. Formal och real demokrati. In: Torgny T:son Segerstedt, Brita Åkerman, Harald Elldin, Gregor Paulsson, Jöran Curman, and Helge Zimdahl: *Inför framtidens demokrati*, pp. 9-36. Stockholm: Kooperativa förbundets bokförlag.

Segerstedt, Torgny T:son; Brita Åkerman; Harald Elldin; Gregor Paulsson; Jöran Curman; and Helge Zimdahl. 1944. *Inför framtidens demokrati.* Stockholm: Kooperativa förbundets bokförlag.

Shore, Cris. 2000. *Building Europe. The Cultural Politics of European Integration.* London: Routledge.

Silberbrandt, Henning. 1993. *Den danske syge. Kritik af dansk selvforståelse.* Højbjerg: Hovedland.

Sjöholm, Carina. 2003. *Gå på bio. Rum för drömmar i folkhemmets Sverige.* Stockholm: Brutus Östlings Bokförlag Symposion.

Skaarup, Jørgen. 1994. Uden respekt for kanonerne. *Skalk* (6): 18-29.

Skarin Frykman, Birgitta. 1983. 23rd Nordic Congress of Ethnologists and Folklorists, May 30 – June 3, 1983, Fuglsø (Århus), Denmark. *Ethnologia Scandinavica*: 160-167.

Smith, Anthony D. 1988 (1986). *The Ethnic Origins of Nations.* New York: Basil Blackwell.

Smith, Anthony D. 1991. *National Identity.* London: Penguin.

Sontag, Susan. 1969. A Letter from Sweden. *Ramparts* (July): 23-38.

Sørensen, Charles. 1880. Gøngefolket og Snaphanerne. Nogle Bidrag til den lille Krigs Historie i Norden. *Historisk Arkiv* II: 253-268.

Sørensen, Jørgen. 1980. *Danmarksfilmen og Danske billeder.* Copenhagen: Gyldendal.

Sørensen, Marie Louise Stig. 1986. "...føie oldtidens Kraft til nutidens Kløgt..." *Stofskifte. Tidsskrift for Antropologi* 13. Danmark: 35-45.

Sörensson, Per. 1916. *Friskyttarna (Snapphanarna) under skånska kriget (1676-79). Deras organisation och militära betydelse.* Lund.

Sörlin, Anton. 1951. *Geografi. Sverige och dess grannländer*, vol. 4. Stockholm: Ehlin.

Spager, Ingeborg. 1935. *Sverige kalder.* Copenhagen: Jespersen og Pios Forlag.

Spager, Ingeborg. 1937. *Se Sverige.* Copenhagen: Jespersen og Pios Forlag.

Sperling, Vibeke. 1990. Er de bare blevet borgerlige? *Information* 20. april 1990: Moderne tider. Copenhagen.

Sprengel, David. 1904. *Från det moderna Danmark. Kåserier och intryck.* Stockholm: Albert Bonniers förlag.

SSF (Stiftelsen Skånsk Framtid; ed.). 1991. *333 års-boken om Skånelandsregionen – historielös – försvarslös – framtidslös.* Örkelljunga: Settern.

Stangerup, Hakon (ed.). 1941. *Det moderne Sverige. Evne, Indsats, Vilje.* Copenhagen: Forlaget af 1939.

Statens Husholdningsråd 1935-1985. Copenhagen: Statens Husholdningsråd (1985).

Stenius, Henrik. 1993. Den politiska kulturen i Nordens ontologi. Modell eller icke? Vara eller icke vara? *TijdSchrift voor Skandinavistiek* 14(1): Godelieve Laureys, Niels Kayser Nielsen, and Johs. Nørregaard Frandsen (eds.): Skandinaviensbilleder: 185-196.

Stocking, George W. (ed.). 1983. *Observers Observed. Essays on Ethnographic Fieldwork.* Madison: University of Wisconsin Press.

Stokes, Martin. 1998. Imagining "the South": Hybridity, Heterotopias and Arabesk on the Turkish-Syrian Border. In: Thomas M. Wilson and Hastings Donnan (eds.): *Borders Identities. Nation and State at International Borders*, pp. 263-288. Cambridge: Cambridge University Press.

Stoklund, Bjarne. 1972. *Bondegård og byggeskik.* Second edition. Copenhagen: Dansk Historisk Fællesforening.

Stoklund, Bjarne. 1994. The Role of the International Exhibitions in the Construction of National Cultures in the 19th Century. *Ethnologia Europaea* 24(1): 35-44.

Stråth, Bo. 2005. The idea of a Scandinavian nation. In: Lars-Folke Landgrén and Pirkko Hautamäki (eds.): *People, Citizen, Nation*, pp. 208-223. Helsinki: Renvall Institute.

Strauss, Erwin S. 1984. *How to Start Your Own Country. How You Can Profit from the Coming Decline of the Nation State* (Second Edition). Port Townsend: Loompanics Unlimited.

Strid, Artur. 1970. Giftas – Resonemangsparti och bondefred. *Ale. Historisk tidskrift för Skåneland* (2): 12-28.

Strid, Artur. 1973. Något om självstyrelsen i Loshults socken. Glimtar ur kyrkoprotokollen 1690-1721. In: Artur Strid (ed.): *En bok om Loshult*, pp. 61-76. Osby: Loshults kommun.

Strunk, Helge 1978. Danmark (PH – 1935). *film-uv* 50, 12(6): 18-19.

Sundback, Susan 1993. (Contribution without title). In: Göran Bexell (ed.): *Värdetraditioner, värdekonflikter och värdegemenskap i de nordiska länderna – Kontinuitet och förändring*. Rapport från ett symposium i Lund den 11-12 mars 1993, pp. 41-44. Lund: Department of Theology.

Sundbärg, Gustav. 1911. *Det svenska folklynnet. Aforismer.* Stockholm: P.A. Norstedt and Söners förlag.

Sundin, Jan. 1986. Bandits and Guerrilla Soldiers. Armed Bands on the Border between Sweden and Denmark in Early Modern Times. In: Gherardo Ortalli (ed.): *Bande armate, banditi, banditismo e repressione di giustizia negli stati europei di antico regime*, pp. 141-166. Roma: Jouvence.

Svensson, Birgitta. 1993. *Bortom all ära och redlighet. Tattarnas spel med rättvisan.* Stockholm: Nordiska museets handlingar 114.

Svensson, Sigfrid. 1958. Danskt och skånskt. *Rig* 41: 33-44.

Sverige-bilden i utlandet. Utdrag ur Svenska Institutets klippbok 1951-57. Stockholm: Svenska Institutet 1958.

Sverige i utländsk press. Stockholm: UD:s Pressbyrå, 1968- (some years with the title: *Sverige i utländska media/medier*).

Swedberg, Sven. 1948. *Våra grannländers näringsliv.* Stockholm: Kooperativa förbundet.

Tägil, Sven (ed.). 1992. *Europa – historiens återkomst.* Hedemora: Gidlunds.

Tägil, Sven (ed.). 1995. *Ethnicity and Nation Building in the Nordic World.* Carbondale: Southern Illinois University Press.

Tägil, Sven (ed.). 2001. *Europe: The Return of History.* Lund: Nordic Academic Press.

Tandrup, Leo. 1971. *Svensk agent ved Sundet. Toldkommisær og agent i Helsingør Anders Svenssons depecher til Gustav II Adolf og Axel Oxenstierna 1621-1626.* Århus: Aarhus University Press.

Tangkjær, Christian. 2000a. *"Åbent Hus". Organiseringen omkring Øresundsregionen.* Copenhagen: Copenhagen Business School.

Tangkjær, Christian. 2000b. Øresund as an Open House. Strategy by Invitation. In: Per Olof Berg, Anders Linde-Laursen, and Orvar Löfgren (eds.): *Invoking a Transnational Metropolis. The Making of the Øresund Region*, pp. 165-190. Lund: Studentlitteratur.

Tangkjær, Christian and Anders Linde-Laursen. 2004. Place-Making in the Global Village. Øresund: a Brand New Future? In: Friedrich Zimmermann and Susanne Janschitz (eds.): *Regional Policies in Europe. Soft Features for Innovative Cross-Border Cooperation*, pp. 9-29. Graz: Leykam.

Thau, Carsten. 1992. Den "danske" arkitekturideologi. In: Uffe Østergård (ed.): *Dansk Identitet?*, pp. 169-209. Kulturstudier 19. Århus: Aarhus University Press.

"The Birth of a Region". *Et strategisk oplæg om Øresundsregionen*. København and Malmö: Öresundskomiteen, Wonderful Copenhagen. Copenhagen Capacity, Skånes Turistråd, Länsstyrelsen i Skåne Län, Malmö stad, and Øresundskonsortiet, October 1997.

Thing, Morten. 1993a. *Kommunismens Kultur 1-2. DKP og de intellektuelle 1918-1960*. Copenhagen: Tiderne Skifter.

Thing, Morten. 1993b. Poul Henningsen – den danske kulturradikalismes pædagog. In: Bertil Nolin (ed.): *Kulturradikalismen. Det moderna genombrottets andra fas*, pp. 223-249. Stehag: Brutus Östlings Bokförlag Symposion.

Thomsen, Knud H. 1989. *Degnen i Kragevig*. Copenhagen: Gyldendal.

Thorkildsen, Dag. 1997. Religious Identity and Nordic Identity. In: Øystein Sørensen and Bo Stråth (eds.): *The Cultural Construction of Norden*, pp. 138-160. Olso: Scandinavian University Press.

Tingsten, Herbert. 1969. *Gud och fosterlandet. Studier i hundra års skolpropaganda*. Stockholm: Norstedt and Söners Förlag.

Tingsten, Herbert. 1979 (1933). National self-examination. *Ethnic and Racial Studies* 2(1): 38-54.

Tolnov Clausen, Laura. 2002. Grønne argumenter og lokal identitet. Kampen om Skjern Å-dalen. *Folk og Kultur* 2002: 48-68.

Tomenius, John. 1961. Karl XI:s anfall mot partisanrörelsen i Örkened 1678. *Göinge hembygdsförenings årsbok*: 21-30.

Tomenius, John. 1984. *Den stora ofärden. Snapphaneskildring från Göinge*. Örkelljunga: Settern.

Trotzig, Birgitta. 1997 (1957). *De utsatta*. Viborg: En bok för alla.

Tunander, Ola (ed.). 1995. *Europa och Muren. Om "den andre", gränslandet och historiens återkomst i 90-talets Europa*. Aalborg: Nordic Summer University.

Tuneld, John. 1960. *Prästrelationerna från Skåne av år 1667 och 1690*. Lund: Gleerups.

Tvede-Jensen, Lars 1985. Clementsfejden. Det sidste bondeoprør i Danmark. In: Jørgen Würtz Sørensen and Lars Tvede-Jensen (eds.): *Til Kamp for Friheden. Sociale oprør i nordisk middelalder*, pp. 196-217. Aalborg: Aalborg University Press.

Tyler, Stephen A. 1986. Post-Modern Ethnography. From Document of the Occult to Occult Document. In: James Clifford and George E. Marcus (eds.): *Writing Culture. The Poetics and Politics of Ethnography*, pp. 122-140. Berkley: University of California Press.

Urry, John. 1990. *The Tourist Gaze. Leisure and Travel in Contemporary Societies*. London: Sage.

Varpio, Yrjö. 1980. Blev "Okänd soldat" censurerad? In: Yrjö Varpio (ed.): *Bilden av ett folk*. En festskrift till Väinö Linna, pp. 82-95. Helsingfors: Holger Schildts förlag.

Velander, Jenny. 1906. *Danmark och danskarna. En skildring*. Svenskt folkbibliotek IV:7. Stockholm: P.A. Norstedt and Söners förlag.

Vila, Pablo (ed.). *Ethnography at the Border.* Minneapolis: University of Minnesota Press.

Vinge, Louise (ed.). 1996-1997. *Skånes litteraturhistoria* I-II. Malmö: Corona.

Vogel-Jörgensen, T. 1943. *Svenskt väsen och danskt.* Stockholm: P.A. Norstedt and Söners Förlag.

Vollertsen, Nils. 1992. Danmark, Island, Nordslesvig. Nationale bevægelser og dansk politik 1848-1920. *Scandia* 58(1): 103-121.

Wagner, Kim A. 2003. Fra Svend Poulsen til Gøngehøvdingen. *Siden Saxo* 20(2): 14-21.

Wägner, S. 1886. *Skånska kommissionen af 1669-1670. Ett bidrag till de skånska landskapens inre historia.* Lund.

Wallerstein, Immanuel. 1974. *The Modern World System. Capitalist Agriculture and the Origins of the European World-Economy in the Sixteenth Century.* New York: Academic Press.

Weibull, Carl Gustav. 1908. *Freden i Roskilde. Aktstycken och framställning.* Särtryck ur Historisk tidskrift för Skåneland, band 3. Lund.

Weibull, Carl Gustav. 1923. *Skånska jordbrukets historia intill 1800-talets början.* Lund: Gleerup.

Wendt, Frantz. 1941. Politik indad og udad. In: Hakon Stangerup (ed.): *Det moderne Sverige. Evne, Indsats, Vilje,* pp. 52-73. Copenhagen: Forlaget af 1939.

Westholm, Sigurd. 1942. *Minimifordringar å storlek av bostadslägenheter i hus avsedda att uppföras med stöd av statligt tertiärlån.* Stockholm: Meddelanden från Statens byggnadslånebyrå.

Wieslander, Anna. 1997. Att bygga Öresundsregionen. Från 1960-talets utvecklingsoptimism till 1990-talets lapptäcksregionalism. In: Sven Tägil, Fredrik Lindström, and Solveig Ståhl (eds.): *Öresundsregionen – visioner och verklighet. Från ett symposium på Lillö,* pp. 77-125. Lund: Lund University Press.

Wiggers-Jeppesen, Lars and Michael Boisen Schmidt. 1981. *Gønger, friskytter og snaphaner. Guerilla-virksomhed i Norden i det 17. århundrede. En bibliografi.* Copenhagen: Det kongelige Garnisonsbibliotek.

Willerslev, Richard. 1983. *Den glemte indvandring. Den svenske indvandring til Danmark 1850-1914.* Copenhagen: Gyldendal.

Wilson, Thomas M. and Hastings Donnan. 1998a. Nation, State and Identity at International Borders. In: Thomas M. Wilson and Hastings Donnan (eds.): *Borders Identities. Nation and State at International Borders,* pp. 1-30. Cambridge: Cambridge University Press.

Wilson, Thomas M. and Hastings Donnan (eds.). 1998b. *Borders Identities. Nation and State at International Borders.* Cambridge: Cambridge University Press.

Wilson, Thomas M. and M. Estellie Smith (eds.). 1993. *Cultural Change and the New Europe. Perspectives on the European Community.* Boulder: Westview Press.

Winge, Vibeke. 1991a. Dansk og tysk i 1700-tallet. In: Ole Feldbæk (ed.): *Dansk identitetshistorie* I, pp. 89-110. Copenhagen: C.A. Reitzels Forlag.

Winge, Vibeke. 1991b. Dansk og tysk 1790-1848. In: Ole Feldbæk (ed.): *Dansk identitetshistorie* II, pp. 110-149. Copenhagen: C.A. Reitzels Forlag.

Würtz Sørensen, Jørgen. 1985. Budstikken går. Bøndernes oprørspraksis i nordisk middelalder. In: Jørgen Würtz Sørensen and Lars Tvede-Jensen (eds.): *Til Kamp for Friheden. Sociale oprør i nordisk middelalder*, pp. 26-40. Aalborg: Aalborg University Press.

Yahil, Leni. 1991. National Pride and Defeat: A Comparison of Danish and German Nationalism. *Journal of Contemporary History* 26: 453-478.

Zabusky, Stacia E. 2000. Boundaries at Work. Discourses and Practices of Belonging in the European Space Agency. In: Irène Bellier and Thomas M. Wilson: *An Anthropology of the European Union. Building, Imagining and Experiencing the New Europe*, pp. 179-200. Oxford: Berg,

Zangenberg, H. 1982 (1925). *Danske Bøndergaarde. Grundplaner og konstruktioner*. Copenhagen: Foreningen Danmarks Folkeminder.

Index